PERIODIC STRUCTURES, GRATINGS, MOIRÉ PATTERNS AND DIFFRACTION PHENOMENA

Volume 240

Contents

I might be very little,
but my **family** is large.
And with so many of us,
we take turns to be in charge!"

DAD'S SHOE

In the reading corner, **Elio** grabs his favourite book.
You'll find your **friends** and **family** in stories if you look.

"My **mummy** is a princess –
she has a silver sword.
She's **stronger** than a giant
and I'm never, ever bored . . .

YE OLDE
BOOK
CART

"We live inside a turret,
high,
high
above
the ground.

We all **read** books together,
even when the dragon's round."

At quiet time, we settle down –
and then I get a shock . . .

Isla's beanbag has **transformed** . . .

into
a
giant
rock!

"My **daddy** is a merman,
and we swim the ocean blue.
We race around the coral
and look for treasure, too.

We're **friends** with great white sharks,
and know the secrets of the sea.
We love that we are just us two –
our little family!"

Our grown-ups come at home time –
HIP HIP HOORAY!
It's like a great big party at the end of every day.

With **astronauts** and **super grans**, **mermen** and **princesses**,
there's so much **magic**, so much fun
(and awesome fancy dresses!).

There are *billions* of families,
a *million* ways to be . . .
but of them all MY favourite is **Mama**, **Mum** and **me**.

Families are interesting,
each different from the rest.
There isn't one that's perfect,
not one kind that's best.

ARLO'S MUM'S GIANT SHOE

ARLO'S DAD'S GIANT SHOE

KAI'S HERO HOUSE

OUR MAGIC NURSERY

GIA'S SPACESHIP

NAOMI'S WIZARD DEN

Proceedings of the Society of Photo-Optical Instrumentation Engineers

Volume 240

Periodic Structures, Gratings, Moiré Patterns, and Diffraction Phenomena

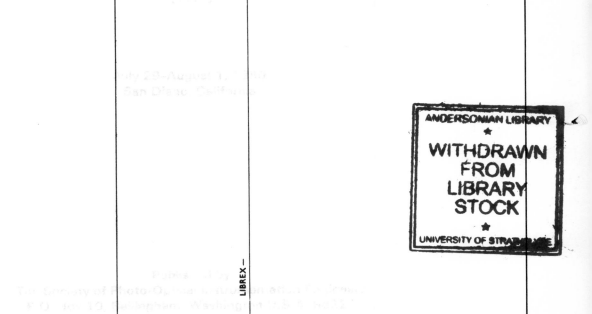

The papers appearing in this book comprise the proceedings of the meeting mentioned on the cover and title page. They reflect the authors' opinions and have not been reviewed, refereed, edited, or copyedited. Their inclusion in this publication does not necessarily constitute endorsement by the editors or by the Society of Photo-Optical Instrumentation Engineers.

Please use the following format to cite material from this book:

Author(s), "Title of Paper," *Periodic structures, gratings, moiré patterns, and diffraction phenomena,* Proc. Soc. Photo-Opt. Instr. Eng. 240, page numbers (1981).

Library of Congress Catalog Card No.: 80-52689

ISBN 0-89252-269-0

PERIODIC STRUCTURES, GRATINGS, MOIRÉ PATTERNS AND DIFFRACTION PHENOMENA

Volume 240

Seminar Committee

Chairmen
Chang H. Chi
Hughes Aircraft Company

Erwin G. Loewen
Bausch and Lomb, Incorporated

Carey L. O'Bryan III
Air Force Weapons Laboratory

Chairman Session 1—Holographic Grating I
Chang H. Chi
Hughes Aircraft Company

Chairman Session 2—Holographic Grating II
Peter L. Misuinas
Air Force Weapons Laboratory

Chairman Session 3—Diffraction Theory I
Allen Mann
Hughes Aircraft Company

Chairman Session 4—Diffraction Theory II
Erwin G. Loewen
Bausch and Lomb, Incorporated

Chairman Session 5—Fabrication Techniques, Applications I
Carey L. O'Bryan III
Air Force Weapons Laboratory

Chairman Session 6—Fabrication Techniques, Applications II
William R. Hunter
Naval Research Laboratory

Chairman Session 7—Fabrication Techniques, Applications III
Terry L. Holcomb
Hughes Aircraft Company

PERIODIC STRUCTURES, GRATINGS, MOIRÉ PATTERNS
AND DIFFRACTION PHENOMENA

Volume 240

INTRODUCTION

The study of physical phenomena originating from the periodic structure is an old science. Yet, the immense range of applications and endless variations of periodic structure types make this technology a fertile ground for scientific and engineering explorations in an ever-increasing degree. Periodic structures include the diffraction gratings, multilayer coatings, corrugated boundary waveguides, periodic mirrors (as in the buried long-period grating), and combinations of different grating types (as in the compound interlaced grating).

In recent years, the tempo of activities in this technology has increased and the significant progress has been achieved on many fronts. There have been advances in grating theory and computational techniques, in fabrication control in the ruling engines and ion etching processes, and in diagnostics and process controls. We have also witnessed the expansion of application areas including optical circuits, high energy laser (HEL) applications, and applications in UV and IR instrumentation.

The symposium which this volume represents was the forum which enabled scientists of this community to come together and evaluate the current status of the technology, exchange ideas, and describe their activities. The conference was attended by more than 300 participants including those from nearly a dozen countries in Europe and Asia, and the participants actively contributed to make this conference an important occasion in the history of grating technology.

Chang H. Chi, Hughes Aircraft Company
Erwin G. Loewen, Bausch and Lomb, Incorporated
Carey L. O'Bryan III, Air Force Weapons Laboratory

ELIO'S TOWER

RAVI'S PIRATE SHIP

ISLA'S OCEAN PALACE

But a **family** *is* **magic**,
no matter who is who.

Real and **fantastical** . . .

ETTIE'S FAIRY GARDEN

. . . they're full of love for
YOU!

PERIODIC STRUCTURES, GRATINGS, MOIRÉ PATTERNS AND DIFFRACTION PHENOMENA

Volume 240

SESSION 1

HOLOGRAPHIC GRATING I

Session Chairman
Chang H. Chi
Hughes Aircraft Company

Wavelength scaling holographic elements

Mark Malin, Howard E. Morrow

Lockheed Palo Alto Research Laboratory, 3251 Hanover Street, Palo Alto, California 94304

Abstract

A simple technique to generate holographic optical elements with diffraction-limited performance at a wavelength different from the construction wavelength has been developed. The method consists of specifying the desired grating performance at the operating wavelength, selecting one construction beam, and utilizing a ray trace program to compute the aberrated construction beam required to create the hologram. A lens capable of focusing this aberrated beam to a point may then be designed using ACCOS-5. A point source used with this lens thus serves to generate the second required construction beam. A sample design which may be used to construct with 0.4579-μm light a grating to operate at 2.8 μm is presented.

Introduction

Many of the adaptive optical techniques now being considered require special-purpose holographic elements. In particular, the need exists for holographic gratings to perform a beam sampling task for outgoing wavefront analysis in a laser beam projector. In general, these elements will be on the primary mirror to obtain a sample of the wavefront after the last element in the optical train and, due to the high fluence levels they will be subjected to, they will generally be ion-etched either into the substrate or into a protective base coat.

In a typical application, the primary may be covered with a number of individually focused, weakly diffracting zone plates. In such a case, as a plane wave is projected to the far field, weak individual samples of the wavefront as it exists on the primary are diffracted and focused through the secondary obscuration. An analysis of these samples essentially forms a Hartmann test and can provide the necessary wavefront phase aberration measurements to correct for a perfect projected beam using an adaptive optical element.

The generation of such gratings is conceptually quite simple. One only needs to coat the substrate with photoresist and record the interference pattern of the nominal illumination beam and the conjugate of the desired sample wavefront in the perfect configuration.

For our sample case, the required construction wavefronts would simply be a point source at the desired zone plate focus and a collimated beam on axis for each grating. The resulting interference pattern recorded may then be etched into the coating or substrate for permanence and, as a parallel beam is projected, the desired diffracted sample is produced.

Unfortunately photoresists are only sensitive to the blue/UV region of the spectrum. Since a number of applications exist for projection systems operating at infrared wavelengths, it was necessary to develop a technique to produce at one wavelength gratings that provide diffraction-limited performance at a different wavelength. For plane surfaces and simple linear gratings this is a trivial matter of changing construction beam angles. For the more complicated optical elements we are concerned with, geometry changes alone would, in general, produce an unacceptably aberrated sample beam. It was thus necessary to design a method of distorting the fringes to remove this sampling aberration. Lockheed has developed a grating system ray trace capability to analyze potential grating sampled system performance under various operational conditions such as segment misalignment and thermal expansion. We chose to use a quick and dirty modification of this analysis to attempt to derive exposure configurations for wavelength scaled holographic elements.

Computation of construction beams

The technique consists of selecting one of the hologram construction beams at the photoresist sensitive wavelength, computing the other beam required to produce the desired interference pattern, and designing the necessary optical elements to produce this beam. We have only been concerned with reflection holographic elements for our projects, but it is clear that the same method presented here can be modified for transmission gratings.

To begin, the surface on which the optical element is to reside is specified (i.e., conic constant, vertex curvature, orientation) and the operation of the grating at the design wavelength is defined. This is done by selecting point locations for the illumination beam and the resultant diffraction spot (i.e., perfect operation).

Next, one of the construction beams is chosen by specifying individual ray coordinates and direction cosines and is propagated to the surface of the element. In practice, to make the lens designer's task easier, the intersection points on the surface are selected and the appropriate rays computed.

At each of these points on the surface a local grating vector and local grating spacing may be computed using standard grating ray trace formulae.[1]

$$d = \frac{\lambda}{\mid (\vec{D} \times \vec{SN}) - (\vec{S} \times \vec{SN}) \mid} \tag{1}$$

$$\vec{GV} = \vec{SN} \times \frac{(\vec{D} \times \vec{SN}) - (\vec{S} \times \vec{SN})}{\lambda / d} \tag{2}$$

Where

d = local grating spacing
λ = operation wavelength
$\vec{D}$ = direction cosines of specified diffraction ray in design configuration
$\vec{SN}$ = surface normal unit vector
$\vec{S}$ = direction cosines of incident ray in design configuration
$\vec{GV}$ = local grating vector

Note that these are fictitious parameters and may not actually exist anywhere on the grating. They merely represent effective grating constants at the particular point of intersection.

The second construction beam required at the exposure wavelength λ' is computed by calculating the diffracted rays from the specified first construction beam.

$$\vec{D'} = \vec{S'} + G\,\vec{SN} + \frac{\lambda'}{d}\,\vec{GV} \tag{3}$$

where S' and D' are the construction and diffracted beams, respectively, and

$$G = \frac{-\vec{S} \cdot \vec{SN}}{\mid \vec{S} \cdot \vec{SN} \mid} \left\{ (\vec{S} \cdot \vec{SN})^2 - \frac{\lambda'}{d}\left(\frac{\lambda'}{d} + 2\vec{S} \cdot \vec{GV}\right) \right\}^{1/2} - \vec{R} \cdot \vec{SN} \tag{4}$$

The construction beam required comprises these diffracted rays reflected from the surface and direction reversed, as

$$\vec{C} = 2(\vec{D'} \cdot \vec{SN})\,\vec{SN} - \vec{D'} \tag{5}$$

These rays are then used to design an optical element which will focus them to a point. Thus, when a point source is used with this optical element, the required aberrated construction beam is produced.

Design of aberration correction element

The algorithm described can produce the direction cosines of any specific ray in the aberrated construction beam. Further, it can give the intersection coordinates of these rays on any specific plane surface displaced from the holographic grating surface. We next describe the selection of seven rays which sample the grating clear aperture in a way suitable for optimization within ACCOS-5.

The optimization phase of ACCOS-5 is prepared by the user by defining data with four subfiles. Within the RAYSET subfile, individual rays are definable with great flexibility. The starting point on the object surface is specified by a statement of the form FOB, Y X Z N where Y, X, and Z are the coordinates on the object surface (surface O) normalized to the object height as specified by the SCY record in the LENS subfile. Once this starting point is given within the RAY-SET subfile, the aiming point for any number of rays is given by statements of the form RAY, Y X N where Y and X are the coordinates on the reference surface (frequently surface 1) normalized to the aperture height as specified by the SAY record in the LENS subfile. Thus for purposes of ACCOS-5, the object surface is simply the curved mirror upon which the holographic grating is desired and the Z value in the FOB record is the sag of the surface at height Y. The reference surface is any agreeable surface for ray aiming. Any ray in the aberrated construction beam is specified by one FOB record followed by one RAY record. We normally set UNITS to mm and set both SCY and SAY to 1 so that all ray coordinates are real mm dimensions. Since current versions of ACCOS-5 are dimensioned to allow a maximum of seven FOB records, only seven rays can be selected for optimization. However, since no field angle in the usual sense is required, we feel this is entirely adequate.

Once these seven rays are defined, the user must create a list of image errors. We always select the ray from the center of the holographic grating and use it as the reference ray. In the DEFINITIONS subfile, the absolute X and Y heights of the intersections of the rays on the image surface are named as preliminary image errors. Then, using the COM "NAME" option, ray deviations are created for each nonzero X and Y difference between the reference ray and each of the other six rays. In normal lens design use, ACCOS-5 automatically traces reference rays which are not applicable in the grating construction problem. Thus the COM option is required to define ray deviations.

The rest is much simpler. The REQUESTS subfile lists those data and targets, with weights, which the user wishes to correct. The VARIABLES subfile lists those constructional parameters which the user will permit to vary in order to produce this correction.

We currently define "correction" to be achieved when all seven rays, after passing through the corrector element, lie within the Airy diameter defined by $D = 2.44 \lambda f/\#$, where $\lambda = 0.4579 \mu m$ and $f/\#$ is found from the image space cone angle of the three rays from the edge of the holographic grating.

Once a correcting configuration has been discovered, a perturbation analysis can be performed to explore the effects of setup errors and certain corrector manufacturing errors. The perturbation (e.g., tilt, thermal effects) is entered via an UPDATE LENS operation, and the new ray coordinates on the image surface plotted out as a mini-spot diagram and compared to the Airy diameter.

Sample design

In Fig. 1 we present a design to generate a 15-cm-diameter focusing grating located on the outer edge of a 4-m, f/1.25 parabola whose vertex is defined as the coordinate system origin. One construction beam was selected 2 m away centered on the grating and the desired grating focus was chosen to be on axis 4 m away from the primary vertex. An operation wavelength of 2.8 μm was specified and the construction wavelength was chosen to be 0.4579 m.

The construction rays required on the surface were generated using the technique described above. (Note that the inherent symmetry of the system about the y axis obviated the need for the rays on both sides, but this will not be the case in general.) A solution found by ACCOS-5 to focus the construction rays to within a diffraction-limited point consists of a tilted spherical mirror as shown in Fig. 1. This turns out to be one solution of an entire solution set of spherical surfaces of varying radii of curvature, each with a unique optimal location, and hence a large number of exposure geometries utilizing relatively simple surfaces are available for this one convenient choice of the first construction beam.

This specific design has not been constructed since 4-m parabolas are somewhat difficult to obtain. The validity of the method has been established by producing a 15-cm zone plate covering an f/6 parabola using an exposure geometry at 0.4579 μm. Reconstruction at the desired operation wavelength of 2.9111 μm has demonstrated the performance expected of the design.

Summary and conclusion

We have demonstrated a technique to design aberration correction lenses to produce holograms in other than the operational wavelength. This method is conceptually and operationally quite simple and only requires access to the ACCOS-5 lens design program. The next step is to integrate the ray generation process into ACCOS-5 so that optimal exposure geometries may be explored and tolerance analysis can be facilitated.

Acknowledgments

The authors wish to acknowledge the work of Tom Pope who was responsible for the laboratory demonstrations of the method and Robert E. Smithson who had overall responsibility for the Lockheed Independent Development program under which this work was performed.

References

1. G. H. Spencer, M.V.R.K. Murty, J. Opt. Soc. Am., 52, 672 (1962).

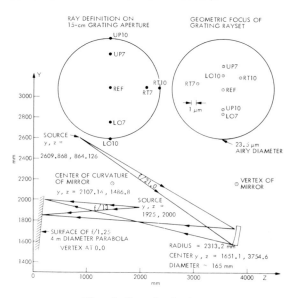

Fig. 1 Sample design.

Blazed holographic gratings—formation by surface waves and replication by metal electroforming

James J. Cowan

Polaroid Corporation, Research Laboratories
750 Main Street, Cambridge, Massachusetts 02139

ABSTRACT

The formation of a blazed holographic grating, using the interference of an ordinary plane wave with a surface wave is considered. The usual holographic recording arrangement has been modified such that a positive photoresist recording layer is coated on one face of a prism; one of the recording beams is totally reflected from the prism such that the surface wave thus formed enters the photoresist from the back side. The other recording beam also enters from the back side, but is an ordinary plane wave. The resulting interference fringes are inclined with respect to the normal. With this arrangement the profile can be contoured more precisely than usual and many variations in the shape of the profile are possible. Modifications of the method described here allow the formation of gratings on curved surfaces and in some cases correction for aberrations. Once a profile has been formed in this fashion in photoresist it is possible to replicate it exactly in a durable metal layer using electroforming techniques. Such gratings are applicable for use in spectrometers where high beam intensities must be taken into account.

INTRODUCTION

In the usual method of forming holographic diffraction gratings, two coherent light beams from a laser interfere in a photosensitive material, generally a positive photoresist, coated on a suitable substrate. After exposure the surface of the photoresist is etched by development to form a periodic profile that can then be overcoated with a metal like Al and used as a reflection grating. Since the etch depth is proportional to the incident intensity, the surface profile in this case would be sinusoidal. Gratings of this type are generally superior to the ruled grating in terms of low light scattering, absence of ghost images, ability to be corrected for aberrations, and ease of fabrication. The ruled grating, on the other hand, can be blazed so that most of the diffracted light falls into a given order. For a holographic grating with a sinusoidal groove profile the diffracted light would be distributed equally at normal incidence between diffracted orders on each side of the grating normal. The holographic grating would thus be improved if a method could be found of fashioning an asymmetric profile.

Some attempts at blazing holographic gratings have been made in the past. One method is that employed by Sheridan[1], whereby two beams are incident on the photoresist layer from opposite sides, forming standing waves within the layer, and the blank is tilted during recording. Upon development the photoresist develops down to nodal planes, and the result is a sawtooth profile. Another method is that used by Schmahl and Rudolph[2], whereby a sawtooth groove is formed by Fourier composition, using interference of several beams of the fundamental and higher order spatial harmonics. In still other embodiments, the photoresist layer is carefully etched so that only the tops of the sinusoid remain on the substrate, then the substrate is etched by ion bombardment to form an asymmetric profile. Recently, blazed holographic gratings have become available from the Jobin-Yvon Company.

BLAZED HOLOGRAPHIC GRATINGS

We present here a blazing method using interference of two beams of light in a photoresist layer whereby one of the beams is a surface wave and the other is an ordinary wave incident on the photoresist layer at a high angle of incidence. We also consider the replication of such structures in a durable nickel layer by metal electroforming[3]. The blazing technique is an extension of an earlier proposal we had made using two beam interference, but in that case an optical guided mode was used instead of a surface wave. We consider two configurations: in the first, the surface wave is evanescent, while in the second, it is propagating but results from total reflection. The evanescent case is considered first.

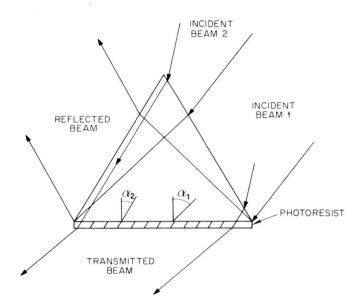

Figure 1. Recording configuration

EVANESCENT SURFACE WAVES

The optical arrangement is illustrated in Fig.1. A positive photoresist (Shipley AZ-1350) is coated onto a glass prism whose index of refraction is higher than that of the resist (n resist = 1.64 at 488nm). One of the recording beams is incident on the prism at an angle greater than that for total reflection at the prism-photoresist interface and the evanescent wave thus formed at the boundary enters the photoresist from the back side. The other recording beam also enters the prism from the back side but at an angle less than that for total reflection at the photoresist-air interface, so it is transmitted through the resist layer as a plane wave. The wave emerges from the top surface of the photoresist, and is refracted away at a large angle from the normal. Interference is between the evanescent surface wave and the transmitted plane wave, and the fringes are inclined at an angle with respect to the normal. The optical arrangement is similar to that used by Nassenstein[4] for recording holographic fringes in a photographic emulsion layer. An alternate recording scheme is to spin coat the photoresist onto a glass plate whose index is the same as that of the prism and then contact it to the prism with an appropriate index-matching fluid.

The mathematical description of the process is as follows: We consider the recording beams to be polarized perpendicular to the plane of incidence (S-polarized) and the boundary between the prism and the photoresist to be the X-Y plane with the positive Z axis extending from this boundary into the photoresist.

From the boundary conditions on an incident wave of the form

$$\vec{E} = A \exp \left[iK \left(x \sin \alpha \ + \ z \cos \alpha \right) - i\omega t \right] \quad (0,1,0)$$

one finds the surface wave can be described by the equation

$$\vec{E}_1 = 2A_1 \exp\left(-i\phi_{21}\right) \cos \phi_{21} \exp\left(-K_\perp z\right) \exp\left[i\left(K_\parallel x - \omega t\right)\right] \quad (0,1,0) \quad (1)$$

and the transmitted wave by the equation

$$\vec{E}_2 = A_2 \exp\left[i\left(n_s K_0 \left[Sx + Cz \right] - \omega t \right)\right] \quad (0,1,0) \quad (2)$$

where $K_0 = 2\pi/\lambda$ = the wave number in free space

with λ = free space wavelength, $K = 2\pi n_p/\lambda$

and $K_\parallel = K_o n_p \sin \alpha_1$ where

$$K_2 = \left(K_\parallel^2 - n_s^2 K_o^2\right)^{1/2}$$

$$K_\perp = \left(n_p^2 K_o^2 - K_\parallel^2\right)^{1/2}$$

$S = \sin\theta$

$C = \cos\theta$

$$\phi_{21} = \tan^{-1}\left(K_2/K_\perp\right)$$

$S = n_p \sin \alpha_2/n_s$

A_1 = amplitude of surface wave

A_2 = amplitude of transmitted wave

n_p = index of refraction of prism

n_s = index of refraction of photoresist

θ = angle that transmitted beam makes with respect to normal within the photoresist

α_1 = angle of incidence of beam no. 1 within the prism

α_2 = angle of incidence of beam no. 2 within the prism

The interference intensity is expressed by

$$I = \left| \vec{E}_1 + \vec{E}_2 \right|^2$$

$$= 4A_1^2 \cos^2 \phi_{21} \exp\left(-2K_\perp z\right) + A_2^2 + 4A_1 A_2 \cos\phi_{21} \exp\left(-K_\perp z\right)\cos\delta \tag{3}$$

where

$$\delta = 2\pi \left\{ \left(\frac{1}{\lambda_s} - \frac{S}{\lambda_m}\right) x - \frac{C}{\lambda_m} z \right\} - \phi_{21}$$

λ_s is the wavelength of the surface wave and $\lambda_m = \lambda/n_s$ is the wavelength of the transmitted beam within the photoresist.

The first term represents the exposure to the photoresist due only to the surface wave and it rapidly dies away in the z-direction. The second term is the exposure due only to the transmitted wave, and this is constant in the z-direction. The third term is the interference intensity between the two beams and this varies exponentially in the z-direction.

From the above equations one can calculate the groove spacing, ℓ, from

$$\ell = \left(\frac{1}{\lambda_s} - \frac{S}{\lambda_m}\right)^{-1} \tag{4}$$

and the blaze angle, γ, from

$$\gamma = \tan^{-1}\left(\frac{\lambda_m}{C\ell}\right) \tag{5}$$

The distinguishing characteristic of this recording scheme over other techniques is that the exposure intensity is inherently larger at the substrate than at the surface of the photoresist. This can be advantageous when the resist is etched to form the profile, because with a low exposure at the surface, the etch rate will be slow initially, but it will increase as the developer reaches the lower layers. One would thus expect a deeper and more pronounced groove that one formed in the usual way where both recording beams enter the photoresist from the top surface. In this treatment we are neglecting absorption in the photoresist layer, but it can become significant for thick layers. Specifically, for a reported[5] absorption coefficient of κ = .02 and an attenuation given by $I/I_o = \exp\left(-\kappa\frac{2\pi}{\lambda}t\right)$, the thickness required to reduce the intensity by one half, at a wavelength of λ = 442 nm, is t = 2.4 μm. One also takes advantage of the non-linear exposure and development characteristics of photoresist in that the etch rate for low exposures is proportionally less than that for higher exposures. The response of the photoresist can be made linear, by employing a uniform pre-exposure prior to the actual recording, and thus some degree of control over the etch rate can be exercised in this way. Another control over the etch rate can be achieved by varying the relative intensities of the surface wave and transmitted beams. As is clear from Fig. 2, the contrast is not constant, but varies from the substrate to the surface, so by varying the relative amplitudes of the two waves, the maximum contrast can be shifted to various depths of the resist layers.

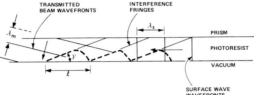

Figure 2a.   Interference fringe structure

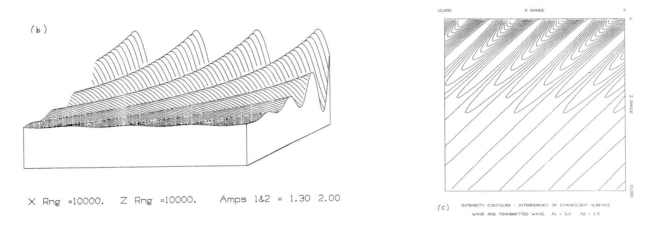

X Rng =10000.   Z Rng =10000.   Amps 1&2 = 1.30 2.00

Figure 2. b&c Equi-intensity contours, evanescent case

The primary disadvantage of this recording scheme is that exposure must be made from the back side of the photoresist layer, so that great care must be taken to eliminate as many aberrations and imperfections as possible in the prism and photoresist layer, and also in the high index plate and the contacting liquid, if they are used.

Another disadvantage is that the blaze angle and the groove spacing are related through Eqn. (5) so that one cannot independently choose these values. Also because the index of photoresist is higher than that of ordinary glass, one must use especially dense glass for prism material to achieve flexibility in recording. For example, the smallest incidence angle in the prism for which an evanescent wave will form within the photoresist is the critical angle, given by $\sin\alpha_c = n_s/n_p$. For dense flint glass with an index n_p = 1.7343 the angle is thus α_c = 71.0°. With high index Schott glass, on the other hand, with n = 1.94, one finds α_c = 57.7°. The largest possible surface wavelength is $\lambda_s = \lambda / \left(n_p \sin\alpha\right)$ and for a recording wavelength of λ = 442 nm is λ_s = 269.4 nm. If the transmitted beam exits at a grazing angle, then S = 1/n_s

and $C = \left(1- \left[1/n_s\right]^2\right)^{1/2}$ from which we get $\ell = 690.6$ nm and $\gamma = 26.20°$. These results represent extreme cases and other angles of incidence would give smaller ℓ values and larger blaze angles. Because of the exponential dependence of the surface wave intensity on depth, care must be taken to use either thin photoresist layers or angles of incidence close to the critical angle.

Figure 2a shows the interference fringe structure for the surface and transmitted wavefronts, where λ_s is the wavelength of the surface wave, λ_m the wavelength of the transmitted wave; and ℓ and γ are the groove spacing and blaze angle, respectively. Figures 2b and 2c show typical computer plots of equal intensity contours for the case where the surface wave is close to its largest possible value at $\lambda_s = \lambda/(n_p \sin \alpha)$ and, for simplicity the transmitted beam is taken to be moving through the photoresist at an angle, $\theta = 0°$.

SURFACE WAVES FORMED IN TOTAL REFLECTION

The restrictions on ℓ and γ can be relaxed if the angle of incidence, α, is reduced so that the evanescent wave is converted into a propagating wave that is totally reflected at the photoresist-air interface. The condition on α is then $(n_s/n_p) > \sin \alpha > (1/n_p)$. Equation (3) would show a propagating wave in the z-direction rather than an exponentially decaying one, and because of total reflection, there would be a similar reflected wave moving in the -z direction. Interference between the incident and reflected beams would cause standing waves to be set up in the photoresist with fringe spacing $d = \pi/K_\perp$. The incident and reflected waves, along with the standing waves, can be regarded as a single surface wave having its propagation vector in the same direction as the evanescent wave. Thus the fringe spacing and blaze angle are defined the same as before. In the extreme case the largest value for λ_s is the free-space wavelength, λ, and for a transmitted beam that is refracted at a grazing angle, ℓ approaches infinity and γ approaches zero. An advantage of this variation is that an ordinary glass prism can be used instead of a high-index one.

The equation of the surface wave for this case becomes

$$\vec{E}_1 = 2A_1 \exp(-i\Psi) \exp\left[i\left(K_\parallel x - \omega t\right)\right]\left(0, \cos\left[\Psi + K_\perp z\right], 0\right)$$

$$= A_1 \exp(-i\Psi)\left[\exp(i\beta) + \exp(-i\beta)\right] \exp\left[i(K_\parallel x - \omega t)\right] \quad (0,1,0) \tag{6}$$

where $\quad \beta = \Psi + K_\perp z \qquad \Psi = \phi_{12} - K_\perp t \qquad \phi_{12} = \tan^{-1}\dfrac{K_1}{K_\perp}$

and $\quad K_\perp = \left(n_s^2 K_o^2 - K_\parallel^2\right)^{1/2} \quad K_1 = \left(K_\parallel^2 - K_o^2\right)^{1/2} \qquad t = $ photoresist layer thickness

and the interference intensity becomes (replacing ϕ^{21} with Ψ in equation 3 for δ)

$$I = 4A_1^2 \cos^2\beta + A_2^2 + 4A_1 A_2 \cos\beta \cos\delta \tag{7}$$

A plot of equi-intensity contours for $\lambda_s \sim \lambda$ and $\alpha_2 = 0$ is given in Fig. 3a - 3c for this equation. One sees that the contours are inclined with respect to the normal, as for the evanescent case, except that here the variation in the z- direction is sinusoidal rather than exponential. This leads to some interesting results especially when the respective amplitudes of the two waves are varied. If the surface wave amplitude A_1 is larger than the transmitted wave amplitude, A_2, (Fig. 3b), one would expect development down to a nodal plane and a resultant shallow, but well-defined sawtooth profile with the above values for blaze angle and groove spacing. If A_2 is larger than A_1 on the other hand, the nodal planes are not well defined, and etching could be quite deep, even to the substrate, but at a steeper blaze angle (Fig. 3c).

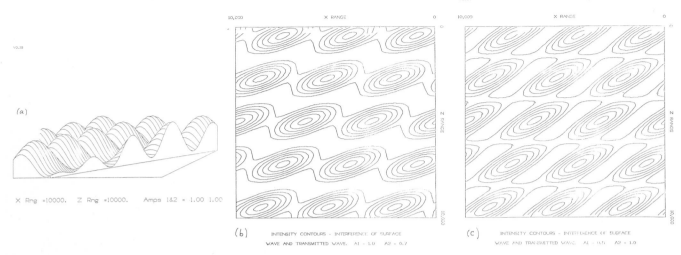

Figure 3.   Equi-intensity contours, totally reflected case
a) $A_1 = A_2$ b) A_1 greater than A_2 c)A_1 less than A_2

EXPERIMENTAL RESULTS

     Figures 4, 5 and 7 show contours that were made by the above technique.   The
profile in Fig. 4 results from recording at a wavelength of 442 nm with a dense flint
prism (n = 1.7432) that of Fig. 5 using a medium density glass prism (n = 1.620)
and a recording wavelength of 459.7 nm, and that of Fig. 7 at a recording wavelength
of 442 nm and a prism with an index of n = 1.5.   The profile depicted in Fig. 5 was
calculated to have ℓ = 1000 nm and γ= 16.59° for an angle of incidence of the surface wave of α_1= 40°, and α_2= 20.7° for the transmitted beam.   The exposure for this
grating was 100 mj with a pre exposure of 60 mj; development time = 30 seconds in AZ developer.   For this exposure A_1 was larger than A_2, thus one expects a profile similar

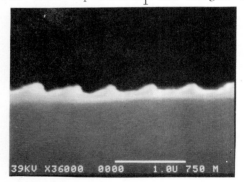

Figure 4. Grating profile, evanexcent case,
recorded on dense prism.

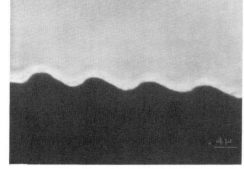

Figure 5.   Shallow profile recorded on medium
density prism, where $A_1 > A_2$

to that depicted in Fig. 3b.   A typical set of efficiency curves is shown in Fig.
6 for the visible spectral region at normal incidence for this grating overcoated
with Al.   The curves are found as the ratio of the diffracted (or reflected) light
intensity to the incident intensity and thus represent the absolute efficiency.
Mirror efficiency (e.g., the diffracted or reflected light intensity divided by the
maximum that is possible for an Al surface) would be higher.   The blaze wavelength
is given by λ_b = 2ℓ sinγ, which for the grating considered here is λ_b = 571.0 nm.

One sees from the figure that the on-blaze intensity (s-polarization) has a broad
peak that is centered at about 550 nm with an absolute efficiency of about 53%.
The dips in the curves for p-polarization at 540 nm are due to the excitation of
the surface plasmon in Al.   The wavelength for this exictation is calculated from
the grating equation at normal incidence including the plasmon interaction, by

$$\lambda_{p1} = \frac{\ell}{m} \sqrt{\frac{\varepsilon_1}{1 + \varepsilon_1}} = 507 \text{ nm}$$

where the integer m is the higher passing-off order responsible for the excitation (in this case, m = 2), and ε_1 = -34.5 is the real part of the dielectric constant of Al (at E = 2.5 eV) obtained from Ref. 6. The discrepancy between measured and calculated values is probably due to the assumption of a flat surface for propagation of the plasmon. Since the blaze angle is so large a more realistic propagation distance would be obtained by dividing ℓ by cos γ so that we would have $\lambda'_{p1_7} = \lambda_{p1}/\cos\gamma$ = 529 nm. The same effect has also been observed for holographic gratings by Hutley[7].

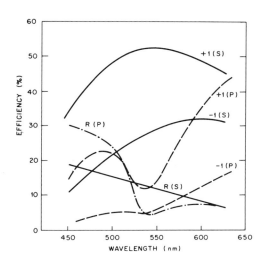

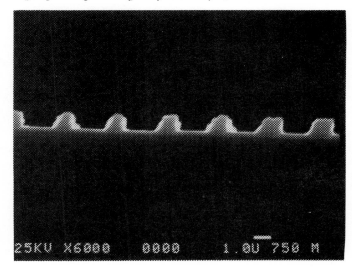

Figure 6. Efficiency curves for grating depicted in Fig. 5

Figure 7. Grating profile, totally reflected case, $A_1 < A_2$

Figure 7 represents a profile obtained for the totally reflected surface wave case when A_2 is larger than A_1. Since the exposure is from the back side, one would expect a slightly smaller intensity at the surface than at the substrate due to absorption. Thus as etching proceeds we see sharp edges, nearly vertical walls, and hardly any etching on the islands. The blaze angle is steeper than for that depicted in Fig. 5 and the standing wave structure is evident at the edges of the profile. Such grating structures might be useful if one wished to subsequently form a blazed profile in the substrate by ion etching techiques.

NICKEL ELECTROFORMING

Ion etched gratings are important in applications where gratings are subjected to high beam intensities, such as for example, their use as dispersing elements in synchrotrons. A grating surface etched into glass and overcoated with a reflecting metal like Al or Au will not deteriorate under high intensities as would a conventional holographic grating that has a reflecting layer coated onto the organic polymer, photoresist, that is coated onto glass. We present here an alternative to ion etching that may have promise for high intensity applications, and that is the electroforming, in nickel, or a replica grating directly from a photoresist master grating. The first requirement is to form the desired profile, whether blazed or otherwise, in photoresist, by holographic techniques. Then a replica is made by conventional electroforming methods. This consists of coating the photoresist master with a conducting layer, immersing in a suitable electrolyte like nickel sulfamate, then allowing the electroforming to continue until the desired thickness of nickel has been achieved. Care must be taken to insure that stresses do not build up in the nickel layer during electroforming, because the surface will be deformed. This can be done if the electroforming proceeds slowly enough and one controls carefully such parameters as the temperature of the bath and the proper amount of additives to insure even coating. Generally a sufficient thickness for a durable nickel layer is .05", but it helps for purposes of rigidity if the grating surface is curved rather than flat. Since the replica formed in this way is the mirror image of the original grating, it is necessary simply to replicate the replica by the same electroforming technique.

The second generation nickel replicas will then have a profile exactly like that of the original photoresist master. Figure 8 shows a profile of a first generation nickel replica made from a blazed grating. Unlike the other profile pictures that were made by shattering the glass onto which the photoresist was coated, the nickel replica was bent, and the scanning electron beam was incident tangentially to the surface. For use in a spectrometer, the nickel grating can be overcoated with a metal like Al or Au.

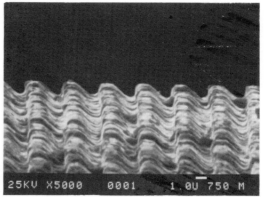

Figure 8. Nickel replica of blazed grating

CURVED SUBSTRATES - ABERRATION CORRECTION

The blazing technique is applicable in certain cases to substrates that are curved. For example, if the substrate were shaped like a plano-concave cylindrical lens, the planar part could be contacted to the prism by an index-matching liquid and then the surface wave could be stimulated by bringing the incident wave through a compensating convex cylindrical lens. With this arrangement the angle of incidence at any point of the cylindrical concave recording surface would be the same.

Under certain limited conditions, correction for aberrations can be made with the surface wave blazing technique. The usual method for correcting aberrations on a spherical blank, as outlined by Seya, Noda, and Namioka[8], requires placing two point sources at specified distances and angles from the blank, so that the resulting grating lines are curved rather than straight, but the profile remains sinusoidal. If correction is desired for gratings used at grazing incidence, in the vacuum UV and soft x-ray regions, one of the recording sources must be placed at an inconveniently large angle of incidence, close to 90°. If the beam from this source could be introduced in the form of a surface wave, by prism coupling as outlined above, the angle of incidence would automatically be 90° and the optical arrangement would be greatly simplified.

Acknowledgements: We appreciate the help of the following people with this study: R. Suitor, for extensive computer programming; C. Giles, for doing the Nickel electroforming; R. Burpee and P. Burnett for carrying out the electron microscopy.

References

1. N. K. Sheridan, "Production of Blazed Holograms", Appl. Physics Letters, 12 (1968), 316; M. C. Hutley, "Blazed Interference Diffraction Gratings for the Ultraviolet", Optica Acta, 22 (1975)1.

2. G. Schmahl and D. Rudolph, "Holographic Diffraction Gratings", Progress in Optics, XIV, 197 (1976).

3. J. J. Cowan, "Blazed and High Line Density Holographic Gratings produced by use of Optical Guided Waves", J. Opt. Soc. Am. 66 (1976) 1114.

4. H. Nassenstein, "Interference, Diffraction and Holography with Surface Waves", Optik, 20 (1969) 597.

5. F. T. Stone and S. Austin, "A Theoretical and Experimental Study of the Effect of Loss on Grating Couplers", IEEE J. of Quantum Electronics, QE-12 (1976) 727.

6. H. J. Hageman, W. Gudat, C. Kunz, "Optical Constants from the Far Infrared to the X-Ray Region" DESY Report SR-7417, May, 1974.

7. M. C. Hutley and V. M. Bird, "A Detailed Experimental Study of the Anomalies of a Sinusoidal Diffraction Grating", Optica acta, 20 (1973) 771.

8. T. Namioka, H. Noda, and M. Seya, "Possibility of Using the Holographic Concave Grating in Vacuum Monochromators", Science of Light, 22 (1973) 77.

Holographic surface grating fabrication techniques

Anson Au, Hugh L. Garvin

Hughes Research Laboratories, Malibu, California 90265

Abstract

Diffraction gratings on the surfaces of optical substrates have been demonstrated to perform functions such as beam sampling and infrared polarization. These surface gratings are defined by holographic exposure and ion beam sputter etched into a gold layer on the substrate. The patterns can be designed with spatially varying periods to provide optical power in the diffracted beam and they can cover large areas on flat or non-flat surfaces. High grating accuracy and edge definition result in improved performance when compared to conventional fabrication techniques.

Introduction

Nearly ten years ago holographic grating pattern generation and ion beam sputter etching were applied to the fabrication of integrated optical compoments including wave guide couplers, distributed feedback solid-state lasers, and wire grid polarizers.[1] These early devices were small (approximately one square centimeter) but they involved submicrometersized pattern generation and etching capabilities that were not possible by prior art techniques. Now these same processes are being used to fabricate high power laser components which approach a square meter in size and there is no fundamental limitation to preclude the scale-up to even larger components.

Holographic grating fabrication

The basic steps in fabricating a holographic grating beam sampler are summarized schematically in Figure 1. The substrate mirror is immersed in a photosensitive resist material and coated uniformly by draining the tank at a controlled rate. After suitable drying the

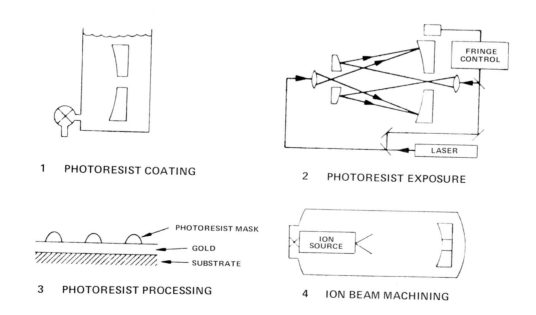

1   PHOTORESIST COATING

2   PHOTORESIST EXPOSURE

3   PHOTORESIST PROCESSING

4   ION BEAM MACHINING

Figure 1.   Grating fabrication steps.

resist is exposed in the interfering region of two laser beams to expose the appropriate grating pattern. The resist is developed in a solvent to remove the exposed portion leaving a masking pattern in areas of less exposure. Ion beam sputter etching is used to controllably remove material from the open areas thus impressing the pattern into the surface film. Finally the remaining photoresist is dissolved away to leave the grating imprint in the mirror surface. The precision with which each of these steps is carried out determines the reproducibility and quality of the optical performance.

The laser sampling mirrors made to date have generally been for infrared applications and gold has been a satisfactory reflective coating. It is readily etched as described in a separate paper[2] by argon ion beam sputtering and is fairly stable against deterioration while in use. Other coating materials such as silver, copper, aluminum, or platinum could also be used if the situation required it. These materials are also reflective to the UV laser exposing radiation (typically at the HeCd laser wave length of 4416 Å or the Kr wavelength at 4131 Å). As a result, standing waves are produced within the photoresist film to give exposure variations up through the thickness of the film. This is overcome by applying a film of compatible material to suppress the reflection at this metallic surface. Figure 2 illustrates the advantage to be gained in edge quality and exposure control which this layer provides.

GOLD SUBSTRATE*   AR COATED GOLD SUBSTRATE

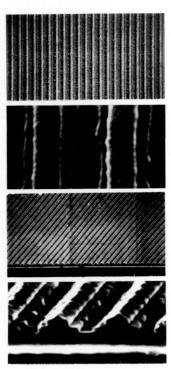 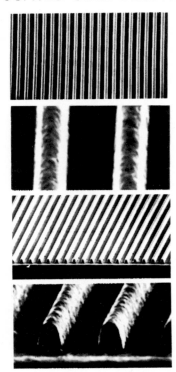

SHIPLEY AZ1350B
(0.4 μm)

AZ DEVELOPER 1:1
(2 MIN)

1 μm/CYCLE

55 mJ/CM^2 EXPOSURE
(4416 Å λ)

*NOTE EXCESS PHOTORESIST IN THE GRATING VALLEYS ON GOLD SUBSTRATE

Figure 2. Reduction of standing-wave effects in holographic exposure.

Photoresist films can be applied to small substrates by spinning or by uniform pulling out of a liquid reservoir. Larger and heavier substrates are coated by immersion into the photoresist liquid and draining the tank at a controlled rate. The thickness of the coating is dependent on the viscosity of the resist, the rate of level dropping along the surface, and the contact angle which the liquid makes to the substrate surface. In holographic exposure the intensity profile is nearly sinusoidal compared with the high contrast of contact exposure through an opaque mask, therefore the resist thickness should be made as uniform as possible if equal grooves and ridges are to be produced in the resist pattern. The coating tank (shown in Figure 3) is isolated from acoustic vibrations and the liquid flow rate is monitored by means of a venturi nozzle which senses the pressure differential and provides an input to a computer control of an electro-pneumatic drain valve. Over complex surfaces of spherical or parabolic mirrors the computer is programmed to provide a uniform coating based on the angle of the mirror surface and the cross sectional displacement of the mirror in the tank. Coatings in the range of 2000 to 5000 Å are made with less than 5% thickness variation over a 35 cm diameter surface.

Holographic pattern construction has significant advantages over contact exposure in several applications: (1) the generation of the interference pattern in the volume of the photoresist provides the capability to produce very high resolution pattern details.

Figure 3. Photoresist deposition system. Tank in background can
coat 35 cm diameter substrates.

Diffraction effects limit replication of master photo patterns to about 1 μm wide lines and spaces. Whereas, holographically exposed grating patterns have been made which are ten times smaller than this and the pattern details are also preserved in the photoresist development and ion beam etching steps. (2) The spatial phase of the hologram is provided by the construction optics and is uniform over the surface whereas a scribed photomaster pattern requires very precise machinery placement to assure uniformity of a pattern phase. Furthermore, even with a perfect master pattern, distortion can later occur in the replication process. (3) The holographic pattern can have optical power when produced by interfering spherical waves. These nonlinear patterns would be very difficult to generate by mechanical scribing; and (4) The holographic interference pattern exists in space and can be exposed on curved or irregularly shaped surfaces to which it would be impossible to apply a contacting master pattern.

Figure 4 shows schematically the manner in which two spherical waves have been used to produce a focussing grating pattern on a flat or curved mirror surface. The placement of the spatial filters is such that the exposed pattern will provide the desired focus of visible or infrared radiation in the final laser optical system. In this case the two converging laser beams have displaced axes and require the angle mirror to bring them into the proper interference. In Figure 1 an alternative exposure arrangement is shown in which the laser beams are coaxial through holes in the centers of the mirror to be exposed and the spherical reference mirror which is used to shorten the length of the stable platform used for the exposure system. As the size of holograms increases the power from a given UV laser

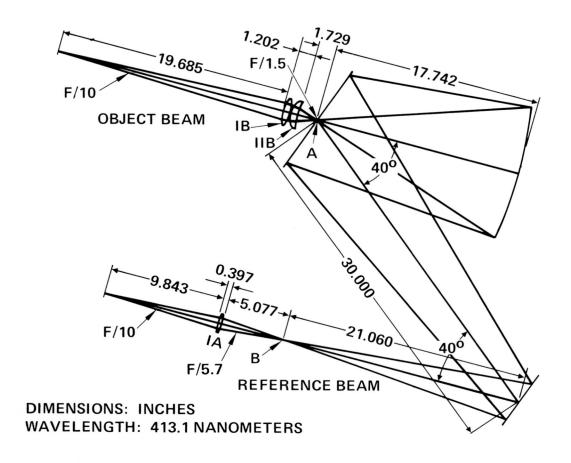

DIMENSIONS: INCHES
WAVELENGTH: 413.1 NANOMETERS

Figure 4. Holographic construction optics for focussing laser beam sampler.

must be expanded with the result that exposures are required which may be an hour or more in length. During this exposing time the interference pattern must not move by half a fringe period (approximately 5 µm) or the exposure will be washed out. To this end the platform which we have used (shown in Figure 5) is a 2 m x 4 m x 1 m thick granite slab on isolation dampers mounted on a floating floor which is separated from the rest of the laboratory flooring. Thermal and acoustic vibrations are reduced by enclosing the system in a padded housing. Long term (1 sec and longer) variations in the length of either laser beam are sensed by an optical monitor set to observe the fringe signal at the edge of exposing pattern. A piezoelectric crystal is used to displace a mirror in one of the laser beam paths and stability of better than 1/8th fringe have been used when making exposures as long as two hours. Figure 6 shows the focussing effect of this type of grating when illuminated with He-Ne laser light. With infrared illumination only one order will be brought to a focus and this can be used for laser beam steering control.

Conclusion

The high resolution pattern generation capability of laser holographic construction was combined with the precise dry etching of ion beam sputtering to produce large area optical beam sampling mirrors which are not possible to be made by other fabrication techniques. Further extension of these techniques will permit fabrication of optical components which are more than a meter in diameter.

Acknowledgement

The authors are pleased to acknowledge the capable assistance of Mr. Klaus Robinson and Mr. Cesar de Anda in performing these fabrication tasks.

References

1. Garvin, H. L., Garmire, E., Somekh, S., Stoll, H., and Yariv, A., "Ion Beam Micromachining of Integrated Optics Components," Appl. Opt., Vol. 12, p 455 1973.
2. Garvin, H. L., Au, A., and Minden, M. L., "Ion Etched Gratings for Laser Applica-

tions," paper presented at SPIE's Annual Technical Symposium, July 28-August 1, 1980, San Diego, California.

Figure 5.   Holographic exposure platform.

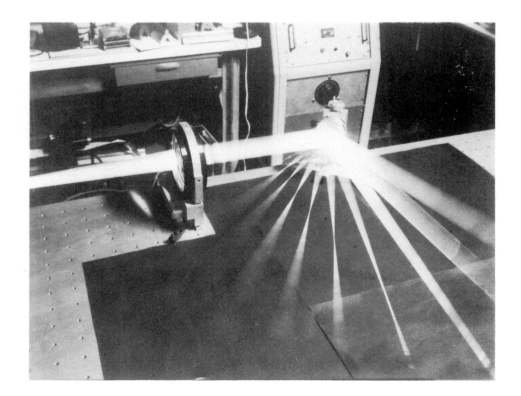

Figure 6.   Diffracted orders of nonlinear holographic grating.

Lithography with metallo-organic resists

A. C. Pastor, R. C. Pastor, M. Braunstein, G. L. Tangonan

Hughes Research Laboratories, 3011 Malibu Canyon Road, Malibu, California 90265

Abstract

Photolithography with metallo-organic resists is a relatively new addition to photo-engraving technology, and involves the chemical incorporation of inorganic constituents into photopolymerizable organic compounds, so that the photoresist functions not merely as a masking material, as in conventional photolithography, but also as the mass transference vehicle itself. The deposition of thin structured films of metal oxides with this method has been accomplished, the metal-doped resist in each case being the metal acrylate in acrylic acid, except in those cases where the metal acrylate was insoluble. Polymerization was effected with uv irradiation. The criteria for depositing other classes of inorganic compounds are outlined.

The conventional photolithographic process

The function of conventional photolithography in the fabrication of electronic and optical microcircuits is to deposit adherent stencil layers on the surface of the substrate material. The process of deposition is carried out by applying a coat of an organic, photopolymerizable liquid, called a photoresist, to the substrate surface, irradiating it through an optical mask, and subsequently fixing the latent image of the mask on the photoresist layer with a selective solvent, i.e., one that will discriminate between the remaining monomeric and the polymerized material. The surface that is now covered with the stencil layer is then ready for the deposition of a film of the desired inorganic material, or alternately, the etching of a thin layer of material off the exposed areas of the substrate surface. After this mass transference operation, the stencil layer remnants are lifted off by mechanical or chemical means.

The characteristics that qualify a photoresist compound for this application are that it must be adherent to the substrate surface in both the monomeric and the polymerized forms and that it must be chemically unreactive with the substrate and the inorganic film materials; obviously, in the case where it is used as an etching mask, it must be impervious to the etchant solution. Furthermore, its fixing solution must also be chemically unreactive with the substrate material.

The significant feature that may be gleaned from this general description of the conventional photolithographic process is that each deposition of a structured inorganic film is accompanied by one other material deposition and two material removal steps. In other words, four mass transference operations are employed to bring about a net effect of one mass transference operation. Surely, such a circuitous procedure cannot be conducive to good lithographic resolution and reproducibility, and when it is considered that the fabrication of some of the most popular microelectronic integrated circuits involves the deposition of six to nine inorganic layers[1], the need for the development of a less cumbersome lithographic process in microfabrication technology becomes apparent.

A modified lithographic process

If the inorganic material which is to be deposited as a structured film were chemically incorporated into the photoresist compound, its lithographic delivery could be accomplished in two mass transference steps, one a deposition and the other a removal step. In a first approximation to such a two-step process, it would be required that the resulting metallo-organic compound retain the photopolymerizability of the parent organic compound, as well as its (single-phase) homogeneity so that it will not tend to scatter the polymerizing radiation. To illustrate the latter point in the negative extreme, lithographic resolution and polymerizable film thickness would be severely limited if the modified photoresist were a turbid emulsion. Further requirements would be that the modified compound must be amenable to some fixing solution and to some post-deposition developing process which will chemically convert in situ the polymerized material to the desired film composition.

Photolithography with metallo-organic* resists (which, for brevity, will be referred to as mo-resists hereafter) for the purpose of depositing thin structured films of oxides of metallic elements has been demonstrated successfully with various elements.[2] Solutions of the acrylates of the metallic elements in acrylic acid were used as the basis for the mo-resist formulation, and their polymerization was effected by uv irradiation. Equations (1), (2), and (3) serve to illustrate the relevant chemical reactions. (See Figure 1.)

The fixing solution used was either water slightly acidified with acetic acid or a light alcohol, such as methanol or ethanol. Developing was carried out by stepwise firing in oxygen, firstly, to pyrolyze the polymerized mo-resist layer, as shown in Equation (4), Figure 1, secondly, to sinter its ash, and, finally, to partly diffuse the film of residue into the substrate surface.

The applicability of this process is confined to those metallic elements which may assume a valency greater than +1 and yet not so high a valency as to result in amphoteric behavior. The reason for the lower limit is that the acrylate of the dopant element must function as a cross-linking agent, as depicted in Equation (3), for the linear polymer that the acrylic acid would otherwise be, as shown in Equation (2), while that for the upper limit is that any tendency on the part of the oxide of the dopant element to display acidic behavior would tend to inhibit the acid-base neutralization reaction between itself and the acrylic acid, as illustrated by Equation (1). The lower limit may be relaxed, however, when a monovalent cation is only a secondary dopant, as in those cases where the desired residue is a binary oxide, e.g., Li_2TiO_3.

A further restriction to this process results from the presence of the metal-oxygen bond through which the metal is incorporated with the photopolymerizable organic matrix material. Because of the generally high degree of stability of this bond, the process becomes useful only in those cases where this bond may be left intact from the point of its formation to the end point of the process. In other words, with few exceptions, such as silver and copper, whose oxides are readily reducible, the acrylate-based process is useful only when the desired residue is the oxide of the metal. Some other organochemical matrix compound must be selected as the basis for non-oxide residues.

Extensions of the modified lithographic process

In order to deposit inorganic constituents in the elementary or alloy form, for example, it would be preferable that the inorganic-to-organic linkage be made through metal-carbon bonds. Such linkages are commonly found in the alkyl or the carbonyl compounds of the metals, which can be caused to yield their inorganic residue by photodissociation.[3] In this case, it is preferred that the organic fraction of the products of photodissociation be volatile, so that they may be removed from the reaction site without difficulty. This would not be likely if the organic component of the mo-resist were a polymer, nor would a polymerizable organic component be desirable, for this would be polymerized by the photodissociating radiation.

Depending on the reduction potential (with respect to hydrogen) at the processing temperature, certain metal oxides can be rendered by hydrogen into the free metal. If the process temperature is less than 830°C, carbon monoxide is an even better reductant.[4] Of course, the latter reductant is not applicable to metals which readily form carbonyls, e.g., nickel.

To deposit chalcogenides of the metal, chemical precursors with the corresponding metal-chalcogen bond would be the practical choice. For example, if the target residue material were the sulfide of the metal, a mo-resist with the M-S bond would be employed. The developing process would be carried out under an atmosphere of carbon disulfide, hydrogen sulfide, or plain sulfur vapor (or, perhaps, some mixture of these).

Free metals and metal-oxide deposits can be converted into their stable sulfide form with either hydrogen sulfide or sulfur. It is kinetically advantageous to carry out the conversion while the residue from the pyrolysis is in its most active form, i.e., prior to sintering. For those metals which are less reducing than hydrogen, hydrogen sulfide is an ineffective reactant; oxidation of the metal to the stable sulfide is best achieved with sulfur. Compared to sulfur, hydrogen sulfide is the more appropriate reactant for the metal-oxide form.

---

*Strictly speaking, metal-to-carbon bonds must be present in a compound for it to fall into the category of a metallo-organic compound as defined by organic chemists; the term metallo-organic is used here loosely, for the sake of convenience.

The acrylate of a bivalent metal forms as follows:

$$2CH_2=CHCOOH + MO \rightarrow CH_2=CHCOO-M-OCOCH=CH_2 + H_2O \qquad (1)$$

Acrylic acid polymerizes according to the reaction,

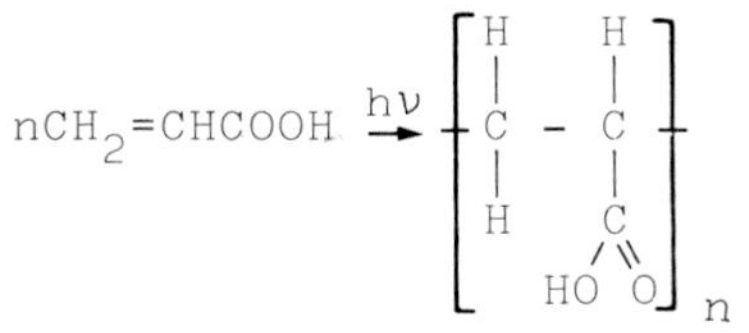

$$nCH_2=CHCOOH \xrightarrow{h\nu} \left[\begin{array}{cc} H & H \\ | & | \\ C & - C \\ | & | \\ H & C \\ & \diagup\!\!\backslash \\ HO & O \end{array} \right]_n \qquad (2)$$

or   or

$$\begin{array}{ccccccc}
H \quad H & & H & H & H & H & H & H \\
\diagdown \; \diagup & & | & | & | & | & | & | \\
C{=}C & & -\;C & -\;C & -\;C & -\;C & -\;C & -\;C \; \text{---}\; \text{to n units}\\
\diagup \; \diagdown & & | & | & | & | & | & | \\
H \quad C{=}O & & H & C & H & C & H & C \\
\quad \diagup & & & \diagup\backslash & & \diagup\backslash & & \diagup\backslash \\
HO & & HO & O & HO & O & HO & O
\end{array}$$

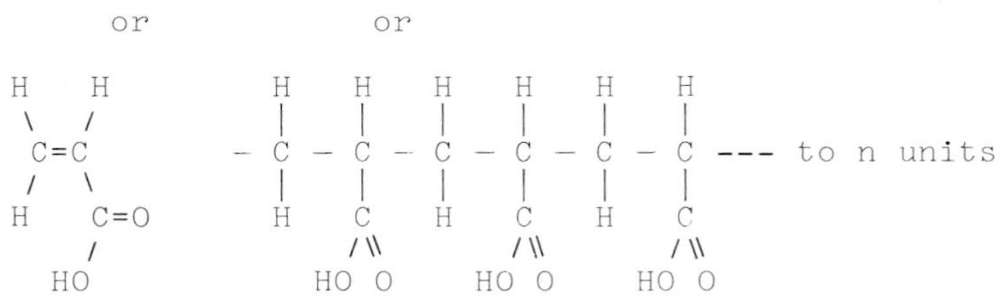

mer, or unit

The acrylate of the bivalent metal acts as a cross-linking agent, as is illustrated in the reaction,

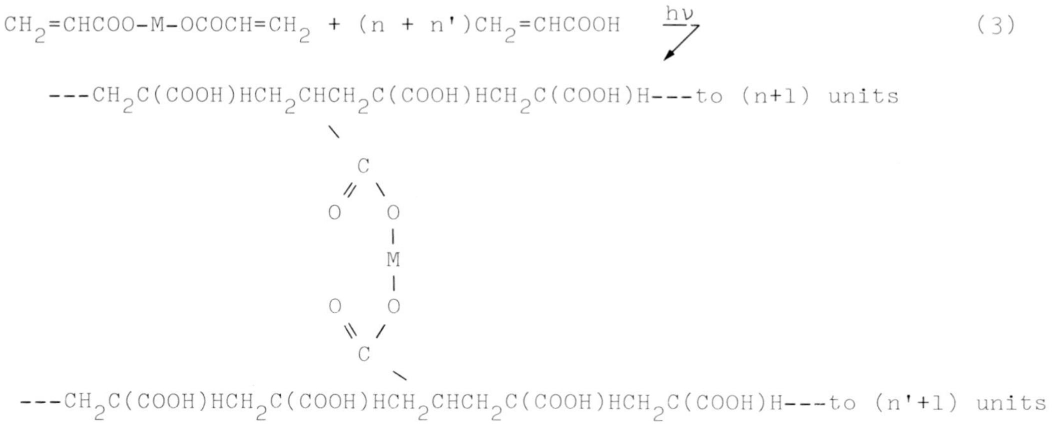

$$CH_2=CHCOO-M-OCOCH=CH_2 + (n + n')CH_2=CHCOOH \xrightarrow{h\nu} \qquad (3)$$

$$\text{---}CH_2C(COOH)HCH_2CHCH_2C(COOH)HCH_2C(COOH)H\text{---to }(n+1)\text{ units}$$

$$\begin{array}{c}
\diagdown \\
C \\
\diagup\!\!\diagup \;\diagdown \\
O \qquad O \\
| \\
M \\
| \\
O \qquad O \\
\diagdown\!\!\diagdown \;\diagup \\
C \\
\diagdown
\end{array}$$

$$\text{---}CH_2C(COOH)HCH_2C(COOH)HCH_2CHCH_2C(COOH)HCH_2C(COOH)H\text{---to }(n'+1)\text{ units}$$

The chemical formula of the product in Equation (3) is written in a different form to illustrate its pyrolysis under oxygen in the following reaction:

$$\left\{ CH_2C\left[COO\left(H_{1-x}M_{x/2}\right)\right]H \right\}_n + 3nO_2 \xrightarrow[\sim400^\circ C]{\Delta} 3nCO_2\nearrow + \frac{n(4-x)}{2}H_2O\nearrow + \underset{\text{ash}}{\frac{nx}{2}MO\downarrow} \qquad (4)$$

Example: For $M = Pb^{+2}$ the value of $2n/x$ in Equation (4) at the room-temperature saturation point of lead acrylate in acrylic acid is approximately equal to 12.

Figure 1. Chemical reactions in the acrylate process.

An analogous lithographic process could be developed for depositing films of compounds of the metals with the elements in the nitrogen group. To deposit compounds of the metals with the halogens, the chemical precursors to use may be almost any of the preceding mo-resists. The metal may be deposited in elementary form and subsequently converted to the halide by developing under an atmosphere of the halogen gas or the hydrogen halide, depending on whether the metal is above hydrogen or not at the process temperature in question. Other precursors may be used, if they are convertible to the halide by some developing process.

Copper and silver are below hydrogen in the emf series. Hence, these metals do not form their respective chlorides at 25°C by the displacement of hydrogen from an aqueous solution of hydrogen chloride. However, at red heat copper and silver react with gaseous hydrogen chloride to form cuprous chloride and silver chloride, respectively.[5]

Compounds of the metals with the Column IV elements pose a quite different problem because the degree of ionicity of those compounds is generally low. It is best to regard them as intermetallic compounds or alloys with specific stoichiometry. Therefore, the lithographic route to follow in this case would be that which has been described, albeit briefly, for the deposition of films of metals or their alloys.

Some comment should be made at this point regarding the requirement of nonpolymerizability of the mo-resist to be used for the deposition of metal films. Irradiation and developing occur concurrently in this process, and therefore fixing comes after developing, as in the usual photographic process. Irradiation in this case is better done with high-intensity focussed beams rather than with low-intensity parallel beams through masks, which is all for the better since it is with the former that the lithographic details of submicron size necessary for very large scale integration are obtainable.

References

1. According to Toombs, D., as paraphrased by Robinson, A. L., in "New Ways to Make Microcircuits Smaller," Science, 208(1980), 1019-22.
2. Pastor, A. C., Tangonan, G. L., Pastor, R. C., Wong, S. Y., and Chew, R. K., Lithography with New Metallo-Organic Photoresists, Thin Solid Films, 67(1980), 9-12.
3. George, P. M., and Beauchamp, J. L., Deposition of metal films by the controlled decomposition of organometallic compounds on surfaces, Thin Solid Films, 67(1980), L25-L28.
4. "Treatise on Inorganic Chemistry," by Remy, H., translated by Anderson, J. S., and edited by Kleinberg, J., (Elsevier, 1956). See page 444 of Vol. I.
5. Op. cit. in Reference 4. See pages 373 and 395 of Vol. II.

Stabilization of the exposing interference pattern in holographic grating production

Lars-Erik Nilsson, Hans Ahlén
Department of Physics II and Institute of Optical Research
The Royal Institute of Technology, S-100 44 Stockholm, Sweden

Abstract

When fabricating holographic gratings the exposing interference pattern must be stationary. Normally this is achieved by working in a very stable interferometric environment where special care has been taken regarding low dependence of temperature, stable mechanical and optical components. Despite considerable precautions in these respects, reproducible results are difficult to achieve, especially considering the fabrication of multiple exposure gratings i.e. Fourier synthesis gratings.

In our laboratory we have developed a technique for fabrication of holographic gratings by using a reference grating and a servosystem to keep the interference pattern stationary. The reference grating monitors both frequency and phase errors. This method allows us to achieve a high outcome in fabrication without utilizing replicas. It is also possible to reproducibly make Fourier synthesized gratings with selectable phase between the different harmonics.

Introduction

One of the most important prerequisites in the process of fabricating holographic diffraction gratings is to keep the exposing interference pattern stationary during the exposure. This is necessary in order to obtain reproducible results. Any movement during exposure will result in a reduction of the effective exposure modulation and thus of the groove depth. When making Fourier synthesized blazed holographic gratings, by subsequent exposure of sine pattern on the same substrate [1,2], it is not only necessary to keep the exposing patterns still but also to locate them in a predetermined way to each other.

In this paper, we want to describe an active servosystem which automatically counteracts any movements of the interference fringes during exposure, and which also permits multiple exposures with arbitrary phase angles between the different exposing patterns.

Instabilities in the exposing system

There are many factors that can cause movement of the interference fringes during exposure. Shifts in the axial output mode from the laser can occur, thermal drift of the components of the two interfering beams, further mechanical disturbances from the surroundings (airborne or propagated through the mechanical structures) can excite the optical components and thus cause the exposing pattern to vibrate.

Several precautions against these factors have to be taken when designing and constructing an interferometric setup. Thus all optical components should be mounted on heavy cast iron mounts that have large mass and are difficult to excite. The entire setup should be placed on a heavy vibration isolated table and thermally isolated from the surroundings. Despite these precautions it is our experience that it is difficult to obtain reproducible results.

Monitoring an interference pattern

There are several methods to monitor the phase of a two-beam interference field. The direct observation of the individual fringes in the interference pattern is difficult. The fringes are very narrow and contain little energy. However, a great number of fringes can be observed simultaneously by the Multiple-Sine-Slit method devised by Biedermann and Johansson [3]. In this method, the fringes are projected on a reference grating of the same or almost the same spatial frequency and angular orientation, and a moiré pattern between the reference grating structure and the structure of the illuminating pattern is observed. The moiré pattern can also (perhaps better) be interpreted as an interference phenomenon. If a grating is illuminated with two monochromatic coherent beams of light, the various diffraction orders will interfere with each other. If two beams intersect at an angle which gives an interference pattern with approximately the same frequency as that of the grating, first order diffraction from one of the beams will interfere with zero order diffraction from the other, with a visible interference pattern as a result. The difference in frequency between the illuminating pattern and the grating appears as fringes of varying intensity. Deviation of angular orientation appears as a skewness of the fringes.

When the frequency and orientation are completely matched the interference pattern consists of an uniformly illuminated zero fringe. The relative intensity of the fringe indicates the phase difference between the grating structure and the illuminating pattern. The result will be the same when the illuminating pattern is two or an integer multiple of the grating period, except that other diffraction orders are interfering.

By providing the holographic setup for making gratings with a reference grating, it is thus possible not only to observe and monitor the fringe movement but also to adjust the exposing pattern relative to the grating to the very high degree of accuracy (10^{-6} to 10^{-7}), which is necessary for multiple exposure gratings.

Fig. 1 shows such a reference grating obtained when the exposing pattern is of slightly different frequency and orientation. In the figure the reference grating is located on the actual substrate of the grating which is to be made. This ensures that the position between the reference and the exposing pattern is constant, over the surface, during the entire manufacturing process. When making single exposure gratings (i.e. gratings with sinusoidal profiles) it is usually sufficient to locate a small reference grating on the substrate holder outside the actual grating substrate. The only requirement on the reference grating is that the diffracted wave fronts should be of interferometric quality. Diffraction efficiency and straylight are of minor importance and the grating need not be coated with a reflecting layer. The grating can also be used in transmission, which is convenient under some circumstances.

Figure 1.   Photograph of a diffraction pattern from a reference grating indicating a slightly different frequency and orientation of the exposing pattern.

In order to be able to adjust the fringe location of the exposing interference pattern, the pathlength difference between the two interfering beams must change. There are several ways to achieve this, and we have found that electromagnetically displacing a small mirror in one of the beams is both simple and reliable up to frequencies of a few hundred Hz.

Active control of the phase of the exposing pattern

Figure 2 shows a schematic view of the phase control system in the holographic setup. Light from the laser passes an electromechanical shutter (SH) and is subsequently divided into two beams by a beamsplitter arrangement. The beam, coming from the beamsplitter BS, is reflected at nearly normal angle at the displacement mirror M1. The two beams are then focused, filtered and collimated in an ordinary way and brought to intersection on the photoresist covered substrate surface (S). The reference grating is situated either on the substrate or beside the substrate.

The light diffracted from (or transmitted through) the reference grating falls onto a silicon photodetector (D1) which continuously monitors the phase between the exposing pattern and the reference grating. The signal from the detector is amplified and fed to a microcomputer. If the exposing pattern moves the detector will see a change in irradiance. The computer checks the degree of movement and reacts by displacing mirror M1 an appropriate amount. The signal from detector D1 is normalized through division with a signal proportional to the total irradiance on the substrate. This signal is provided by a second detector D2. In this way a change in laser power will not influence the location of the

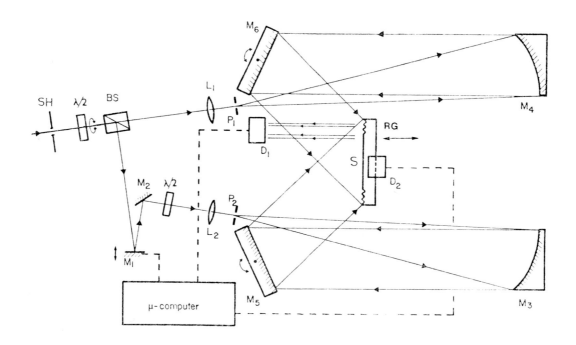

Figure 2.   A schematic view of the phase control
             system.

exposing pattern. The total exposure is controlled by integration of the signal from detector D2 to a predetermined level.

    Before exposure, data regarding exposing wave length, total energy required and phase
angle between exposing pattern and reference grating, are fed into the computer. At the
start command the computer opens the shutter and starts to oscillate the moveable mirror
M1 back and forth so that the exposure pattern moves several periods over the reference.
The intensity at D1 will then pass several maxima and minima. When the mirror oscillates
the computer samples and calculates the mean-value of the maxima and the mean-value of
the minima from the detector D1. From these values the computer determines what the
irradiance on D1 should be in order to correspond to the demanded phase angle. After this
initial phase, which takes about 0.5 s, the computer goes into control mode keeping the
pattern stationary until the exposure is terminated when the computer closes the shutter.
In this manner, the exposing pattern can be kept stationary for any length of time re-
gardless of temperature fluctuations and mechanical vibrations. The system can compensate for phase variations of any amplitude and with frequencies up to about 10 Hz. Some
residual phase variations remain, however, due to the finite response time of the control
system.

    These variations, which occur at high frequencies, however, are of minor importance
since they do not exceed one tenth of the grating period (such a phaseshift would, even
if it were present at fifty percent of the exposure time, not decrease the fringe modulation more than five precent).

    Fig. 3 shows the electronic control system used in our laboratory for making hologra-
phic gratings. The instrument in the middle of the right row is the described computer.
The keys to the right are used for control and input of exposure parameters (phase, total
exposure,exposing wavelength etc.). The display shows the phase shift corrections fed to
the moving mirror which is a measure of the stability in the setup.

Frequency and orientation control

    The described system above efficiently compensatesany phase variation of the inter-
ference pattern during exposure. The exposing pattern is always stationary in phase. In
the fabrication of Fourier synthesized gratings two additional factors are important for
a successful result, namely the frequency of the harmonics and their angular orientation
relative to the fundamental frequency (the lowest frequency of the final Fourier blazed
grating). In order to ensure that the groove profile is uniform over the grating surface,

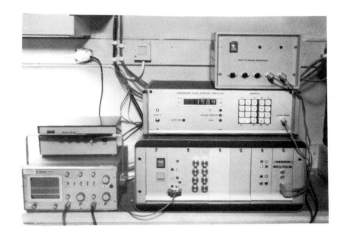

Figure 3. A photograph of some of the electronics in the exposure control system.

the harmonics must be defined to less than one tenth of the grating constant over the entire surface. For a grating of 1000 grooves/mm, with a diameter of 100 mm, this implies an angular accuracy between the two interfering beams of the order of few tenths of a microradian. That is an angular accuracy of one millimeter in ten kilometers. The angular orientation of the exposing pattern relative the reference pattern, is in the order of two microradians.

With the aid of a system similar to the one above, it is possible to electronically adjust the different exposing patterns to this high degree of accuracy. In this case the reference grating is a narrow rim on the substrate from which the multiple exposure grating will be formed. In the diffraction pattern from this reference, four detectors are orientated symmetrically.

The two collimated interfering beams are reflected from the two high quality plane mirrors M5 and M6, one of which can be rotated through an axis parallel to the fringes and the other one through an axis perpendicular to the fringes. For coarse adjustments the mirrors can be moved mechanically, while they are moved with electromagnets for fine corrections. In this way, the two plane mirrors are adjusted manually for "zerofringe" from the reference grating. A change in irradiance on the detectors can be attributed to three factors.

1) a phase shift between the interfering beams.
2) a change in frequency of the interference pattern due to an angular shift between the interfering beams perpendicular to the grooves of the reference grating.
3) as 2) but with a shift parallel to the grooves.

The signals from the four detectors are decoded by the associated electronics and resolved into two signals corresponding to 2) and 3) above. Here we discard the information about phaseshift since here phase changes are taken care of by the phase control described earlier. The mirrors are displaced, by means of electromagnets fed from the electronics in a manner which counteracts any change of frequency and orientation in the exposing pattern.

The system is designed to keep the patterns located with a deviation of less than one tenth of the grating period. Correction of the pattern is performed every second.

Results and discussion

The phase control system drastically changed the effective interferometric stability of the holographic setup. At first, due to unavoidable thermal and mechanical disturbances in the laboratory, the interference pattern moved up to two fringes per hour. We also then encountered a permanent shift of the interference pattern which is difficult to prevent. These factors made the manufacture of holographic gratings often a game of chance. With the incorporation of phase control these factors were efficiently removed. Consequently now production, of single exposure gratings, runs with some 80 % yield (nearly 100 % in the exposure part of the production). Earlier the yield was only some 20-25 %. Our phase control system can also be adapted to other interferometric systems.

A prerequisite for a systematic and reproducible exploitation of the Fourier blaze technique [4], however, is a system for monitoring stability of frequency and angular orientation. Before the introduction of this system the making of Fourier synthesized gratings was an extremely delicate and sensitive process.

We have thus demonstrated that it is possible to make a large quantity of high quality single exposure holographic gratings with very high yield. It still remains to determine the yield for production runs of multiple exposure blazed gratings. Our results, however, appear promising in this respect.

References

1.   Johansson, S., Nilsson, L.-E., Biedermann, K. and Kleveby, K. "Holographic Diffraction Gratings with Assymetrical Groove Profiles", Applications of Holography and Optical Data Processing, Edited by E. Marom, A.A. Friesem and E. Wiener-Avnear, Pergamon Press 1977, p. 521
2.   Breidne, M. and Johansson, S. "Blazed Holographic Gratings", Optica Acta, Vol. 26, p. 1427, 1979.
3.   Johansson, S. and Biedermann, K. "Multiple-Sine-Slit Microdensitometer and MTF Evaluation for High Resolution Emulsions", Appl. Opt., Vol. 13, p. 2280, 1974.
4.   Breidne, M. and Maystre, D. "Optimization of Fourier Gratings". In this issue.

Triangular and sinusoidal grooves in holographic gratings—manufacture and test results

Erwin G. Loewen, Leon Bartle

Bausch & Lomb, Inc., 820 Linden Avenue, Rochester, New York 14625

Abstract

Photographic recording of interference fringe fields has long been recognized as a method of producing diffraction gratings, with the potential of avoiding residual mechanical effects that occasionally inhibit the performance of ruled gratings. Practical realization required the development of suitable light sources (in the form of ion lasers), and a photographic medium capable of accepting smooth, fine pitch modulation (in the form of photoresist). However, to expand from this base and produce high grade diffraction gratings requires much more than standard holographic techniques. Some of the techniques are described. Since reduction of stray light is one of the prime reasons for adopting a direct interferometric approach to making gratings, it follows that a corresponding figure of merit ought to be adopted, capable of being determined by a simple procedure. One such approach is suggested and results given. It indicates no difference between the performance of a master and high grade replicas derived from a chain of several generations, but does indicate a difference between holographic and mechanically ruled gratings.

Introduction

Holographic gratings have established themselves over the last decade as an interesting and useful alternative to mechanically ruled gratings. In the form of plane gratings they offer, as a principal advantage, reduced amounts of in-plane stray light. In the form of concave gratings they offer, in addition, possibilities for reducing imaging aberrations that have long detracted from their usefulness.

The concept of making diffraction gratings by photographic recording of standing wave fields is quite old. Both Lord Rayleigh[1] and Michelson[2] suggested the possibilities, the former actually producing some that still exist[3]. There was no follow up for the simple reason that even the finest grained photographic emulsion is much too coarse to lead to a clean, efficient grating. In addition, for full sized gratings one requires a very intense, highly monochromatic and coherent source of light. Argon (or Krypton) ion lasers, which became available in the mid 60's, represent the first fully qualified source. Photoresists, developed for the microelectronic industry, provided the second key, in the form of a molecular rather than a grain type of recording structure. This provides almost infinite resolution.

The field of holography began with other lasers and with HR emulsions. Holograms have always been thought of as a special kind of diffraction grating, which may explain the origin of the term "holographic" used for gratings made by techniques borrowed only in part from holography. A plane grating can be described as a hologram of a pinhole at infinity[4]. A case could be made, but presumably too late, to use instead more descriptive terms such as interference, photoresist or photomechanical, to identify this type of grating.

Since equipment for making holograms is available in many labs, there is sometimes the impression that producing diffraction gratings is so simple a procedure that it can be accomplished in a few days. As will be detailed below, the degree of perfection demanded of instrument grade diffraction gratings is so great that normal holographic techniques are just a starting point. Techniques for measuring departures from perfection are so well developed that problems observed in minutes may take months to overcome.

Early holograms achieved the necessary phase modulation in transmitted light mainly through variations in refractive index in photo emulsions rather than by surface modulation. If that were still a limitation today, holographic diffraction gratings would be restricted to the extremely inconvenient Bragg mount, familiar from x-ray crystal spectrometers.

Fortunately methods were developed that led to the required periodic surface modulation. The groove shape most readily attained is a close approximation to a sinusoid. While very useful, it imposes some significant restrictions on diffraction efficiency behavior, compared to the traingular groove shape typical of mechanical ruling.

However, several methods have since been developed that lead to interferometric generation of gratings with triangular groove shape, with a spectral efficiency performance that, in some cases, is virtually the same as that of a ruled grating.

Optical systems

To generate high resolution sinusoidal groove gratings requires a stationary interference fringe field, with nodal planes evenly spaced, normal to the face of the photoresist coated master. After a suitable exposure, followed by a chemical development that produces a corresponding surface modulation, only a metal overcoating is required to convert this into a reflection grating. However, to achieve a quality grating requires attention to a great many details. It is necessary to start with the optical system.

The required uniform fringe field results from the intersection of two perfectly collimated beams of laser light, and in its simplest version is achieved by the optical schematic of Figure 1. The simplicity is based on the absolute minimum of critical optical components, i.e. two, in the form of the collimating mirrors. Since they are used off-axis they need to be figured as off-axis paraboloids. To achieve the desirable wavefront perfection of $\lambda/4$ (at 400nm) on a, say, 200mm diameter paraboloid is not a simple undertaking. To make matters more difficult for the optician he must simultaneously obtain a very fine polish. Since the grating is a hologram it will faithfully record on its surface everything it "sees". This includes not only polishing marks left on the mirrors, but every particle of dust on their surface. Furthermore, unless mirror edges are properly masked, Fraunhofer diffraction from the edges is readily included. All such effects end up as some form of stray light, the very factor these gratings are intended to minimize. It is remarkable easy to observe such disturbances qualitatively by eye. All it takes is laser illumination of the finished grating from a distant pinhole. The observer can see the holographic reconstruction of any part of the optical system that is not perceived as completely black, simply by placing his eye near any of the diffracted orders. Needless to say, a low power laser is recommended for safety reasons. It has been suggested by Hutley[5] that this property of holograms could be used to convey any appropriate message, e.g. logos, into the background of a grating. In a sense all this extra information can be considered to be a family of auxilliary gratings, super-imposed on the desired gratings. Fortunately, their contribution as light at the exist slit of a spectrometric system can, with care, be held to very low levels. However, an optical accident can easily lead to levels high enough to end up with classical Rowland ghosts. In addition, light scattered from the edges or the rear surface of a blank will quickly degrade any grating, and must be completely suppressed.

An important point is that perfect fringe fields can also be obtained from a pair of imperfect wavefronts, provided that their errors match in phase space. Schmahl and Rudolph[6] have developed an elegant technique for doing this that depends on generating a pair of special holographic mirrors, which can lead to excellent fringe fields from commercial quality paraboloids, although a good polish is still necessary.

Since the groove spacing d of a holographic grating is given by $d = \lambda_0/2 \sin \theta$, where λ_0 is the wavelength of the laser light and θ the half angle between the intersecting beams, the construction of Figure 1 implies some limitations. The lowest groove frequency is determined by the closest approach between the mirrors and the highest by the maximum usable value of θ (around 60°) and possibly by the size of the optical bench. It is therefore, desirable to sometimes add additional beam steering mirrors to the optical system, as in Figure 2. Compared to Figure 1, this approach uses a single, relatively large collimating mirror instead of a pair of off-axis ones.

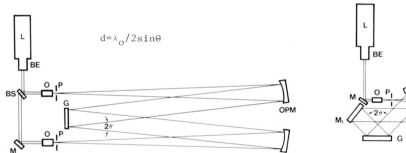
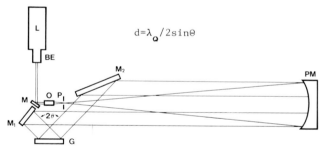

Figure 1. Optical schematic of a 2-collimator interferometer for generating holographic gratings. L = Laser, BE = Beam Expander, O = Micro Objective, M = Mirror, P = Pinhole Spatial Filter, G = Grating Blank, OPM = Off Axis Parabolic Mirror.

Figure 2. Optical schematic of a single collimator system with beam stearing mirrors M_1 and M_2 for generating holographic gratings. PM = Parabolic Mirror.

Several techniques are available for adjusting θ to obtain a specific groove spacing. The most convenient is to place a mechanically ruled grating of the desired pitch into the grating mount and adjust until the resulting moiré fringe pattern is fluffed out. Moiré fringes will also occur when frequencies are even multiples of each other. Such a test will also reveal improper collimation, i.e. if the pinholes are not located at the focal point of the collimating mirrors, one of the critical adjustments.

Air path lengths for a grating type optical bench will be quite long, typically 2m or greater. Therefore, turbulence or temperature stratification can easily upset optical stability. One solution is to shield the entire system with a cover, but good results have been achieved without, provided the laboratory is properly insulated and temperature controlled.

Like any multicomponent interferometric system, this one will be sensitive to mechanical vibrations from the outside world, especially over the 2 to 100 Hz range. Interference fringe field motion of d/2 is sufficient to completely eliminate useful modulation and d/10 is about the maximum acceptable. For 3600g/mm gratings this criterion allows only 28nm (1.1µ-inch) fringe motion during an entire exposure. It is clear that vibration isolation of the optical bench is an important feature, and requires a low natural frequency. Air cushions are the most common approach.

It is hardly surpirsing that mechanical stability, and temperature and vibration environment demands are about as stringent as those required for mechanical grating ruling engines, except that constancy need not be provided for so long a period.

Lasers

Most holographic grating work has been done with Argon or Krypton ion lasers and Helium-Cadmium types. The 457.9nm Argon line is most frequently used because it has the best convolution of output intensity with photoresist sensitivity. The equation above shows that at this wavelength a 111° included angle between the beams leads to 3600g/mm frequency. The 350.7nm line of Krypton allows groove frequencies 30% greater, which means that gratings up to 5000/mm are just about attainable. Finer pitches require still shorter wavelengths, which can be obtained by frequency doubling of the 514.5 Argon line for example. If the gratings required are small enough to make them on the face of a special prism (Figure 3) then, as Shank and Schmidt have shown[7], the effect is to reduce wavelengths by a ratio of 1.5. By this approach one can obtain acceptable modulations at frequencies up to 10,000/mm, using the 325nm line of a He-Cd laser.

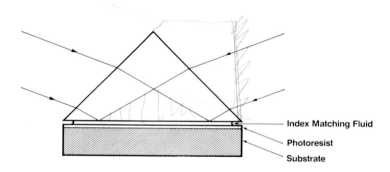

Figure 3. Generating maximum frequency diffraction gratings in a high index medium.

The laser must maintain a very high degree of absolute frequency stability, at least 200MHz, which is readily attainable, but only with the use of a stabilized Fabry-Perot etalon in the cavity. In addition, mode hopping must not be permitted, and this may require operating the laser at less than full power. Coherence length will be so great that no attention need be paid to maintaining equal path lengths.

Spatial filtering is essential, as it would be for ordinary holography, and follows the same rules.

Photoresist

There are two classes of photoresist, negative and positive. With negative resists the portion exposed to light becomes insoluble under the action of a developer and with positive resist the opposite takes place. Both have been used to make gratings, but, in general, positive resists of the diazo type are much more useful. Their resolution and uniformity is better. Like all resists, they were formulated originally to act as masks for microcircuit patterns, which call for developing down to the substrate. While this approach can also be applied directly to gratings, it leads to grooves approximately trapezoidal in shape, hardly a desirable result, because it is difficult to get uniformity over a large area when the groove spacing is fine enough to be interesting, (i.e. 600/mm or finer). Also efficiency behavior is not ideal. Fortunately, underexposure of positive resist, compared to its normal use, leads to a uniform surface modulation that is close enough to being sinusoidal for most practical purposes. The characteristic curve of positive resist, i.e. the amount removed during a standardized development cycle as a function of exposure, is roughly exponential in shape. It starts off so slowly as to imply a threshold, then increases to roughly linear behavior. An appropriate pre-exposure of the resist can take it up to the beginning of the linear range, so that a good sinusoidal groove shape is attained from the $\sin^2$ intensity field of the interferometer. In most practical instances sinusoidal grooves are more useful than those containing higher harmonics. But there are cases where lamellar groove shapes are desirable. They can be derived from resist patterns developed to the substrate.

Applying the photoresist is a critical step, since any departure from uniformity will be reflected in performance of the grating. The method most often mentioned is rotary spinning. Equipment can be "borrowed" from the microelectronic industry, but the quality required in terms of smoothness exceeds their standards. Details are rarely described in print, partly for proprietary reasons, partly because everyone has to develop his own techniques. An obvious problem arises when square or odd shaped blanks need to be coated. Surface tension discontinuities in corners will disturb flatness in their vicinity, no matter how careful the spinning. If gratings are to be generated on large, heavy blanks, spinning becomes even less attractive. The obvious alternative is dipping. This calls for withdrawing the blank at a slow but very steady rate from a resist filled tank, or, in the case of large blanks, emptying the tank at a controlled rate. A high degree of cleanliness is called for at every step, the blank surface, the resist itself and the environment.

Additional processing steps are baking the resist before exposure (to remove any solvent) and again after development (to improve adherence). Resist suppliers typically offer a choice of two developers, to which is added a choice in the degree of dilution. Once standardized, together with time and temperature, most of the combinations work equally well, which may explain why many authors publish firm recommendations, no two of which are alike. An important and useful difference between resists and photographic emulsions is that they are used in layers only 1/10 the thickness or less, and contain no gelatin that swells or contracts.

Blazed holographic gratings

Sinusoidal groove shapes achieve first order diffraction efficiencies comparable to ruled gratings only when diffraction angles exceed about 20°, a region where polarization factors are quite strong. For some applications, such as dye laser tuning or Raman spectroscopy, the resulting short useful tuning range (typically 2:1) is entirely adequate. Longer tuning ranges are attainable, but only by sacrificing diffraction efficiency. However, in many monochromator and spectrophotometers the wavelength range of 3 or 4:1 must have a high efficiency peak, usually near the low end. Such performance is attainable only with triangular groove shapes, and explains the desirability of generating them by direct interference techniques in order to have the best of both worlds.

Several approaches have been tried, some of them very elaborate. The simplest one by far is to record a standing wavefront system[5],[8] with the blank inclined at a relatively small angle, as shown in Figure 4. The concept is a very old one, analogous to Wiener's 1890 classical experiment in which he used photographic emulsion and was concerned not with the resulting grating, but with settling the question of which electromagnetic vector was photographically active. Again the laser serves to provide good coherence and high intensity while the photoresist serves as a clean recording medium. Developing the resist, as indicated in Figure 4b, will lead to the desired triangular groove shape. This approach works only with positive resists, as it depends on development to stop whenever it reaches the unexposed planes that correspond to the fringe field antinodes.

It is obvious from Figure 4 that the inclination angle θ of the blank, for a given laser wavelength λ_o, determines the groove spacing d: $d = \lambda_o/2 \sin \theta$.

Groove frequencies as coarse as 100/mm (d=10μm) are readily attainable, but in the fine pitch direction an upper limit of 3000 to 3600/mm is set both by the angle at which total internal reflection prevents light being passed through the blank and by the shortest usable wavelength from a laser.

Not immediately obvious is that all gratings made by this arrangement have the same blaze wavelength. The spacing of the antinodal planes within the photoresist will be $\lambda_o/2n$ where n is the refractive index of the resist. The first order Littrow blaze wavelength λ_B is always twice the groove depth, i.e. $\lambda_B = \lambda_o/n$. Thus for a typical resist index of 1.7 and a 457.9nm laser wavelength the gratings will be blazed around 270nm, with slight non-linearities leading to actual figures 10% less. Shorter blaze wavelengths are attainable with shorter wavelength lasers and with some control over the developing cycle. Thus this process leads to gratings well matched to the needs of UV-VIS spectrophotometers. For blazing at longer wavelengths the limit is set not by available lasers but by the very rapid drop in the sensitivity of all know photoresists to wavelengths exceeding 500nm.

It should be noted from Figure 4 that in contrast to Figure 1 the blank is now an important part of the optical system, and for full resolution must not deform the wavefront more than $\lambda/4$, nor have significant surface defects on either face. Unless special measures are undertaken to reduce the spatial coherence of the laser light it is also important that the rear surface have a high efficiency anti-reflection coating.

It is possible to replace the simple optics of Figure 4 by the more complex system of Figure 5. Variation in the direction of the beam entering from the rear leads to only minor changes in the blaze angle unless the edge of the blank is polished to a steep angle prism. In such a case it becomes possible to bring in the rear beam at more nearly grazing angles, the effect of which is to increase, up to a factor of two, the blaze angle for a given groove frequency[9].

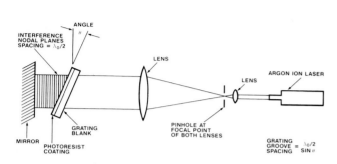

PRODUCTION OF BLAZED HOLOGRAPHIC GRATING

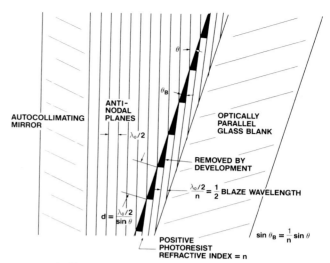

INTERACTION OF NODAL PLANES
TO DETERMINE BLAZE ANGLE AND WAVELENGTH

Figure 4a.   Optical schematic of inclined wavefront recording system for generating blazed gratings in photoresist.

Figure 4b.   Enlarged view of nodal plane recording system. θ_B is the blaze angle.

The necessary opto-mechanical stability is the same as that required to produce sinusoidal gratings. A reference ruled grating, in transmissions form, can again be used to check set-ups. The moiré fringes that are generated from the interaction of the interference fringe field and the grating grooves correspond each to one groove spacing. An observer can detect 1/20 of a moiré fringe by eye, which corresponds to an optical instability of 0.04μm (1.5μ-inches). This level must be maintained during the exposure cycle, that may take from 1 to 30 minutes, depending on the laser and size of the grating. It is an excellent technique for monitoring stability.

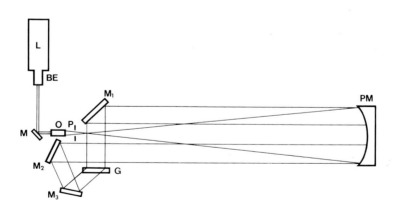

Figure 5.   Two beam recording system for generating blazed photoresist gratings. M_1, M_2, M_3 are the recording beam steering mirrors.

An example of the type of blazed grating that can be achieved by the above methods can be seen in the electron-micrograph of Figure 6, taken of a 1200g/mm grating. The shadow cast by a small asbestos fiber shows an excellent triangular groove shape, although theory predicts[10], and efficiency measurements confirm, that there is a minute flattening of the resist at the tip of the grooves. It is also interesting to compare the surface roughness with that of a 1200g/mm diamond ruled grating produced in a good aluminum film, Figure 7; they seem to have similar values.

Figure 6. Electron Micrograph of a blazed 1200/mm holographic type grating. The shadow cast by a fine asbestos fiber makes clear the nature of the triangular groove shape. Groove angle is 8.5°.

Figure 7. Electron Micrograph of a blazed grating ruled with a diamond tool in an aluminum coating. Blaze angle is 1°.

The desire to blaze at other wavelengths by interference methods has led to two additional approaches, one purely optical the other more mechanical. Both have limitations, especially in the size of the gratings attainable.

The optical approach is based on the fact that a triangle can, to a reasonable approximation, be synthesized by the first two terms of a Fourier sine series, with the second term ½ the amplitude of the first and the phase between the two properly controlled. Thus if one could generate a matched set of sinusoidal double exposures in photoresist, the result would be a roughly triangular groove shape. Despite the difficulties involved in matching groove frequencies to at least one part in 10^5 and phase location to 1/10 the period, successful gratings have been produced experimentally. Schmahl and Rudolph[6] used a transmission grating to produce the two matching frequencies, taking advantage of the fact that the first and second diffracted orders automatically give the exact ratios. A limitation is that a 100mm transmission grating is required to produce a 25mm blazed grating, and the incident light must be very well collimated. Johansson et al[11] made 50mm gratings by interferometric double exposures. The optical alignment specifications are so tight that they are impossible to maintain by conventional means for the several minutes required per exposure. The solution was photoelectric sensing system providing an active optical feedback to one arm of the generating interferometer. It appears difficult to scale up such a system to produce large gratings.

The optomechanical approach is to generate an appropriate resist pattern on the surface of the grating and then subject it to slant bombardment of a beam of Argon ions. The substrate material must be removed at a rate sufficiently faster than the resist if triangular grooves are to result. Aoyagi and Namba[12] have shown that properly oriented single crystal GaAs can lead to 1000g/mm gratings with blaze angles from 7 to 27°. An alternate recording material in the form of a thin layer of PMMA is perhaps more useful. The gratings produced so far are only 20 x 20mm. An interesting question is whether gratings produced by a multi-step procedure can retain the high degree of spectral purity obtainable by one-step methods.

Stray light

Stray light, the composite of misplaced spectral energy, has always been a problem area with diffraction gratings. Perhaps the principal difficulty comes from the lack of an accepted physical definition, which in turn is related to the fact that different spectrometric systems impose quite different demands. In some, the grating is the principal source of stray light, in others merely an equal participant. In some it is not even a problem. The spectral intensity distribution of the light source and the spectral response of the detector often play a key role. For example, a solar blind detector has no sensitivity to stray light from the visible. To complicate the picture there are several separate attributes of a grating that give rise to stray light and do so to a degree that sometimes varies with wavelength. Yet an experienced observer can tell in seconds how to classify a grating, armed with nothing more than a flashlight. What he looks for is the amount of light visible between orders. It is suggested below how, with the help of a laser, this simple test can be made quantitative and thereby provide a very simple figure of merit by which gratings can be categorized.

Classification of stray light sources

Periodic errors

Periodic errors in groove spacing give rise to well defined spurious line images termed ghosts. When the period is traceable to the ruling engine lead screw, i.e. is a long one, the lines are close to the parent line and termed Rowland ghosts. Modern ruling engines, especially with interferometric feedback control, keep error amplitudes so low that only in special applications, such as Raman spectrometry, do they seriously detract from performance. Lyman ghosts arise from compound drive effects and have much shorter periods. This locates them in fractional order positions, far from the parent line, and are thus more difficult to identify. They not only contribute to integrated stray light but can lead to misidentification of lines, and must therefore be kept to very low levels. Again, interferometric control, properly applied, will do this quite well. A good photoresist grating will be entirely free of both ghosts, which represents one of its advantages.

Random groove misplacements. Small irregular groove spacing gives rise to a more or less continuous scatter between orders, often called grass because of its appearance as a bright line when Hg 546nm light is used for examination. It is particularly easy to observe when a grating is directly illuminated by a laser beam and the diffracted orders picked up by a screen. Sharpe and Irish [13] have done a careful analysis of this effect, based on scalar diffraction theory that is quite applicable to most gratings where stray light is of special concern, i.e. UV gratings of relatively shallow groove depth. One important conclusion is that the random groove based scatter is not a strong function of wavelength.

Groove depth and surface roughness. Variations in depths of grooves as well as surface roughness, on the other hand, produce stray light that is an inverse function of wavelength. A typical result in a ruled grating is negligible amounts at 500nm but each producing a roughly equal contributions to random groove displacement effects at 230nm. In fact, the variation with wavelength becomes a tool for identifying the origins of stray light. The effects of groove depth variation and surface roughness can be separated by noting that the former directs light in the diffraction plane while the latter leads to diffuse scatter over a 2π solid angle.

Holographic gratings will be very low in grass, but have diffuse scatter of the same order as a good ruled grating. As pointed out by Hutley[5] and Verrill[14] it is important to appreciate that in a spectrometer the amount of grass transmitted is proportional to slit width and the amount of diffuse scatter is proportional to the area of the slit. This illustrates why it is difficult to make general statements concerning the stray light performance ratios of different gratings. In general, one can say that the larger the slits and the shorter the wavelength the greater the dominance of surface roughness and therefore the less the advantage of a holographic grating.

Halo effects. Several authors, including Verrill[14] and Flamand[15], have noted that holographic gratings sometimes display a diffuse halo around the laser image, that changes orientation depending on where the grating is illuminated. This is the result of a micro-modulation of the resist surface generated during the spinning of the photoresist, unless the greatest of care is excersised.

Measurement of stray light. The classical approach to measuring stray light[14, 16, 17] is to supply a spectrometer with high intensity monochromatic light and move the detector arm across the diffraction plane while recording the output of its photomultiplier tube. This is not a completely straightforward undertaking; it involves some arbitrary decisions. Of concern are not only slit dimensions, but there is conflict between a desire to use an incoherent source and the fact that no source is as convenient or intense as a laser. It is important to underfill the grating to lessen Fraunhofer edge diffraction of the grating aperture. Neutral density filters are needed to calibrate the system over the large dynamic range to be covered (10^6 or more). The optical system must be of a quality such that even a good grating contributes the bulk of the output stray light.

If diffuse scatter is to be monitored, data taking becomes more complex because the instrument must have provision for moving the detector out of the diffraction plane.

The output of such tests is presented in curves, usually with semi-log or log-log scale, with zero or first order output normalized to 1.0 as the ordinate, and angles or wavelength as abscissa. This provides a great deal of information, but does not readily produce a figure of merit.

Simple figure of merit stray light test for gratings

A simple test fixture was recently constructed that enables an inspector to obtain a useful figure of merit in less than a minute. As one can see in the optical schematic of Figure 8, the grating is illuminated by the direct beam of a small 633nm He-Ne laser, equipped with a red filter to remove light from its plasma. A 78mm diameter collecting lens with a 116mm focal length is positioned to cover 95% of the angular span between zero and first orders. It must be translated along its axis for gratings of different groove frequencies, and the laser must be repositioned for the same reason. Fixed stops for standard frequencies make this easy. A simple Densichron®️ photocell detector has just enough dynamic range (10^5) to be effective, but this could easily be increased with a neutral density filter. A horizontal mask 3mm wide restricts the acceptance angle to the diffraction plane, so that by simple difference from the total one obtains a rough estimate of the diffuse scatter. In order to have results independent of the blaze wavelength the 100% setting is made with a mirror of the grating material directing the laser beam to the detector, rather than referencing the first or zero order of the grating.

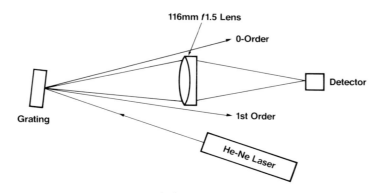

Figure 8. Optical schematic of simple integrated stray light measuring system.

On a sample set of gratings, both ruled and holographic, the data obtained with the 633nm light of the He-Ne laser did not differ much from that obtained when the 458nm blue light of an Argon ion laser was substituted. The only advantage of the shorter wavelength is that the range of attainable groove frequencies is slightly increased, from 2400 to 3600/mm, because red light is not diffracted by a 3600/mm grating. In the other direction mechanical interference limits coarse spacings to 600/mm, which also conveniently represents the level below which holographic gratings show up to advantage only in special instances.

A set of 1200/mm gratings, blazed at different wavelengths, produced holographically and by ruling on a mechanical and a small interferometrically controlled ruling engine, were evaluated by the method above, with results as shown in Figure 9.

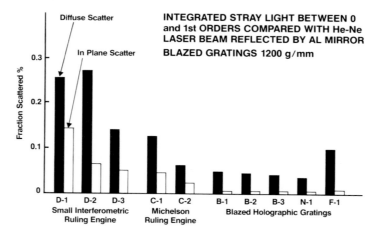

Figure 9. Integrated in-plane and out-of-plane scattered light measured between zero and first order of a family of 1200g/mm gratings.

That there are significant differences is immediately obvious. The interferometric engine has higher stray light levels compared to the mechanical (Michelson) engine, despite its reduced ghost intensities. It is not as rigid a machine. Yet gratings from both engines perform well in most applications. The holographic gratings have the same nominal specifications as grating D-3, but have only about 1/10 the amount of in-plane stray light. That good results are not inherent with all holographic gratings is evident from the grating labelled F-1. Its stray light levels are more than double the others. It is also evident that out-of-plane scattered light on a holographic grating is not much less than that of a good ruled grating.

Replication

There seems to be some concern about the maintenance of quality, particularly regarding maintenance of stray light qualities, when going from masters to replicas, that are typically three or four generations removed[17]. Such fears appear to be groundless, at least based on our experience. For example, grating C-2 is a second generation replica from the 250th replica made from the original master. Its low stray light level is obvious.

Grating B-1 is a holographic master of which B-2 is a third generation replica, while B-3 is a fourth generation replica of another master similar to B-1. They are substantially alike. The generation replication of gratings N-1 and F-1, from different sources, is not known. The good performance of grating N-1 is supported by the smooth surface structure evident in Figure 10. The poor performance of F-1 seems to be traceable, at least in part, to its relatively rough groove edges, as seen in the electron photomicrograph of Figure 11.

Figure 10. Electron photomicrograph of blazed holographic grating N-1.

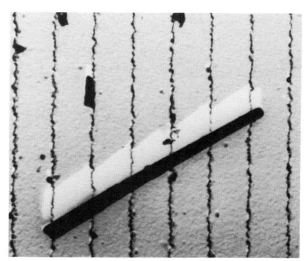

Figure 11. Electron photomicrograph of blazed holographic grating F-1.

Diffraction efficiencies

Since the diffraction efficiency behavior of gratings with sinusoidal grooves differs from those with triangular ones, some comments are in order. It is not possible in the limited space here to cover this topic in detail, as has been done in previous publications[18, 19, 20].

A good example is a 1200g/mm sinusoidal grating with a depth modulation, or ratio of groove depth to spacing, of 0.31. This is considered a medium high modulation, by no means an upper limit. Its first order near Littrow efficiency performance is plotted in Figure 12, with S and P polarization data shown separately as well as that of unpolarized light. The values are very close to theoretical calculations for such a grating. It is typical that high efficiency is attainable in such deeply modulated gratings only when $\lambda/d>0.7$, the region where no higher orders can diffract, and that values are quite similar to those observed with triangular grooves with a depth of about 46% of the spacing (27° blaze angle). If 20% efficiency is acceptable at the low end, this grating is seen to be usable over a spectral band from 300 to 1500nm.

It is a basic characteristic of sinusoidal modulation to suffer by comparison with ruled gratings when they are used in orders higher than the first. A hint in this direction is given in Figure 13, which shows the second order efficiency curve for the same grating as Figure 12. A corresponding ruled grating would have about 50% higher efficiency peaks and also reduced polarization effects. Successive higher orders show even greater differences. However, especially with maximum modulations, holographic gratings have been successfully used in orders up to four.

An entirely different type of efficiency curve is shown in Figure 14. It represents the blazed holographic grating whose surface was shown in Figure 6. Due to its low blaze angle (8.5°) it has near scalar or non-polarizing behavior near its blaze region and exhibits Wood's anomalies at 550 and 650nm that are relatively weak. Efficiency values are almost identical to those of a good, mechanically ruled grating.

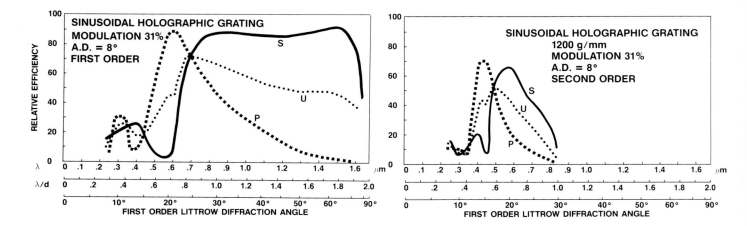

Figure 12. Relative first order efficiency of an aluminum replica grating of a 31% modulated 1200g/mm sinusoidal grating. Angular deviation between incident and diffracted beams is 8°. S and P plane designate polarized light with electron vector perpendicular and parallel to the grooves respectively. Unpolarized light, U, is the average.

Figure 13. Same as Figure 12, but measured in second order.

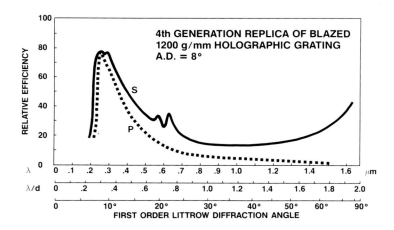

Figure 14. Relative first order efficiency of a fourth generation aluminum replica of a 1200g/mm blazed holographic grating. A.D. = 8°.

Conclusion

The ability of holographically recorded diffraction gratings to produce clean spectra, of superior high spectral purity, stands without question. Replicas without measurable stray light degradation can be produced and verified with a very simple test.

Acknowledgements

The authors wish to thank Victor Shay and Alison Palumbo for making the efficiency measurements and John Ziarko for the electron photomicrographs. William Morgan was invaluable in establishing many of the process controls and Ralph Dakin was responsible for most of the optical system designs. Ian Wilson, of the University of Tasmania, was a major contributor during a C.S.I.R.O. post doctoral fellowship in 1978.

Bibliography

1. Lord Rayleigh, On the Manufacture and Theory of Diffraction Gratings, Phil. Mag., 4, 47, p. 81., 1874.
2. Michelson, A. A., The Ruling and Performance of a Ten Inch Diffraction Grating, Proc. Am. Phil. Soc., 54, p. 137. 1914. Also: Studies in Optics, University of Chicago Press. 1927.
3. Learner, R. C. M., Personal Communication.
4. Palmer, E. W., Hutley, M. C., Franks, A., Verrill, J. F., Gale, B., Diffraction Gratings, Rep. Progr. Phys., 38, pp. 975-1048. 1975.
5. Hutley, M. C., Interference (Holographic) Diffraction Gratings, Jl. Phys. E, Scientific Instruments, 9 pp. 513-520. 1976.
6. Schmahl, G., Rudolph, D., Holographic Diffraction Gratings, Progress in Optics, Vol. XIV, Academic Press. 1976.
7. Shank, C. V., Schmidt, R. V., Optical Technique for Producing 0.1μm Periodic Surface Structure, Appl. Phys. Letters, 23, pp. 154-155. 1973.
8. Nagata, H., Kishi, M., Production of Blazed Holographic Gratings by a Simple Optical System, Japan, J. of Appl. Phys., 14, pp. 181-186. 1975.
9. Obermayer, H. A., Interferometric Methods for Generating Reflection Diffraction Gratings, Dr. Ing. Thesis, University of Stuttgart. 1976.
10. Brown, B. J., Wilson, I. J., A Numerical Study of Blazed Holographic Gratings, Dept. of Physics, University of Tasmania, Hobart, Australia. 1978.
11. Breidne, M., Johansson, S., Nilsson, L. E., Ahlen, H., Blazed Holographic Gratings, Optica Acta, 26, pp. 1401-1441. 1979.
12. Aoyagi, Y., Sano, K., Namba, S., High Spectroscopic Qualities in Blazed Ion Etched Holographic Gratings, Optics Communication, 29, pp. 253-255. 1979.
13. Sharpe, M. R., Irish, D., Stray Light in Diffraction Grating Monochromators, Optica Acta, 25, pp. 861-893. 1978.
14. Verrill, J. F., The Specification and Measurement of Scattered Light from Diffraction Gratings, Optica Acta, 25, pp. 531-547. 1978.
15. Flamand, J., Design, Production and Test of V.U.V. Holographic Gratings, Vth Intern. Conf. on Vac. Ultraviolet Rad. Physics, Abstracts, Montpelier, Paper 2. 1977.
16. Brown, S., Tarrant, A. W. S., Scattered Light in Monochromators, Optica Acta, 25, pp. 1175-1186. 1978.
17. Francis, R., Performance Details of Original and Replicated Holographic Gratings, 29th Pittsburgh Conference on Anal. Chem. and Appl. Spectroscopy. 1978.
18. Maystre, D., Petit, R., Duban, M., Gilewicz, Theoretical Study of the Efficiency of Metallic Gratings in the Near Ultraviolet, Nouv. Rev. Opt., 5, pp. 79-85. 1974.
19. Loewen, E. G., Neviere, M., Maystre, D., Grating Efficiency Theory as it Applies to Blazed and Holographic Gratings, J. Appl. Opt., 16, pp. 2711-2721. 1977.
20. Loewen, E. G. Neviere, M., Simple Selection Rules for VUV and XUV Diffraction Gratings, 17, pp. 1087-1092. 1978.

Holographic grating development at Lockheed Palo Alto Research Laboratory

R. C. Smithson, T. P. Pope

Lockheed Missiles and Space Company, Incorporated, Lockheed Palo Alto Research Laboratory
Department 52-10, Building 202, 3251 Hanover Street, Palo Alto, California 94304

Abstract

During the past several years the Lockheed Palo Alto Research Laboratory has developed a capability to produce holographic optical elements using a variety of production techniques. Gratings produced have included reflective gratings using both overcoated photoresist and ion etched substrates up to 15 cm. in diameter. Conventional bleached photographic transmission gratings have also been produced for some applications. Methods in use at Lockheed are capable of producing extremely rugged gratings which can withstand both physical abuse and high incident flux levels. Photoresist coating techniques have been developed which are sufficiently precise that optical quality obtained with reflective gratings is limited primarily by the quality of the construction optics. Both high and low efficiency gratings have been produced for a variety of applications. This paper presents an overview of Lockheed holographic grating technology.

Introduction

The general problem of the production and use of holographic optical elements has received considerable attention in the past few years. Holography allows the production of high performance diffraction gratings of conventional design without the need for elaborate ruling engines. Perhaps more important, however, is the possibility of producing optical elements with unique qualities difficult or impossible to achieve by conventional means. Gratings generated holographically on a reflective surface can be used to diffract a small portion of incident light to a convenient spot for analysis of the quality of the wavefront reflected from that surface. Gratings so generated can be made to have negligible effect on the quality of the mirror on which they are placed. Thus, they can serve as a sampling element in an adaptive optical system. Finally, they can easily be generated on a curved surface.

For some time, Lockheed Palo Alto Research Laboratory has been producing holographic gratings for use with various optical systems. Gratings have been made which are both rugged and of high optical quality. Figure 1 shows a Lockheed grating in operation. A beam of light normally incident on the mirror is diffracted to a point where the wavefront quality can be analyzed. A microscope projection of this sampled beam (about 20% of the incident energy) shows a high quality diffraction pattern.

Figure 2 shows a photograph of a 15 cm. diameter parabola with 18 holographic gratings applied. The diffracted samples from the gratings come to a common focus to produce the diffraction pattern shown on the left in Figure 3. On the right is the diffraction spot from a single grating.

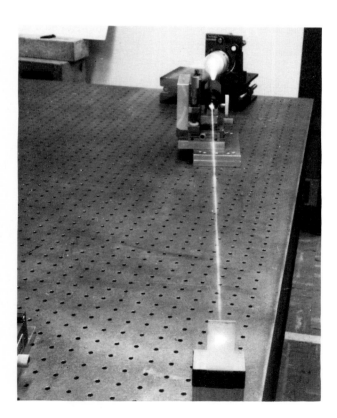

Figure 1. A Lockheed holographic optical element in operation.

Grating fabrication techniques

The area of greatest interest in grating technology is the production of reflective or transmissive holographic optical elements on large, stable substrates. This discussion will primarily concern the fabrication of reflective holographic gratings on parabolic or other curved glass substrates. Although the Palo Alto Research Laboratory has the capability to produce gratings using bleached photographic emulsions, this technology will not be further discussed here.

Production of holographic gratings on glass substrates involves the sensitization of the substrate, exposure and development of the sensitive layer, possibly the ion etching of the grating into the substrate, and finally the application of the final reflective coating.

The first step in the fabrication process (sensitization of the substrate) has been approached at Lockheed in three different ways: conventional spin coating with photoresist, dip coating, and evaporative coating. All three methods have been found to produce satisfactory coatings. The spin coating technique has been used to produce coatings capable of near diffraction limited performance on substrates up to 15 cm. in diameter. Dip coating has been used to provide similar performance on larger substrates. Evaporative coating methods, which involve the deposition of photoresist by the evaporation of dilute solutions, have also been used at Lockheed to produce optical quality photoresist films. In short, the problem of producing an adequate photoresist coating is amenable to a variety of solutions.

The chief difficulties in the exposure of the sensitized surface lie in the fact that photoresist is a relatively insensitive material that responds only to violet or shorter wavelengths. The insensitivity gives rise to long exposure times and attendant long-term stability problems when large gratings are made, and the limited useful wavelength range of photoresist sensitivity requires the production of carefully distorted ("scaled") exposure beams in order to produce gratings for use at wavelengths other than violet or ultraviolet. Since many applications for holographic gratings involve wavelengths in the red or infrared, scaling (i.e., the production of the required construction beams) can be a major problem, especially if the gratings are to be constructed on curved substrates. This subject is treated in detail in another paper[1] and will not be discussed further here.

Figure 2. Holographic gratings applied to a 15 cm. parabola.

After development of the photoresist, a straightforward process which leaves a pattern of grooves etched in the photoresist coating, two options exist. The first is to leave the photoresist layer on the substrate and simply overcoat the grooves with an appropriate reflective surface. The second is to ion etch the grating directly into the substrate, and then to apply the reflective coating. Both methods have been used successfully at Lockheed.

Overcoated photoresist gratings eliminate the ion etching step, and thus allow a simpler fabrication process. However, a stable coating over a photoresist layer is not easy to achieve, especially if the grating is to be used at high incident flux levels. Lockheed has developed techniques for producing overcoated photoresist gratings which are stable for most applications, and this technique is routinely used at Palo Alto for the production of prototype gratings for use in various optical test systems. We feel, however, that ion etched gratings are vastly superior to overcoated photoresist for most uses.

The ion etched gratings produced at Lockheed are etched directly into the glass substrate of the mirror. Figure 4 shows a scanning electron microgram of a photoresist grating compared with an ion etched glass grating. No loss in surface smoothness or optical quality of the substrate is produced by the ion etching process. Since the ion etched grating is simply a glass mirror with shallow grooves etched into it, there are no problems with coating adhesion or substrate stability. The gratings so produced are extremely rugged, and are more resistant to mechanical abrasion than either photoresist or conventional ruled metal gratings.

Performance of ion etched glass holographic gratings has been excellent, both in terms of optical quality and uniformity of grating efficiency. Figure 5 shows the diffracted sample spots at focus and at various distances outside focus for the grating shown in Figure 3. As can be seen, the uniformity of the spots is excellent. Zone plates covering 15 cm. diameter parabolic mirrors have been fabricated which also show excellent uniformity.

Figure 6 shows a comparison between an actual diffraction pattern measured at the focus of the first order diffraction pattern of the gratings shown in Figure 3, and a computer simulation of a perfect system. Differences are commensurate with the quality of the mirror substrate used.

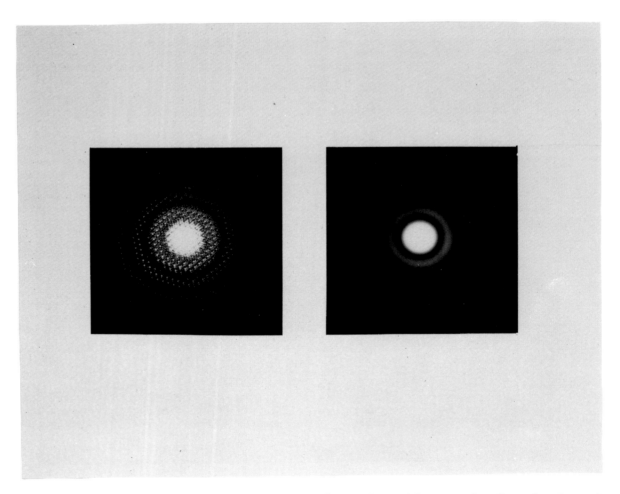

Figure 3. The pattern on the left is produced by 18 superimposed beam samples from the mirror in Figure 2. On the right is the diffraction pattern from a single grating.

In summary, we believe that ion etched glass holographic gratings have been shown to be a mature technology which can be used routinely for a variety of applications.

8,300X
GRATING PATTERN IN PHOTORESIST, GLASS SUBSTRATE

8,300X
GRATING PATTERN ION ETCHED, THRU RESIST, INTO GLASS

Figure 4.

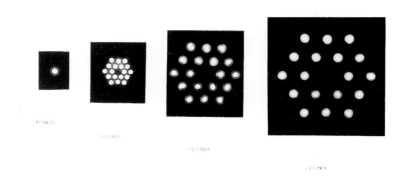

SIX INCH PARABOLIC MIRROR
WITH 18 HOLOGRAPHIC GRATINGS

Figure 5.

DIFFRACTION SPOT
18 HARTMANN ZONES
AT FOCUS

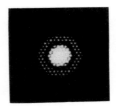

ACTUAL   COMPUTER SIMULATION

Figure 6.

<u>References</u>

1. Malin, M., and H. Morrow, Wavelength Scaling Holographic Elements, published elsewhere in these proceedings.

PERIODIC STRUCTURES, GRATINGS, MOIRÉ PATTERNS AND DIFFRACTION PHENOMENA

Volume 240

SESSION 2

HOLOGRAPHIC GRATING II

Session Chairman
Peter L. Misuinas
Air Force Weapons Laboratory

Holographic gratings at the Institute of Optical Research, Stockholm

Stefan Johansson, Klaus Biedermann
Department of Physics II and Institute of Optical Research
The Royal Institute of Technology, S-100 44 Stockholm, Sweden

Abstract

Based on long experience in the field of holography and recording materials, research on holographic gratings has been pursued at the Institute of Optical Research for a number of years. Grating production techniques have been developed including an interferometric setup with microcomputer control of exposure and fringe stabilization as well as methods for grating performance evaluation. A specialty is the synthesis of blazed grating profiles from successive exposures of their Fourier components. Theoretical studies based on electromagnetic theory and experimental investigations are carried on aiming at the optimization of these Fourier blazed gratings. Finally, as an example for the design flexibility of holographic gratings, a multiple grating flame photometer for the simultaneous determination of five elements is presented.

Introduction

Together with spectroscopy, grating manufacture has some tradition in Sweden. Manne Siegbahn, Nobel Prize Winner in 1924, designed and built ruling machines to make the high-quality gratings he needed for his research in X-ray spectroscopy [1]. The latest one is still in good condition but has not been in use for grating manufacture for the last fifteen years or so.

At the Institute of Optical Research, interest in grating manufacture developed from the work in holography, which had started in 1966. Hologram interferometry and high resolution recording materials were the main areas of research. In the latter field, methods and instruments for characterizing and investigating light-sensitive materials were developed [2]. In particular, high resolution photographic materials were studied by exposing them to two-beam interference and measuring their response by diffraction and/or by irradiating and scanning the sample with an identical interference pattern. This principle, which we called the "Multiple-Sine-Slit-Microdensitometer" [3] can be interpreted either according to its name, but also in terms of a scale-down of high spatial frequencies to an easily detectable moiré-pattern, a cross-correlation, or Fourier transform operation, or a superposition of coherent fields diffracted at a Fourier series of complex gratings [4]. In this way, also the phase modulation in photographic layers, caused by internal refractive index variation and gelatin surface relief, could be studied separately from the absorption modulation [5].

Early work on holographic gratings in photoresist by Rudolph and Schmahl [6,7] presented at the Meetings of the German Society for Applied Optics suggested to us to transfer the experience and methods elaborated at the Institute of Optical Research to a project on holographic gratings.

Since we had experience in analysing the Fourier components of gratings [4], the synthesis of blazed gratings from Fourier components [8] was to become one very challenging topic in the project. Our method of synthesizing uncommon holographic grating profiles relies upon sequential addition of sinusoidal irradiance distributions with integer spatial frequency ratios and suitable phase relations [9,10]. Extreme accuracy in adjusting and maintaining the spatial frequency, orientation and phase of the exposing interference patterns, is one precondition.

This paper is intended to give a general view of the holographic grating work at the Institute of Optical Research, while four other contributions report on some specific topics of the experimental [11], instrumental [12] and theoretical [13,19] work connected with this project.

The exposing instrument

For a holographic grating, the "ruling" accuracy is determined by the accuracy of the two interfering waves and, hence, of the system used to expose the gratings. An instrument, determined to yield high accuracy, reproducibility and easy handling, should have the following properties:

1. Exposing fields plane to better than $\lambda/10$ at 500 nm.
2. Small variation of irradiance over the substrate with less than 5 % fall-off towards the edges.
3. Simple adjustment of irradiance ratio in the fields.
4. Fast and simple adjustment of arbitrary spatial frequencies or groove densities.
5. Stable setup with active stabilization of the interference pattern.

Fig. 1 shows the setup schematically. The beam from a Model 170 Spectra Physics Ar-laser is divided by the beam-splitter BS, a Glan-Thomson prism, giving two perpendicularly polarized beams. Their intensity ratio depends on the polarisation of the beam incident on BS and can easily be adjusted by rotating the $\lambda/2$-plate in front of the prism. The second $\lambda/2$-plate is necessary to turn the reflected beam by $90°$ for

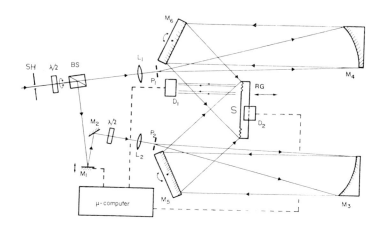

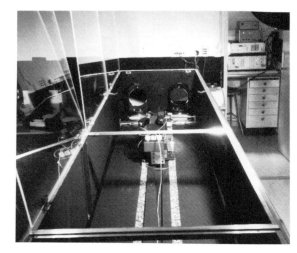

Figure 1.   Arrangement for exposing holographic gratings

Figure 2.   A photographic view of the exposing arrangement showing the plane
mirrors and the substrate holder.
In the background, electronics for
fringe stabilization.

vertical polarisation. P_1 and P_2 are point sources, pinholes of 5 µm diameter, onto which the laser beams are focused by means of the microscope lenses L_1 and L_2. Of special interest is the mirror M_1. It forms part of the adaptive fringe stabilization system described elsewhere in this volume [12]. A variation in optical pathlength between a fraction of a wavelength and about 10 wavelengths is achieved by moving this mirror electromagnetically. M_3 and M_4 are collimating devices. They are off-axis paraboloidal mirrors with a focal length of 3.5 m polished to an accuracy of $\lambda/20$ over the diameter of 130 mm. The low relative aperture of the mirrors assures an even irradiance distribution in the reflected field since only the central part of the fields arriving from the pinholes is used. As a consequence, the duration of the exposures is extended, however, the active stabilization of the fringes ensures constant modulation in the time integrated exposure pattern irrespective of exposure time. According to the purpose of the machine to provide fast access to arbitrary spatial frequencies, beam collimation and angular adjustment of spatial frequency are assigned to two separate mirrors in each beam. M_5 and M_6 are plane mirrors that can be turned simultaneously in opposite directions for fast and simple adjustment of groove density. For the same purpose, the substrate holder can be moved on air bearings along the bisector of the interfering fields. The whole instrument is assembled on a block of granite with all parts rigidly mounted.

When making blazed gratings by Fourier synthesis of subsequently exposed interference patterns, the mirrors M_5 and M_6 have to be adjusted with an accuracy comparable to that for the mirrors in an interferometer in order to achieve the fringe orientation and spatial frequency required. These fine adjustments can be done by means of micrometer screws as well as by electronic remote control [12].

Fig. 2 is a photographic view of part of the machine taken from the direction of the parabolic mirrors to the plane mirrors and the substrate holder.

Techniques for evaluation of gratings

The properties of importance for the quality of a grating are in the first place

1. Diffraction efficiency as a function of wavelength.
2. Stray light level.
3. Absence of false spectral lines.
4. Spectral resolution

1. Diffraction efficiency

By efficiency we usually mean relative efficiency, i.e. the ratio of the light flux diffracted into a given order to the light flux reflected from a mirror with the same reflective coating as the grating. The absolute efficiency may be obtained by multiplication by the reflectance of this coating.

The efficiency curve (efficiency versus wavelength) can be calculated from the groove profile, as to be discussed in the next section. The groove profile depends on the parameters of the exposure and the developing process. It is often sufficient to measure efficiency at some laser wavelengths. By matching these measured values to a suitable calculated efficiency curve it is possible to obtain the whole efficiency curve in this way. For prototype gratings and for experiments with gratings with special groove shapes it is necessary to measure the η (λ)-curve directly.

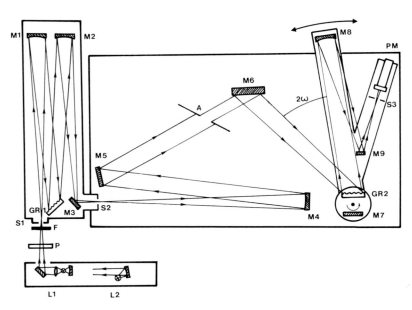

Figure 3. An instrument for measurement of efficiency and straylight of gratings.

The principle of an instrument developed for this purpose is outlined in Fig. 3. The spectral range from 200 to 800 nm is covered by two light sources, a deuterium lamp and an ordinary incandescent halogen lamp. There is a 1 m Czerny-Turner type pre-monochromator. The test monochromator has two spherical mirrors M_4 and M_8 with identical focal lengths. In between the two plane deviating mirrors M_5 and M_6 an aperture A may be inserted at an angle yielding the projected area over which the grating is used at the wavelength in question. The grating under test GR_2 is inserted on a turntable on which also the reference mirror M_7 is mounted. Each of them is easily turned into the measuring beam in order to determine diffraction efficiency and reflectance, respectively. The focusing mirror M_8, the exit slit S_3, and the detector comprising a photomultiplier tube are mounted on the same arm for easy adjustment of the Ebert angle ω. To be able to detect low efficiencies and low stray light levels we use a photon counting detection system.

2. Scattered light

The level of scattered light from a grating is difficult to define in a general way since it depends on a great many factors, especially those of geometry. For an emission line of a certain wavelength in a certain order we can quantify scattered light as the light flux observed in a different spectral band relative to the light flux at the locus of the line fed through the system. For these measurements the photon counting detection unit was conceived. Most of the sources of stray light inherent to ruled gratings are not present in holographic gratings. Besides to the possibility of unwanted contributions to the exposing fields in the coherent optical system, attention has to be paid to the photoresist developing process which may cause some surface roughness.

3. False spectral lines

In spite of the absence of ghosts and satellites in holographic gratings false spectral lines may be detected with a strictly monochromatic (laser) light source. They indicate that the set-up has to be checked for reflections or other secondary sources.

4. Resolution

Resolution is determined in a Twyman-Green interferometer. With high groove densities we found it often practical to use a moiré technique where the grating is compared with the exposing pattern. Strictly speaking, this method just checks the exposing pattern against itself. However, by turning the grating upside down, at least assymmetrical errors become visible.

Calculation of diffraction efficiency

Through our collaboration with the Laboratoire d'Optique Electromagnetique of Marseille, France, the project has very much benefitted from the theoretical experience of that group. For the calculations, we use a computer program based on the integral formalism [15]. It can calculate the efficiency of sinusoidal, lamellar and echelette gratings for both polarisation directions and for finitely as well as infinitely conducting gratings, with or without dielectric coatings. An additional subroutine makes it possible for us to calculate the efficiency for any Fourier blazed grating. Furthermore, we have added to the program different photoresist response curves which enable us to study the influence of nonlinearities in the photoresist process on the efficiency. This makes it possible to test different resist response theories theoretically and thereafter make comparisons with experimentally obtained data for efficiency [11].

In the following, we want to discuss the two most studied types of holographic gratings, sinusoidal and Fourier blazed gratings.

Sinusoidal gratings

The interference pattern has a sinusoidal irradiance distribution. We normally use Shipley AZ 1350 photoresist and develop in AZ 303 developer diluted 10 times. This combination yields a fairly linear exposure v. groove depth dependence, at least for moderate groove depths. (This developer is also used for Fourier gratings where linearity is important.) This means that we produce gratings with sinusoidal groove shapes. When $\frac{2}{3} < \lambda \cdot \nu < 2$, where λ is the wavelength of the light incident on the grating and ν is its groove density, only 2 orders propagate in Littrow configuration. In Fig. 4, showing ν v. λ, the region where this relation holds is dashed. For this region, Breidne and Maystre [14] have shown that a sinusoidal grating has the same efficiency curve as any ruled grating if the groove depth is appropriately chosen. An asymmetrical groove profile does not improve efficiency.

In spectrographs, where the space for the spectra is limited, the dispersion of gratings relying on this sinusoidal blaze would be too high. Therefore, lower groove densities with asymmetrical groove profiles are desirable. In Ref. 10, we discuss and compare different blaze techniques for holographic gratings.

The Fourier blaze technique covering the dotted region in Fig. 4 turns out to have the greatest design flexibility.

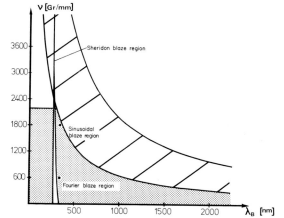

Figure 4. Regions in the wavelength-groove-density plane covered by different blaze techniques. Recording wavelength 458 nm.

Fourier blazed gratings

Any periodic structure can be synthesized from its Fourier components. Obviously, by multiple sinusoidal exposures any groove profile may be composed. A theoretical suggestion had been made by McPhedran et al [16], experimental results have been published by Schmahl [8] and later on by Johansson et al [9]. However, the technical problems of superimposing even only two exposures on a grating of reasonable extent are severe. The alignment of the exposing patterns has to be parallel to within 0.02 mrad, the accuracy in their frequency has to be within $< 10^{-6}$. Moreover, their relative phase has to be maintained constant to within $< 1/10$ of a grating period. Our solution to this problem is based on the principle of comparing exposing patterns to a reference grating in the same location. Deviations will be visible in the form of moiré fringes. By observation of these moiré fringes, the exposing patterns can be adjusted manually or electronically with the required accuracy [12].

Design of Fourier blazed gratings

We may start with the assumption that a grating with a triangular groove shape yields maximum efficiency. The Fourier expansion of such a profile with an apex angle of 90°, blaze angle α and grating period d is

$$f(x) = \Sigma \, b_n \sin nKx, \quad K = 2\pi/d$$

$$b_n = \frac{d}{(n\pi)^2} [\tan \alpha + \cot \alpha] \sin [n\pi \cos^2 \alpha] \tag{1}$$

For practical reasons, only the fundamental and first harmonic in the expansion are used. Their coefficients can be found by means of Eq. (1). The result $b_1 \approx 2 \, b_2$ means that the first harmonic shall be given half the exposure of the fundamental.

Is this optimisation still true when only two exposures are applied and what efficiency can we expect? In the scalar domain we may make the following simplified discussion [9]. Let us regard the Fourier blazed grating as a superposition of two gratings having groove densities ν_o and $2\nu_o$ with the complex amplitude transmittance along the x-coordinate

$$\tau(x) \quad = e^{i\phi_1(x)} \cdot e^{i\phi_2(x)} \quad \text{where}$$

$$\phi_1(x) \quad = \frac{m_1}{2} \sin 2\pi\nu_o x$$

$$\phi_2(x) \quad = \frac{m_2}{2} \sin (2\pi 2\nu_o x + \beta)$$

β describes the relative position of the gratings, and the parameters m_1 and m_2 describe peak-to-peak excursions of the phase delay and are proportional to the groove depth of the gratings. The diffracted field from the first sinusoidal grating can be expressed by a sum of Bessel functions [17]. Considering only zero and first order diffraction, $J_o\left(\frac{m_1}{2}\right), J_1\left(\frac{m_1}{2}\right),$ and $J_{-1}\left(\frac{m_1}{2}\right)$ are proportional to the amplitude, and $J_o^2\left(\frac{m_1}{2}\right),$ $J_1^2\left(\frac{m_1}{2}\right),$ and $J_{-1}^2\left(\frac{m_1}{2}\right)$ are proportional to the efficiencies in zero, first and minus first diffraction order, respectively. $J_1^2\left(\frac{m_1}{2}\right) = J_{-1}^2\left(\frac{m_1}{2}\right) = 0.34$ is the well-known maximum for the efficiency of a sinusoidal grating, which appears for $m_1/2 = 1.8$. However, if, in our simplified model, these three diffracted fields fall on the second grating, they are again diffracted, as sketched in Fig. 5.

The first order term $J_1\left(\frac{m_1}{2}\right)$ from the first grating is attenuated to $J_1\left(\frac{m_1}{2}\right) \cdot J_o\left(\frac{m_2}{2}\right)$ On the other hand, an additional field in the same direction appears, namely $J_{-1}\left(\frac{m_1}{2}\right) \cdot J_1\left(\frac{m_2}{2}\right)$

By a proper choice of the relative position β of the two gratings, namely $\beta = 0$ in this simple model, the fields can be added in phase resulting in the efficiency,

$$\eta = J_1^2\left(\frac{m_1}{2}\right) \cdot \left[J_o^2\left(\frac{m_2}{2}\right) + J_1^2\left(\frac{m_2}{2}\right) + \right.$$
$$\left. + 2 J_1\left(\frac{m_2}{2}\right) \cdot J_o\left(\frac{m_2}{2}\right) \right]$$

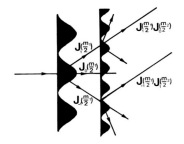

Figure 5.   Illustration of the contributing diffractions into the first order from two superimposed sinusoidal gratings.

The factor within the bracket is an amplification factor applied to the single grating. It reaches the value 1.5 when $m_2/2 = 0.8$. Hence, superimposing two gratings gives the maximum efficiency in the scalar region, $\eta = 1.5 \cdot 0.34$, which occurs when $m_1 = 2.3 \, m_2$. Also this model arrives at the result that the fundamental shall have about twice the exposure of the first harmonic.

The restriction to the scalar region, however, is a severe one, since most gratings are to be used in the electromagnetic domain. Hence, electromagnetic theory has to be used to find optimum design also with regard to bandwidth, a topic discussed in a separate paper in this issue [13].

Production of Fourier blazed gratings

We want to exemplify our process of making Fourier blazed gratings by the following description of the production of a 1800 gr/mm grating.

A substrate is coated with photoresist by the usual spinning technique. As a first exposure, at its rim, a 900 gr/mm reference grating is recorded. The unexposed areas remain unaffected by the developer, due to the low concentration we use.

When we reinsert the ring shaped reference grating into the exposure set-up the two plane wavefronts give rise to a number of diffracted orders from the reference grating. When first order diffractions from the two fields are brought close together a visible interference pattern is projected on a screen (Fig. 6 a). With this interference pattern as a guidance, fine adjustments are made until the interference fringe is considerably larger than the substrate (Fig. 6 b). In the central part of the substrate, surrounded by the reference, we now can make our first exposure, which is 1800 gr/mm, exactly twice the groove density of the reference grating. Subsequently, we repeat the procedure with two second order diffractions brought together (Fig. 6 c). The fringe density becomes 3600 gr/mm, exactly four times that of the reference grating. With this adjustment the first harmonic is exposed. During both these exposures the active phase control system stabilizes the phase in a predetermined position.

After development, we obtain a 1800 gr/mm Fourier blazed grating in the center while the reference grating

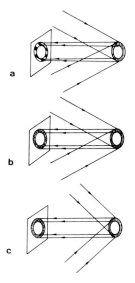

Figure 6. Illustration of adjustment of fundamental
and first harmonic exposure by means of
a reference grating.

    a. Imperfect adjustment of both fringe den-
sity and angular orientation. A skew interference fringe can be seen.
    b. Coincidence of first order diffractions
gives the fundamental frequency.
    c. Coincidence of second order diffractions
gives the first harmonic.

Figure 7. Electron micrograph showing the
profile of a Fourier blazed holographic grating with 600 grooves/mm.

becomes overexposed and disappears.

Experimental results

In order to confirm our theories we made Fourier blaze on 600 gr/mm gratings which fall in the scalar
domain and on 1800 gr/mm gratings in the electromagnetic domain. In Fig. 7, an electron microscope picture
shows the groove profile determined on such a grating. The asymmetry is obvious. The energy distribution
among the different diffraction orders for a grating blazed for the wavelength λ = 340 nm is shown in Fig. 8.

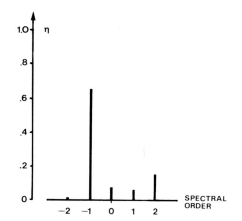

Figure 8. Diffraction efficiency distribution
into various orders of a Fourier
blazed grating with 600 grooves/mm.

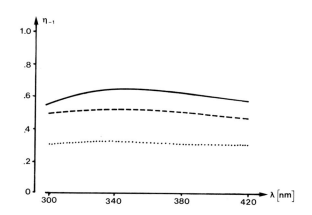

Figure 9. Comparison of theoretical and measured
efficiency values for a 600 gr/mm grating.
Continuous line, Fourier blazed grating
measured; dashed, Fourier blazed grating
calculated; and dotted, sinusoidal grating.

Fig. 9 displays the efficiency curve of the same grating together with theoretical curves for a simple sinusoidal grating and a Fourier blazed grating. The same kind of efficiency curves are seen in Fig. 10. However, in the electromagnetic domain the $E_{\parallel}$ and $H_{\parallel}$ polarisation directions have to be treated separately.

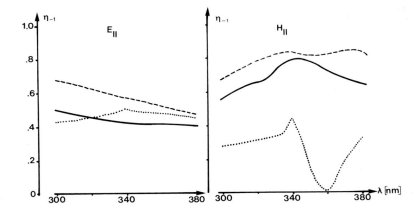

Figure 10.   Comparison of theoretical and measured efficiency values of two polarisation directions for a 1800 gr/mm grating. Continuous lines, Fourier blazed gratings measured; dashed, Fourier blazed gratings calculated; and dotted, sinusoidal grating.

The agreement between theoretical and practical results may be called fairly good. There are tolerances in such effects as nonperfect exposures, nonlinearity effects in the photoresist as well as in the reflective coating process which altogether have a non-negligible influence on the practical results. All the curves shown were calculated and measured with Al coating and for first-order Littrow configuration with an angular deviation of 10°. So far, our studies deal with the principle of Fourier blaze, the gratings shown were not intended for practical use. The diameter of these experimental gratings was 50 mm.

An example of instrument design

We may conclude this review on holographic gratings at the Institute of Optical Research with an example of an instrument design where the free choice of groove density has been essential.

In collaboration with the Department of Analytical Chemistry at the Royal Institute of Technology we developed a flame emission spectrometer for simultaneous determination of five elements [18]. The five elements (for serum analysis) are listed in table 1 together with a typical wavelength for each element used for spectroscopic analysis. It is desirable that these wavelengths can be brought together on the same fixed exit slit in the same spectrophotometer without rotation of the grating. This requirement can be fulfilled by using gratings with ruling characteristics as listed in table 1. Owing to the holographic production tech-

Table 1

Analyte	Wavelength (Å)	Groove density (mm^{-1})
Mg	2852	2500
Ca	4228	1686
Na	5895	1209
Li	6708	1063
K	7660	931

nique these odd groove densities can easily be realized and, moreover, they can be obtained on the same substrate, as is shown in Fig. 11.   The flame emission spectrometer built with this type of grating was of a Czerny-Turner type but with a cylindrical second mirror in order to separate the five wavelengths along the exit slit of the instrument, Fig. 12, where they were picked up by fiber optics.

The result was a stable and simple instrument that is easy to handle and with a performance comparable to other multichannel instruments.

Conclusions

We have given a survey on the current work on holographic diffraction gratings at the Institute of Optical Research. The principles of the instruments for exposing and testing holographic gratings have been outlined. It is shown that blazed gratings for a wide range of wavelengths and groove densities are feasible by synthesis of the profile from its Fourier components. It is, e.g., possible to produce gratings blazed for the visible region in which, to our knowledge, the frequently used Sheridon technique has not been shown to be applicable. The exceptional experimental demands have to a great extent been fulfilled by the precision of the instrument design and microcomputer controlled fringe adjustment and stabilization. Current research includes studies of electromagnetic diffraction theory for the calculation of optimum grating profiles and experimental investigations for the realization of high-efficiency, broad-band Fourier blazed holographic gratings.

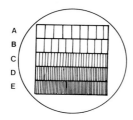

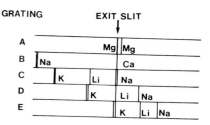

Figure 11. Schematic diagram of a composite holographic grating comprising 5 gratings with odd numbered groove densities.

Figure 12. Spectra obtained in the vicinity of the exit slit from the composite grating.

References

1. Siegbahn, M., "Från ljuset till röntgenstrålningen". Kosmos, vol. 8, pp. 164-196, 1930 and "Studies in the extreme UV and the very soft X-ray region". Proc. Phys. Soc., London, vol. 45, pp. 689-698, 1933.
2. Biedermann, K., "Silver Halide Materials" in Holographic Recording Materials, H. M. Smith, Ed., Topics in Applied Physics, vol. 20, pp. 21-74. Springer-Verlag 1977.
3. Johansson, S. and Biedermann, K., "A Microdensitometer without Lens Optics for Film MTF Evaluations". Optical Engineering, vol. 13, pp. 553-556, 1974.
4. Johansson, S. and Biedermann, K., " Multiple Sine-Slit Microdensitometer and MTF Evaluation for High Resolution Emulsions". Appl. Opt., vol. 13, pp. 2280-2291, 1974.
5. Biedermann, K. and Johansson, S., "A Method for Evaluating MTF and Phase Structures of Photographic Materials". J. Appl. Phys., (Japan), vol. 14, suppl. 14-1, pp. 241-246, 1974.
6. Rudolph, D. and Schmahl, G., "Spektroskopische Beugungsgitter hoher Teilungsgenauigkeit erzeugt mit Hilfe von Laserlicht und Photoresistschichten", Optik, vol. 30, pp. 475-487, 1970.
7. Schmahl, G. and Rudolph, D., "Holographic Diffraction Gratings". Progress in Optics, E. Wolf, Ed., vol. XIV, pp. 195-244, 1976.
8. Schmahl, G., " Holographically made Diffraction Gratings for the Visible UV and Soft X-Ray Region. J. Spectrosc. Soc. Japan, vol. 23, suppl. 1, pp. 3-11, 1974.
9. Johansson, S., Nilsson, L.-E., Biedermann, K. and Kleveby, K., "Holographic Diffraction Gratings with Asymmetrical Groove Profiles". Applications of Holography and Optical Data Processing, Edited by E. Marom, A. A. Friesem and E. Wiener-Avnear, Pergamon Press 1977, pp. 521-530.

10. Breidne, M., Johansson, S., Nilsson, L.E., and Åhlén, H., "Blazed Holographic Gratings". Optica Acta, vol. 26, pp. 1427-1441, 1979.
11. Lindau, S., "A Simple Method for Determining the Groove Depth of Holographic Gratings". This volume.
12. Nilsson, L.E., and Åhlén H., "Stabilization of the Exposing Interference Pattern in Holographic Grating Production". This volume.
13. Breidne, M., and Maystre, D., "Optimization of Fourier Gratings". This volume.
14. Breidne, M., and Maystre, D., "Equivalence of ruled, holographic and lamellar gratings in constant deviation mountings". Appl. Opt., vol. 19, pp. 1812-1821, 1980.
15. Maystre D., "A New Integral Theory for Dielectric Coated Gratings". J. Opt. Soc. Am., vol. 68, pp. 490-495, 1978.
16. McPhedran, R.C., Wilson I.J., and Waterworth, M.D., "Profile Formation in Holographic Diffraction Gratings". Optics Laser Technol., vol. 5, p. 166, 1973.
17. Goodman J.W., " Introduction to Fourier Optics". McGraw Hill, 1968.
18. Johansson, A., and Nilsson, L.E., "A Multiple Grating Flame Photometer for simultaneous Determination of Five Elements". Spectrochem. Acta, vol. 31 B, pp. 419-428, 1976.
19. Breidne, M., and Maystre, D., "100 % Efficiency Gratings in Non-Littrow Configurations". This volume.

Soft x-ray transmission gratings

E. T. Arakawa, P. J. Caldwell, M. W. Williams

Health and Safety Research Division, Oak Ridge National Laboratory, Oak Ridge, Tennessee 37830

Abstract

A technique has been developed for producing transmission diffraction gratings suitable for use in the soft x-ray region. Thin self-supporting films of a transparent material are overlaid with several thousand opaque metallic strips per mm. Gratings with 2100, 2400, and 5600 l/mm have been produced and tested. Representative spectra over the wavelength range from 17.2 to 40.0 nm are given for a grating consisting of a 120-nm-thick Al support layer overlaid with 2400, 34-nm-thick, Ag strips/mm. The absolute transmittance is ~13% at 30 nm, and the efficiency in the first order is ~16%. The observed resolution of ~2Å is acceptable for many of the potential applications. These gratings have several advantages over the two presently available alternatives in the soft x-ray region (i.e., reflection gratings used at grazing incidence and free-standing metallic wire transmission gratings). Fabrication is relatively quick, simple, and cheap. The support layer can also serve as a filter and help conduct excessive heat away. Higher line densities and hence higher resolutions are possible, and when used at normal incidence the spectra are aberration free. Suitable materials, component thicknesses, and line densities can be chosen to produce a grating of optimum characteristics for a particular application.

Introduction

In the soft x-ray region, reflection gratings must be used at grazing incidence because of the low normal incidence reflectance of all materials at these wavelengths. Grazing incidence geometries are cumbersome, efficiencies are relatively low, and astigmatic imaging properties are a problem. On the other hand a transmission grating can be used at normal incidence, resulting in a very simple geometry with reasonable transmittance, efficiencies, resolution, and stigmatic imaging properties.

Interferometrically formed, electrodeposited, free-standing gold transmission gratings supported by a randomized auxiliary gold grid system[1,2] have been produced by a complex process. Later, gratings produced by the same techniques have been developed with a regular support structure[3,4] with a resultant reduction in scattering compared with gratings supported by a randomized structure. The highest line density reported is 1000 l/mm, the limiting factor being the mechanical rigidity of the free-standing gold strips.

We have recently developed a new method for the production of transmission gratings suitable for use in the soft x-ray region. In this paper, we present some representative spectra obtained using one of our gratings, and compare and contrast these gratings and their use with other gratings available for use in the soft x-ray region.

Grating production

Our method for producing holographic transmission gratings for use in the soft x-ray region has been described previously.[5,6] Photoresist is spin-coated onto a glass substrate and then exposed to the interference pattern created by two coherent point sources. The sinusoidal relief pattern which is formed after development is coated at normal incidence with a suitably transparent material by vacuum evaporation. An opaque metal is then added obliquely (i.e., to one side of the sinusoidal profile) to provide the necessary wavefront amplitude modulation. Gratings are then removed from the substrate by dissolving the underlying photoresist.

So far we have fabricated gratings with support layers of carbon or aluminum, chosen to be transparent to the wavelength region of interest and metallic strips of silver, copper, or gold chosen to be opaque over the same wavelength region. Line densities of these gratings are 2100 and 2400 l/mm. A grating with 5600 l/mm has also been produced by passing the interfering beams through a prism optically coupled to the surface of the photoresist, increasing the line density by a factor of the index of refraction of the prism.

Test results

Our transmission gratings were tested in the system illustrated in Fig. 1. Radiation from a condensed spark source is limited by a slit before striking the grating at an angle of incidence, ϕ. The diffracted spectra are then recorded as a function of the angle of diffraction, θ, by a channel electron multiplier. Figures 2 and 3 show the first order spectra over the wavelength range from 17.2 to 40.0 nm for $\phi=0°$ and $\phi=50°$, respectively. The grating consisted of a 120-nm-thick Al support layer overlaid with 2400, 34-nm-thick, Ag strips/mm. The size of the entrance slit was 70μ and the size of the exit slit was 150μ. The absolute transmittance for this grating is ~13% compared with a theoretical 35% at 30 nm. This latter figure is calculated on the assumption that the "transparent" regions and opaque strips are of equal sizes and then allowing for the known absorption in the Al "transparent" regions.[7] The efficiency in the first order was found to be 16% of the 0 order, whereas an ideal, half-open, symmetric transmission grating has a theoretical efficiency in the first order of 40% of the 0 order.[2] The observed resolution is ~3.5Å at normal incidence (see Fig. 2) and ~2Å at a 50° angle of incidence (see Fig. 3). In the present case, the 2Å resolution is not limited by the grating but by the single slit diffraction pattern of the entrance slit and by the non-focussing entrance and exit slit geometry used in the experiment. Using the grating at a 50° angle of incidence spreads the spectrum out over a larger angular range but, in the example shown, does not alter the efficiency significantly, as might be expected if the fabrication techniques had produced a "blaze" on the grating.

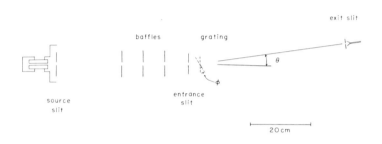

Figure 1. Soft x-ray transmission monochromator.

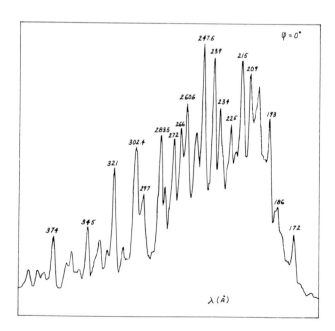

Figure 2. First order spectrum of condensed spark source produced by a transmission grating of 2400 1/mm used at normal incidence.

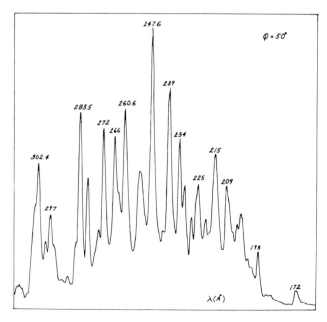

Figure 3. First order spectrum of condensed spark source produced by the same grating as for Fig. 2 used at a 50° angle of incidence.

For comparison with the spectra produced using our holographic transmission grating, Fig. 4 shows the spectrum for the same condensed spark source dispersed by a 2.2 m grazing-incidence monochromator employing an aluminum reflection grating with 300 l/mm. In this case the size of the entrance slit was 100µ and that of the exit slit was 250µ.

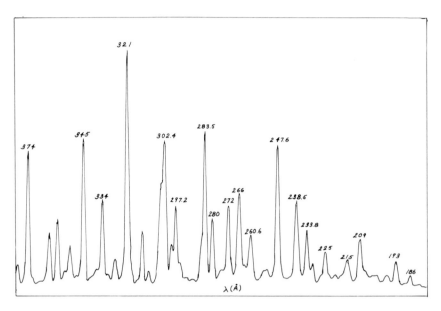

Figure 4. First order spectrum of condensed spark source produced by a 2.2 meter grazing incidence monochromator.

Figures 2 and 3 are typical of the spectra obtained with 2400 l/mm gratings. The grating with 5600 l/mm gave spectra which were similar except that the efficiency was lower.

Discussion

A comparison of the spectra obtained using the holographic transmission monochromator with that obtained using the grazing incidence reflection monochromator shows that, as expected, the positions of the structures in the spectra agree. The intensities of individual spectral lines depend on many factors including the pressure of the gas in the source discharge tube, the blaze on the grating, and, for the transmission grating, the absorption characteristics of the "transparent" support layer. The resolution observed in the recorded spectra also depends on many factors, including the width of the entrance slit, the rate of scanning, and the characteristics of the grating. Generally the resolving power improves as the grating line density increases. Since, in our holographic gratings, the support layer is continuous and thick enough to be self-supporting, the mechanical rigidity of the metal strips is not the limiting factor in the line density which can be achieved. For practical purposes, at the moment, we are limited to gratings which can be produced with the 457.9 nm emission line from an Argon ion CW laser. The highest line density which we have produced with this laser source is 5600 l/mm. Thus it is seen that these new holographically produced gratings should be capable of higher resolutions than other gratings used in the soft x-ray region. The demonstrated resolution of ∿2Å is acceptable for many of the potential applications for these gratings; their unique advantages offsetting the higher resolutions presently attainable with a grazing-incidence reflection monochromator.

The gratings are light-weight transmission gratings which can be used in a simple monochromator geometry with stigmatic imaging properties. The support layer can be used as a filter, transmitting a chosen wavelength range while absorbing other wavelengths. In this connection, it can be used to sort out the interfering multiple orders encountered in existing spectrometers used at synchrotron radiation storage rings. In addition, these gratings could be used in new, simple, designs of soft x-ray spectrometers, and as the monochromatic window for space telescopes. The support layer can also serve to help conduct heat away from the grating, thus lengthening the useful life of a grating used with high intensity, short wavelength sources, such as synchrotron radiation.

Acknowledgments

One of the authors (P. J. Caldwell) is also with the Department of Physics and Astronomy, University of Tennessee, Knoxville, Tennessee 37916. This research was sponsored by the Office of Health and Environmental Research, U.S. Department of Energy, under contract W-7405-eng-26 with the Union Carbide Corporation.

References

1. Schnopper, H. W., Van Speybroeck, L. P., Delvaille, J. P., Epstein, A., Källne, E., Bachrach, R. Z., Dijkstra, J., and Lantward, L., "Diffraction Grating Transmission Efficiencies for XUV and Soft X-Rays," Appl. Opt., Vol. 16, pp. 1088-1091. 1977.

2. Källne, E., Schnopper, H. W., Delvaille, J. P., Van Speybroeck, L. P., and Bachrach, R. Z., "Properties and Prospects of Holographic Transmission Gratings," Nucl. Instrum. Methods, Vol. 152, pp. 103-107. 1978.

3. Delvaille, J. P., Schnopper, H. W., Källne, E., Lindau, I., Tatchyn, R., Gutcheck, R. A., and Bachrach, R. Z., "Diffraction Efficiencies of Holographic Transmission Gratings in the Region 80-1300 eV," Nucl. Instrum. Methods (in press).

4. Bräuninger, H., Predehl, P., and Beuermann, K. P., "Transmission Grating Efficiencies for Wavelengths Between 5.4Å and 44.8Å," Appl. Opt., Vol. 18, pp. 368-373. 1979.

5. Arakawa, E. T. and Caldwell, P. J., "Holographically Produced Transmission Diffraction Gratings for Soft X-Rays," Nucl. Instrum. Methods (in press).

6. Caldwell, P. J. and Arakawa, E. T., "Transmission Diffraction Gratings for the 20 to 200 eV Region," Proc. VIth Int. Conf. Vacuum UV Radiation Physics, Charlottesville, Virginia, June 2-6, 1980.

7. Hagemann, H.-J., Gudat, W., and Kunz, C., "Optical Constants from the Far Infrared to the X-Ray Region: Mg, Al, Cu, Ag, Au, Bi, C, and Al_2O_3," DESY Report SR-74/7 (May 1974).

Design of holographic gratings on primary mirrors for beam sampling

R. J. Withrington, N. Wu, H. M. Spencer

Hughes Aircraft Company, Culver City, California 90230

Abstract

Nonlinear holographic gratings are of interest in laser systems for sampling beams from the primary mirror of a large beam expander. The sampled beam is then used for wavefront control purposes. This paper discusses the optical design and geometrical properties of the grating and its construction optics. The design and performance of a 30-cm aperture breadboard covering a 4-mrad field of view with an 8X, f/1.5 beam expander are then discussed. The wavefront aberration differences between the sampled and laser beams are corrected to within a maximum error of 0.3 μm.

Introduction

Nonlinear holographic gratings are of considerable interest in high-energy laser (HEL) systems for sampling the beam from the primary mirror of a large beam expander since in such systems it is extremely difficult to obtain a beam that is close to diffraction-limited without using adaptive optics. Also, because the primary mirror is the last element in the optical train of the outgoing beam, wavefront errors in this beam are imprinted on the sampled beam. This beam can then in principle be used for wavefront error diagnostic and control purposes. It should be noted that rather than sampling the high-energy laser beam itself, the general requirement is to sample an alignment beam (typically at 6328Å) which traverses the same optical path as the HEL beam. In this paper, it is assumed that methods of inserting this alignment beam at a suitable location in the optical train are indeed available because such insertion involves a different technological area.

The type of optical system discussed is shown in Figure 1, and the typical range of optical characteristics of interest is shown in Table 1. The input laser beam traverses a

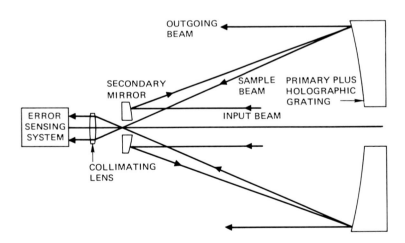

Figure 1 Optical schematic of a holographic beam-sampling system.

Table 1. General range of optical characteristics of interest
in primary mirror holographic beam-sampling systems

Parameter	Value
Primary mirror f-number	1.25 to 2
Beam expander magnification	6 to 20
Outgoing beam FOV, mrad	2 to 4
Sampled beam magnification	6 to 30
Sampled beam wavelength, μm	0.6328
Hologram construction wavelength, μm	0.4131

pair of confocal parabolic mirrors that constitute the main beam expander. For compactness, the f-number of the primary mirror lies between f/1.2 and f/2, and the afocal magnification ratio is between 6X and 20X. The sampled beam is diffracted by a holographic grating on the primary mirror and is focused through a hole in the secondary mirror. This beam is subsequently collimated by a collimating lens before it reaches a sensor package containing a wavefront error diagnostic system. The collimating lens corrects for aberrations in the hologram, and also images the primary mirror (aperture stop) onto the error sensor entrance pupil. The afocal magnification ratio between the outgoing beam space and the error sensor space is at least 6X and is preferably 30X or more in order to provide a small sensor package. Although the laser points in only one direction at a time, there is a field of view (FOV) requirement that the beam be capable of being pointed off the optical axis of the beam expander. Referred to the outgoing beam space, this FOV is about 2 to 4 mrad, and hence implies a field of view of up to 6° or so at the error sensor.

The following section describes optical design considerations in the hologram construction optics, and includes a brief discussion of the aberrations in the hologram. This description is followed by a design example of the sampled beam optical path that ends at the error sensor entrance pupil. The performance of the design is discussed in terms of the wavefront aberrations that occur across the FOV of interest. The design is that for a 30-cm breadboard that is being fabricated.

Hologram construction optics design

The steps in fabricating the holographic grating are to coat the primary mirror with photo-resist, expose this coating to the holographically generated grating pattern, develop the photo-resist coating, and then etch the grating into the mirror surface by means of ion machining. Because of the blue sensitivity of the phot-resist, the exposure wavelength is the 4131Å line of a krypton laser. This line is also advantageous in that 2 to 3 watts of power can be obtained, although the exposure times for sizable primary mirrors are still unduly long. The resultant difference between the hologram construction and playback wavelengths must be taken into account in establishing the construction point source locations in order to obtain the desired focal length and to minimize the hologram aberrations. It should also be noted that since the primary mirror is in effect the system aperture stop, nothing can be done in the hologram construction optics to minimize field aberrations.

The initial design of the construction optics therefore centers around minimizing spherical aberration. The sources used in the construction optics are at distances R_1 and R_2, respectively, from the hologram; see Figure 2. The relationship between R_1 and R_2 that

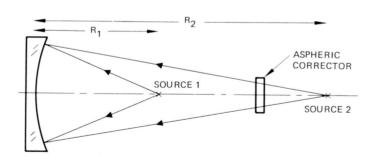

Figure 2 Hologram construction optics conjugates. An aspheric can be used to correct residual spherical aberration.

is needed to give the desired focal length is illustrated in Figure 3 for a 1-meter focal length primary mirror with a hologram focal length of 90 cm. Also shown is the variation in spherical aberration with the long conjugate distance R_2. The pair of distances at which the spherical aberration passes through zero provides a natural design starting point; this occurs when R_1 is 97.7 cm and R_2 is 335 cm. However, residual higher-order spherical aberration remains (see Figure 4) and must be corrected in the integrated design of the construction optics and main system. As indicated in Figure 2, an aspheric in one of the construction optics beams will, of course, provide this correction, but it is preferable to obtain the correction desired by using only spherical surfaces.

The next stage in the construction optics design is to minimize the obscuration associated with source 1 and to take care to eliminate as much scattered and diffracted light as possible. The latter requirement is of particular concern because any coherent stray radiation causes spurious interference patterns to be recorded. These patterns can, in fact, be so severe as to result in an unacceptably poor hologram. One approach used at Hughes

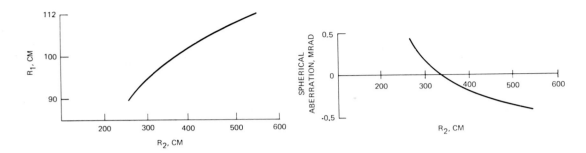

Figure 3. Relationship between point-source distances from primary mirror and spherical aberration in hologram

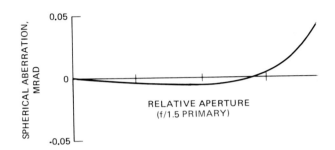

Figure 4. Residual high-order spherical aberration in sampled beam at hologram focus when constructed with spherical wavefronts. The aberration is expressed in milliradians referred to outgoing beam space.

Aircraft Company in the fabrication of 30-cm-diameter test holograms is shown in Figure 5. A large folding mirror separates the two beams from each other. With beam 1 passing through a small hole in the folding mirror, the obscuration of beam 2 is minimal and can be less than the natural obscuration of the beam expander. Figure 5 also shows the use of positive

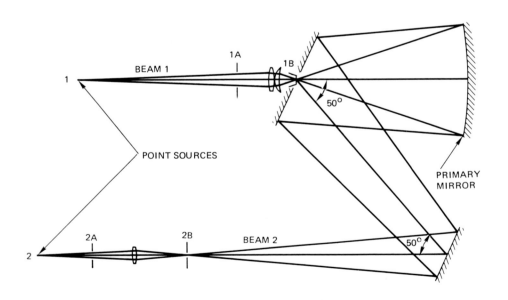

Figure 5. Hologram construction optics for minimum obscuration of beam 2 by beam 1. Stops 1,A,1B,2A, and 2B are used to control stray radiation.

lenses to cause the two construction beams to diverge to the desired f-numbers in order to fill the primary mirror aperture. This arrangement allows f/10 beams to be used at the pinholes. Also, it permits aperture stops to cut down stray radiation very effectively. Stops 1A and 2A, which prevent illumination of the lens edges etc., are positioned in such a way that they are imaged at the primary mirror by the lenses. Consequently, radiation scattered or diffracted from the edges of the stops does not spread across the desired hologram aperture. Stops 1B and 2B are the equivalent of oversized field stops and effectively keep most of the remaining stray radiation from reaching the primary mirror. Lastly, it should be noted that all of the lenses have spherical surfaces and are bent to assist in correcting for higher-order spherical aberration.

A difficulty encountered with the construction optics shown in Figure 5 is the length of the long conjugate. For mirrors of modest sizes, this is no real problem, but this length can reach several meters in larger systems. One way to alleviate this difficulty is shown in Figure 6. The large folding flat is replaced by a large sphere in order that the source for the long conjugate beam may be close to the vertex of the primary mirror. In this way the overall length of the construction optics is shortened to approximately the short conjugate distance, and spherical aberration can still be correctly balanced. Although the construction optics are now reduced to more manageable dimensions, problems of stray radiation are compounded because of the possibility of multiple reflections between the two mirrors. The

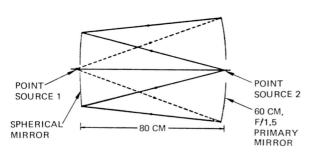

Figure 6. Compact hologram construction optics using a spherical mirror in long conjugate beam path.

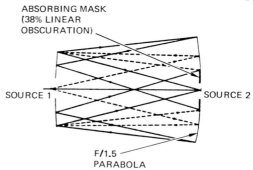

Figure 7. Multiple reflections between mirrors can be blocked by a mask that covers the central linear 38% of primary mirror.

result is a linear obscuration of about 38 percent for an f/1.5 primary mirror (see Figure 7), and a good absorbent mask must be used to cover the central area to prevent further multiple reflections. As with the configuration shown in Figure 5, positive lenses and stops are used in the configuration shown in Figure 6 to cause the two laser beams to diverge and to minimize stray radiation.

Field aberrations at sampled beam focus

The combination of the primary mirror and holographic grating is optically equivalent to the parabolic primary mirror followed by a transmission holographic grating. Viewed in this way, the sampled beam first traverses the beam expander which has the well known property of being corrected for all third-order aberrations except for field curvature. The aberrations at the sampled beam focus are therefore due to the transmission hologram which is, in effect, a single element being operated at a speed of between f/1 and f/2 and consequently has large off-axis aberrations. By far the dominant aberration is coma, which is shown in Figure 8 for a 1-meter focal length primary mirror and 90-cm focal length hologram. For ease of scaling the aberration has been expressed in milliradians by dividing by the hologram focal length. In different applications, the ratio between the focal length of the hologram and that of the primary mirror can vary somewhat from the value of 0.9 assumed in Figure 8, but such variation does not greatly affect the amount of coma present. Some astigmatism and field curvature are also present, but they are relatively negligible in comparison with the coma.

The diffraction limit of the outgoing beam can vary from at most 30 mrad down to a few microradians, depending on the infrared HEL laser wavelength and the diameter of the primary mirror. Hence it is clear that the coma in the hologram must be balanced in the collimating lens to provide correction over the desired 2- to 4-mrad FOV. It should also be noted that this coma implies that the collimating lens must be well aligned in relation to the primary mirror since a small decentration introduces a corresponding amount of coma on axis.

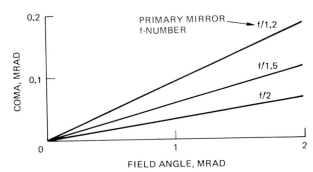

Figure 8. Coma as a function of field angle and f-number at sampled beam focus. Coma is expressed in milliradians referred to outgoing beam space.

Design of a 30-cm breadboard system

A breadboard of a holographic beam sampling system of the type discussed in the previous sections is being built to illustrate the principle involved. It uses an 8X beam expander with an f/1.5 parabolic primary mirror. The sampled beam leaving the collimating lens is also 8X in comparison with the outgoing beam, the criterion for the breadboard being to provide a beam of a convenient size for wavefront measurement in the laboratory rather than to ensure compactness. A ray trace of the system is shown in Figure 9, and the optical characteristics are summarized in Table 2.

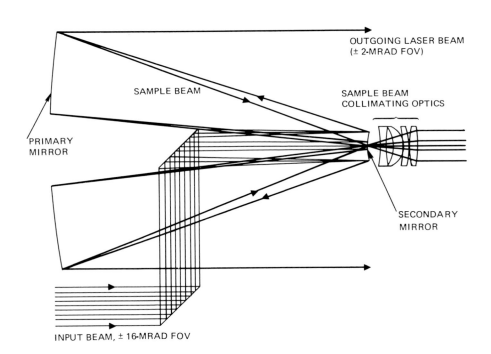

Figure 9. Ray trace of 30-cm breadboard optical system.

Table 2. Optical characteristics of breadboard beam-sampling design

Parameter	Value
Aperture diameter, cm	30
Primary f-number	1.5
Beam expander magnification	8
Sampled beam magnification	8
Field of view (outgoing beam), mrad	4

The collimating lens consists of a negative element followed by a closely spaced group of three positive elements. This configuration makes it possible to introduce sufficient coma to the collimating lens to balance the coma in the hologram. Also, the negative lens assists in matching the field curvature of the sampled beam with that of the outgoing beam. The objective here is to ensure that the sampled beam wavefront error matches that of the outgoing beam as closely as possible over all field angles.

The sampled beam wavefront aberrations are shown in Figures 10 and 11, respectively; it is seen that they differ from those of the outgoing beam by at most 0.3 μm, the off-axis wavefront error in the outgoing beam being pure field curvature. Whereas this level of correction is somewhat large compared with the 0.6328-μm sampled wavelength, it is acceptably small in relation to an actual HEL wavelength. Also, over most of the aperture and field, the actual match between the sampled and the main beam wavefronts is considerably below 0.3 μm. The spherical aberration is almost completely corrected, as is also the field curvature and astigmatism. The principal source of residual aberration is the residual coma evident in the wavefront aberration curves in Figure 10 and in the contour plots in Figure 11.

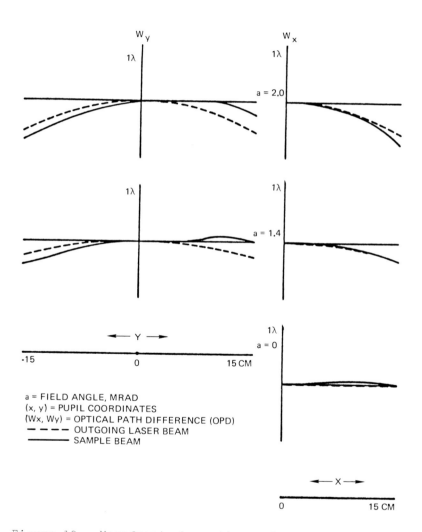

Figure 10. Wavefront aberrations of sampled and outgoing beams (λ = 0.6328 μm).

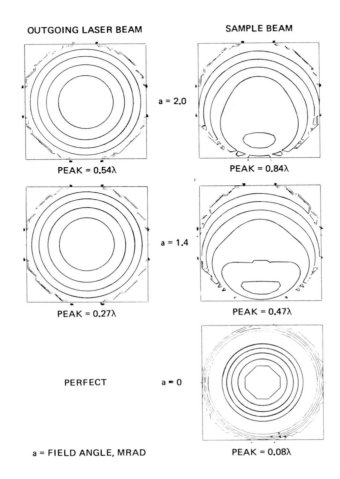

OUTGOING LASER BEAM   SAMPLE BEAM

a = 2.0

PEAK = 0.54λ   PEAK = 0.84λ

a = 1.4

PEAK = 0.27λ   PEAK = 0.47λ

PERFECT   a = 0

a = FIELD ANGLE, MRAD   PEAK = 0.08λ

Figure 11. Wavefront contour maps of outgoing and sampled beams

Conclusions

Some considerations involved in primary mirror holographic beam-sampling optical design have been discussed, including the hologram construction optics. It has been shown that the aberrations in the hologram can be readily corrected in a breadboard design of modest size. For reference purposes, a similar design for a 60-cm aperture has also undergone a preliminary phase. This system has a sampled beam magnification of 30X and a 3-mrad FOV. The current level of wavefront correction is about 0.6 μm peak and can be increased, a fact that indicates that the better performance required of intermediate-sized systems may be achievable, but that for larger systems there is still a considerable optical design challenge to be faced.

Acknowledgments

The work presented in this paper was primarily supported by the Holographic Grating Study and Holographic Alignment Brassboard Programs funded by DARPA through the Rome Air Development Center.

Ion-etched gratings for laser applications

Hugh L. Garvin, Anson Au, Monica L. Minden
Hughes Research Laboratories, Malibu, California 90265

Abstract

Ion beam sputter etching has proven to be a superior technique for producing grating sampling mirrors for large optical systems. The patterns to be etched are defined by a photoresist masking film on the mirror surface. Grating patterns have been produced on laser mirrors by replication of diamond-scribed master patterns, while holographic construction has been used to produce linear and nonlinear gratings. The microscopic details of ion etched grating profiles show that the process is capable of high resolution pattern delineation and large area device fabrication.

Introduction

In the early 1970s, it was demonstrated that ion beam sputter etching was a useful technique for the production of optical components. Ion sources that had been developed as prime electric propulsion devices on outer space missions were adapted to perform high-resolution etching of microelectronic and IR optical devices.[1-3] These sources also provided uniform and controllable etching over large surface areas of high-power laser sampling mirrors.[4] In the sputtering process, ions are produced from inert gas (usually argon) in the discharge chamber of the ion source. These ions are accelerated to pass through a vacuum region and (as shown in Figure 1) impinge on the surface to be etched. Momentum transfer from the incident ions causes material to be ejected from the surface; since this process is very repeatable and atomic in scale, it is capable of delineating very fine details in the etching process. The pattern to be etched is applied to the surface by a photo- or electron-sensitive masking film; this film survives the etching process and is generally dissolved away when the etching is complete. The resolution of patterns achieved by the ion beam etching processes has been shown to be limited only by the ability to produce the masking pattern on the surface, and these patterns have been produced by several methods. (1) Conventional contact photo masking with UV exposure has been used to produce patterns with details as small as 1.5 μm. Original master patterns have been generated by photo reduction of original art or by diamond scribing of grating patterns into metallic films on glass substrates. (2) Electron beams, X rays, and ion beams have been used for exposing certain masking films, and device geometries have been produced and replicated with dimensions down to 0.1 μm. (3) Large-area (30 to 60 cm in diameter) optical diffraction gratings have been produced by laser holography. In these patterns, the grating spacings may be linear or they may be designed for focusing a particular coherent radiation. In either case, the optical phase of the pattern must be maintained over the large surface, and this can be done better holographically than by the other pattern generation processes.

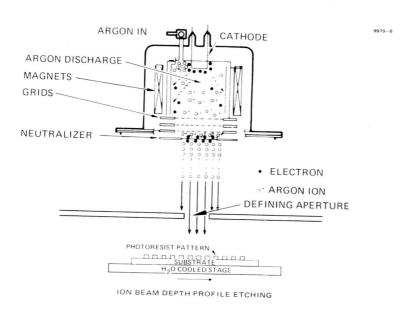

Figure 1. Schematic arrangement for ion beam sputter etching.

Ion beam etching process

In the ion beam etching process, the depth of etch is controlled by the number of incident ions and their energy. Commercially available ion sources produce nearly constant beam distribution and particle energy; therefore, to achieve a uniform etch depth requires only that the time exposure of the ion beam to the surfaces be uniform. One approach to etching a large mirror surface is to sweep the ion beam over the surface in a raster fashion. Ion beams (typically 1 mA/cm²) are not easily deflected by electrostatic fields; however, the task can be accomplished by mechanically moving the ion source to sweep the ion beam across the substrate. Stabilization of the beam and monitoring the resultant etching are somewhat difficult when using this arrangement. An alternative approach (illustrated in Figure 1), which has proven satisfactory in several programs, is to pass the mirror substrate behind a masking aperture which may be shaped to correct for nonuniformities in the incident ion beam density. When depth variations are required (as in variable efficiency feedback gratings or in reflective array surface wave acoustic devices), the speed of travel behind the aperture is programmed to produce the desired depth profile.

Figure 2 shows the manner in which a grating beam sampler diffracts a portion of a high power laser beam. This diffracted signal can be analyzed to determine the intensity distribution and wavefront characteristics of the main (high power) laser beam, or it may be used to provide steering signals for controlling the direction of the main output beam. Figure 3 is a photograph of a 35 cm x 30 cm rectangular beam sampler made by the ion beam etching technique. Half of the shaped masking aperture has been removed to show the mirror, but the left side of the mask illustrates the shape required to compensate for the decrease in ion beam intensity at the ends. When radiation detection instruments are to be used, the intensity of the diffracted signal generally needs to be very low compared to the main laser beam. As an example, the grating depth for a rectangular-groove-shaped grating used at 10.6 μm radiation will be about 950 Å for 1% sampling of the main beam. In some applications, the diffracted beam is still too intense and it must be sampled by a second low-efficiency grating to reduce the final signal to an acceptable detector level.

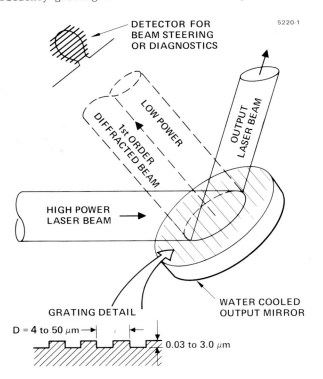

Figure 2. Grating laser beam sampler.

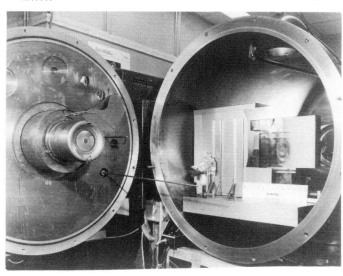

Figure 3. Laser beam sampling mirror in ion etching system.

The theory of diffraction from gratings of various depths and shapes has been treated analytically by Loewen et al.,[5] and our results for ion beam etched gratings have been in good agreement with this theory. Figure 4 compares theoretical and experimental results for groove depth variations as studied by laser radiation at 3.39 μm. When only one diffracted order is allowed or when symmetric orders are added, diffraction efficiencies in excess of 95% are observed for grating depths near λ/4. Efficiency is also influenced by the shape of the grating profile (as illustrated by the high efficiency of blazed gratings used for laser wavelength tuning). In addition, the diffraction efficiencies and phase relationships of radiation polarized parallel to (TE) or perpendicular to (TM) the grating lines are dependent on the relationship of the groove width to the grating period. Detailed results of these studies are discussed in a separate paper.[6]

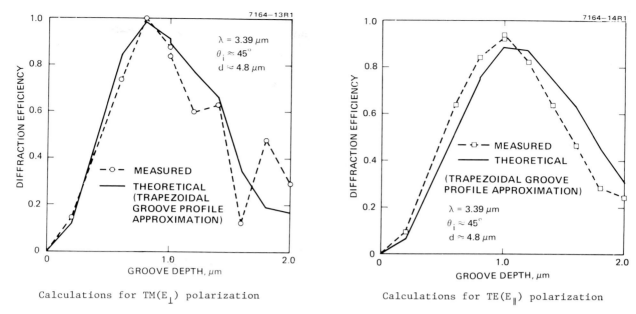

Calculations for TM($E_\perp$) polarization   Calculations for TE($E_\parallel$) polarization

Figure 4.   Ten high-efficiency samples.

Nonlinear grating applications

Several laser applications have required diffracting mirror surfaces made by the techniques described above. One example was the use of a highly efficient circular grating to serve as an axicon, replacing a conical mirror in a laser beam. Figure 5 shows schematically the manner in which the replacement is possible. The circular pattern for this grating was made by a computer-driven pattern generator, and then the photoresist mask was used while ion beam etching 2.0 μm deep into the gold mirror surface. Figure 6 shows an SEM photograph of the resultant grating.

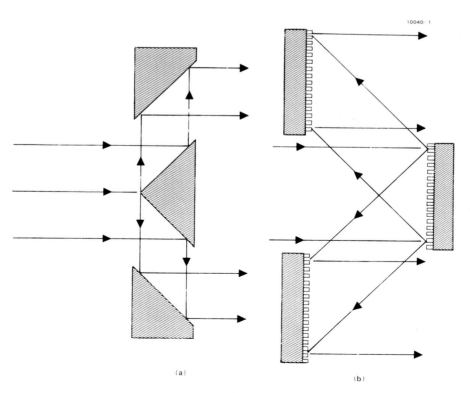

(a)   (b)

Figure 5.   Schematic arrangement for (a) conical mirror and (b) grating axicon.

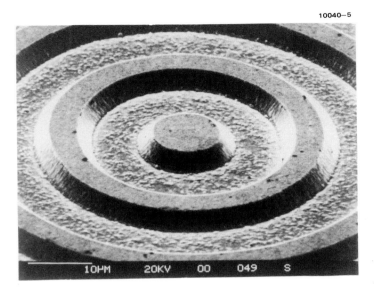

Figure 6. Scanning electron microscope (SEM)
photograph of high efficiency circular
grating.

In other applications, laser holography was used to produce nonlinear diffraction patterns on flat and non-flat mirror surfaces. For these applications, point source diverging laser beams were used to construct a focusing grating pattern which was ion beam etched into the surface of the flat mirror. The resultant etched pattern provides a focused diffracted beam (as shown in Figure 7); the position of this focused spot varies with the wavelength of the incident radiation. Figure 8 shows a photograph of the diffracted orders produced by this optical element.

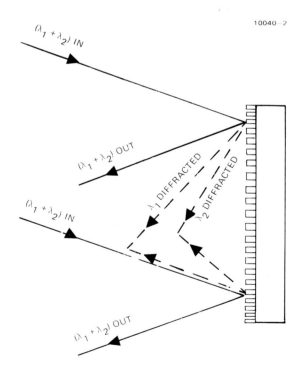

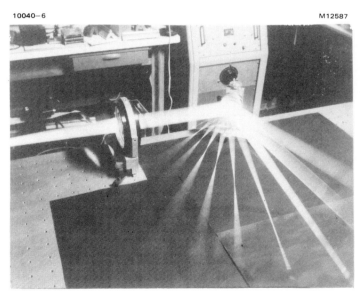

Figure 7. Focusing effect of nonlinear
holographic grating.

Figure 8. Diffracted orders of nonlinear
holographic grating.

In larger laser beam systems now under construction, ion etching will be used with holographic pattern generation techniques to provide multiple wavelength operation in non-flat mirror substrates as shown in Figure 9. While the parabolic mirror substrate provides long focal length operation at one wavelength, λ_1, a steering beam at λ_2 is simultaneously focused onto sensors to actuate servo controls.

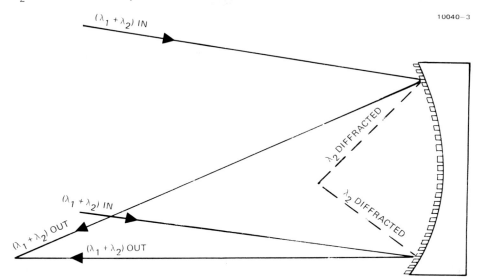

Figure 9. Focusing effect of nonlinear grating on non-flat mirror surface.

Wire grid polarizer fabrication

Wire grid polarizers are another example of an ion beam etched component that has proven useful in IR laser systems. The operation of the grid structure shown in Figure 10 was predicted by electromagnetic theory in early designs of antenna arrays for microwave radiation.[7] Photolithographic techniques made these devices available for long-wavelength IR spectrometers, and the application of holographic and ion etching techniques has pushed the performance nearly to the visible range. Figure 11 summarizes the performance of wire grid polarizers made on IR-transparent substrates. The etched grid with a period of 1.1 μm (Figure 12) gives a value of λ/d of 10 for 10.6-μm radiation. Extinction ratios (ERs) of 100 to 1000 are typical for this device. When finer gratings (with a 0.4-μm period, as shown in Figure 13) were produced and ion etched, the polarizers exhibited ERs of 2000 to 1 for 3.39-μm radiation and greater than 50,000 to 1 for 10.6 μm. These devices plus the grating diffraction devices described above have resulted from the application of the high-resolution pattern generation of laser holography plus the etching capability of ion beam sputtering. These techniques of fabrication are well suited to high-resolution (submicrometer) device features, and, at the same time, they are scalable to the production of very large (meter diameter) optical components. In the rapidly advancing field of microfabrication, very few processing technologies have such breadth of capabilities.

Acknowledgments

The authors acknowledge the technical fabrication skills of Mr. K. Robinson in preparing these laser components and the system analysis and consultation of Dr. G.M. Janney.

References

1. Garvin, H.L., "High Resolution Fabrication by Ion Beam Sputtering," Solid State Technology, Vol. 16, No. 11, p. 31 (November 1973).
2. Garvin, H.L., Garmire, E., Somekh, S., Stoll, H., and Yariv, A., "Ion Beam Micromachining of Integrated Optics Components," Appl. Opt., Vol. 12, p. 455 (1973).
3. Garvin, H.L. and Kiefer, J.E., "Wire-Grid Polarizers for 10.6 μm Radiation," Proceedings of the IEEE/OSA Conference on Laser Engineering and Applications, May 30 - June 1, 1973, Washington, D.C., p. 100.
4. Janney, G.M. and Garvin, H.L., "Diffraction Gratings for High-Power Laser Applications," Proceedings of the IEEE/OSA Conference on Laser Engineering and Applications, May 30 - June 1, 1973, Washington, D.C., p. 100.
5. Loewen, E.G., Neviere, M., and Maystre, D., "Grating Efficiency Theory as It Applies to Blazed and Holographic Gratings," Appl. Opt., Vol. 16, p. 271 (1977).
6. Minden, M.L. and Dunning, G.J., "Polarization-Dependent Phase Shift in High Efficiency Gratings," paper presented at SPIE's Annual Technical Symposium, July 28 - August 1, 1980, San Diego, California.
7. Auton, J.P., "Infrared Transmission Polarizers by Photolithography," Appl. Opt., Vol. 6, p. 1023 (1967).

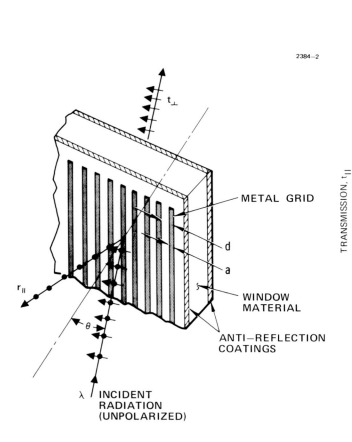

Figure 10. Schematic arrangement of wire grid
            polarizer.

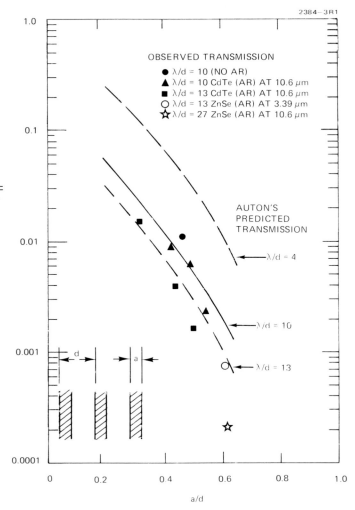

Figure 11. Extinction ratios observed for wire grid
            polarizers.

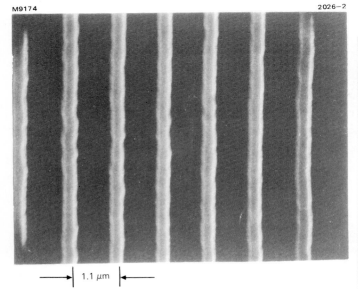

Figure 12. SEM photograph of wire grid polarizer
            with 1.1 μm period.

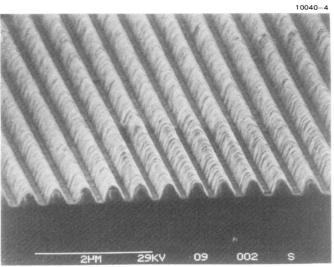

Figure 13. SEM photograph of wire grid polarizer
            with 0.4 μm period.

High-spectroscopic qualities in blazed ion-etched holographic gratings

Yoshinobu Aoyagi, *Kazuo Sano, Susumu Namba

The Insitute of Physical and Chemical Research, Wakoshi, Saitama, 351, Japan
*Shimazu Seisakusho, Ltd., Nishinokyo, Nakagyo, Kyoto, 604, Japan

Abstract

Blazed ion-etched holographic gratings with a high absolute diffraction efficiency, good diffracted wavefront, high resolution, and low stray light have been produced. The performances were proved to be the same as, or superior to, commercially available blazed gratings produced by a ruling technique.

Introduction

The holographic technique used to produce gratings has various advantages over the conventional ruling technique, eg. low cost, high groove positioning accuracy, and capability of producing high groove densities. However, with the holographic technique, the groove profile can not be controlled as readily as with a ruling engine. Recently, an ion etching technique was successfully applied to produce blazed holographic gratings[1] in which clear blazing properties in the diffraction efficiency were obtained. From a spectroscopic point of view, the resolution, stray light and wavefront quality are also important performance factors for gratings. In this paper we report the successful production of blazed ion-etched holographic gratings with high diffraction efficiency, low stray light and high resolution. It is emphasized that the ion-etched holographic gratings have practical applicability in monochromators and spectrophotometers.

Experiments

Necessary conditions for the substrate to produce ion-etched holographic gratings with good spectroscopic performance are a high etching rate for an ion beam, and high thermal stability of the grating blank. A glass blank spin-coated with polymethyl methacrylate(PMMA) satisfies these requirements[2]; it has a higher etching rate than the photoresists which are used as mask gratings for ion etching, and the thermal stability is the same as the polished glass blank itself. In our experiment we used spin-coated PMMA on the glass blank- with a dimension of 30×30 mm^2 - as the substrate. AZ 1350 photoresist was coated on the substrate and the grating was made by the conventional holographic technique with a groove density of 1200 lines/mm. The PMMA was ion-etched through the grating with the beam directed at an angle to the normal of the substrate, to make a triangular groove profile. The blaze angle can be controlled from 9 to 22 degrees by varying the beam angle as shown in Fig.1. Details of the technique of making the ion-etched holographic gratings have been reported in Ref.1. Ruled gratings which are compared with our gratings are commercially available gratings with the same size.

Results and Discussion

Fig.2 shows a typical cross section of the holographic grating blazed by our ion etching technique. The beam angle of the ion etching is 65 degrees and the blaze angle is about 16 degrees. To get gratings uniformly ion-etched the stage of the ion beam apparatus must be rotated while keeping a constant beam angle with reference to the groove direction.

Fig.3 shows typical diffraction efficiencies of the holographic gratings etched at various ion beam angles and of a ruled grating with the blazed wavelength of 500 nm. These efficiency curves were measured by a conventional double monochrometer as shown in Fig.4. The absolute efficiency of the grating blazed by the ion etching is 73% at

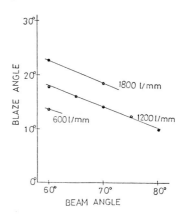

Fig. 1. Blaze angle vs. ion-beam angle for groove numbers.

Fig.2. The cross section of a blazed ion-etched holographic grating observed by a scanning electron microscope. The groove density is 1200 lines/mm and the blaze angle is 16 degrees.

440 nm which is comparable to that of ruled gratings and is much higher than that of conventional holographic gratings in which the diffraction efficiency is about 40%[2]. By increasing the ion beam angle from 65 degrees to 80 degrees, the wavelength of the maximum diffraction efficiency decreases from 440 nm to 280 nm. The anomaly of the ion-etched holographic gratings was smaller than that of the ruled grating.

Fig.3 shows a typical stray light performance for the ion etched grating, unetched holographic grating and commercially available ruled grating at the wavelengths of 410 nm and 240 nm. The stray light of the ion-etched grating is much lower than that of the ruled grating but is higher than that of the holographic grating. The stray light performance of the ion-etched holographic grating depended on the quality of the initial holographic grating and the etching condition.

Fig.5 shows the light scattered by the ion-etched holographic grating and commercially available ruled grating. It is proved to be the scattering of light of our grating is much less than that of the ruled grating.

Fig. 6 shows the resolution of gratings produced by ion etching and ruling technique, which were observed in the same spectrometer with the same experimental condition. This figure shows spectral lines of 265.512 nm, 265.368 nm and 265.204 nm from a Hg lamp, which are clearly resolved, and the resolution of the ion-etched holographic grating is the same as that of the ruled grating. Under the wide slit width condition of the monochromator the resolution limit $\delta\lambda$ is given by $(WD/f\lambda)\Delta\lambda$, where W is the slit width, D is the dimension of the aperture, f is the focal length of monochromator, $\Delta\lambda$ is the theoretical resolution limit of the grating and λ is the wavelength used. In our experimental condition with W=0.1 mm, D=25 mm, f=400 mm and $\Delta\lambda=7\times10^{-3}$ nm, $\delta\lambda=0.015$ nm at 265 nm, and this calculated resolution limit is obtained with our gratings. The wavefront quality

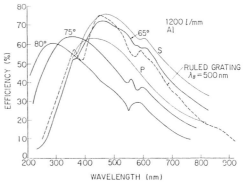

Fig.3. Absolute diffraction efficiencies of the holographic gratings ion-etched at various beam angles. The dotted line shows a diffraction efficiency of the ruled grating blazed at 500 nm. S and P show the diffraction efficiencies of the grating ion-etched at 65 degrees for the S and P polarized light, where S and P polarizations show perpendicular and parallel polarization to the groove direction, respectively.

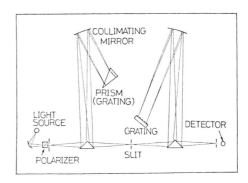

Fig. 4 Optical arrangement of Littrow mount double monochromator.

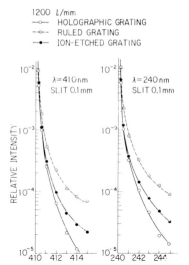

1200 ℓ/mm
—○— HOLOGRAPHIC GRATING
-⊶- RULED GRATING
—•— ION-ETCHED GRATING

Fig.5 Stray light performances for the blazed
ion-etched holographic grating, unetched holographic grating and ruled
grating at the wavelengths of 410 nm
and 240 nm.

of both gratings are almost the same, as
shown in the inset of the figure. For the
grating with the groove density of much more
than 1200 lines/mm, it was easy to controle
the groove profile by this technique.

Conclusion

    The ion etching technique of controlling
the groove profile of holographic gratings
is a simple and useful technique. The
diffraction efficiency, stray light, resolution and wavefront quality obtained by this
technique are the same as or superior to
those of the conventional grating produced by
a ruling engine and it has higher efficiency
than conventional holographic gratings.
It is emphasized that ion etched holographic
gratings have a good enough performance to
use in practical monochromators and spectrophotometers.

Fig.6 Comparison of scattered light between
ruled (bottom) and ion-etched (top)
grating.

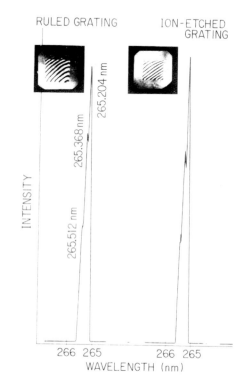

Fig.7 Resolution of the blazed ion-etched
holographic grating and ruled grating.

References
1. Aoyagi, Y. and Namba, S., "Blazed Ion-etched Holographic Gratings", Opt. Acta, 23. 701
   (1976).
2. Sano, K., Aoyagi, Y. and Namba, S., "Production of Blazed Holographic Gratings by Ion-
   etching Technique", J. Spectroscopic Soc. Japan 26, 327 (1977).

Aberration corrected holographically recorded diffraction gratings

J. M. Lerner

Instruments SA, Inc., 173 Essex Avenue, Metuchen, New Jersey 08840

J. Flamand, J. P. Laude, G. Passereau, A. Thevenon

Jobin Yvon, 16-18 Rue du Canal, 91160 Longjumeau, France

Abstract

This paper outlines how ray trace techniques can be used to predetermine new configurations of monochromator and spectrograph gratings. Examples of Rowland Circle aberrations vs. aberration corrected configurations are given in which it has been seen that astigmatism can sometimes be reduced by over a factor of 10. When non-Rowland Circle configurations are used, astigmatism can often be reduced to zero. This is demonstrated in a new type of grating that acts as a scanning flat field spectrograph operating at better than F3 from 600-1200nm.

2. Introduction

In most optical systems the design of the components are typically optimized by the use of computerized ray trace techniques.

In an optical system, such as a monochromator or spectrograph, a diffraction grating is often used as the dispersing element. Until fairly recently, plane gratings have been used in conjunction with mirrors or lenses to permit focussing of the radiation.

However, the needs of the spectroscopist have been growing dramatically as new solid state detectors have become available. In addition, there are many instances where very high aperture systems (better than F/3) are required in order that the expanding state-of-the-art in technologies such as fibre optics can be utilized. Even new light sources are becoming available emitting all the way from X-ray to the infra-red, requiring new diffractive focussing instruments.

The dilemma of conventional spectrometer systems, such as Czerny-Turner and Ebert designs, is that they do not work well when high resolution plus numerically low apertures are required. In addition, such systems are typically bulky, require at least three optical surfaces, each of which can create stray light and, not least, are expensive.

In this paper it is described how concave gratings can be used in new configurations in which aberrations normally present can be significantly reduced. Also described is how this is accomplished using as an example a new "monograph" grating that acts as a scanning spectrograph with an aperture of F 2.95 and a flat spectral field.

The manufacturing techniques used to produce these concave gratings are proprietary; however, principle and theory is briefly described both here and in detail in the literature. [1,2,3,4]

This paper reviews the steps used in determining how a grating will perform by the use of new ray trace techniques developed within Instruments SA.

3. Classical concave gratings

In the case of a classically ruled concave grating, the spectrum falls upon the Rowland Circle as shown in Figure 1. The grooves are parallel and the grating, if ruled on a ruling engine, may have been produced in up to seven zones to optimize blaze angles over the whole surface of the grating. Rarely are these zones phase matched, with the result that the maximum possible resolution is reduced by a factor equivalent to the number of non-phased zones present.

In order to eliminate this and other problems, it is possible to produce gratings "holographically". This technique requires two intersecting beams from a single longitudinal mode laser to produce interference fringes in a photosensitive material deposited on an optically flat substrate--typically glass.

The interference fringes thus recorded in the photosensitive material (a photoresist) are then processed creating grooves in relief.

The groove spacing is controlled by adjusting the angle α between the two laser beams. The laser wavelength is shown as λ_0 in Figure 2.

The resulting grating can be efficient, ghost free, and show very high stray light rejection. Figure 2 shows the production of a holographic concave grating with parallel grooves which is, therefore, a ruled grating equivalent (usually called a Type I holographic grating). In that the grating possesses continuous grooves in only one zone, the resolution obtained may be close to theoretical.

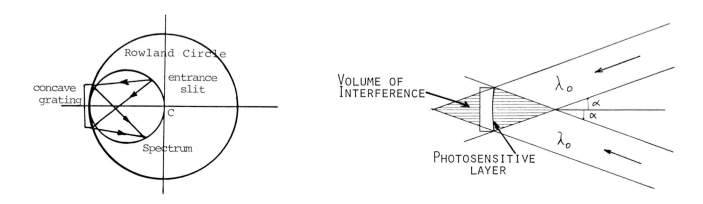

Figure 1. Rowland Circle configuration   Figure 2. Recording a Type I holographic grating

3.1   Rowland Circle aberrations - Figure 3.

On the basis of the Fermat Principle, if A is the object source, B the image of A at wavelength λ (where A and B are in the XY plane), the Fermat Principle could be written

$$MA + MB = IA + IB + kn\lambda + \Delta(M) \tag{1}$$

$\Delta(M)$ = aberrant optical path
 M = any point on the surface of the grating
 k = an integer (the order of the diffracted image at wavelength λ)
 n = number of grooves between I and M
 I = center of the grating

Therefore, if we express $\Delta(M)$ in terms of M (Y,Z), then one obtains

$$\Delta(M) = TY + \frac{DY^2}{2R} + A\frac{Z^2}{2R} + C_1\frac{Y^3}{2R^2} + C_2\frac{YZ^2}{2R^2} + S_1Y^4 + S_2Y^2Z^2 + S_3Z^4 \tag{2}$$

where R = radius of curvature of the blank
    T = tangential focus (when T = O the grating equation is obtained)
    D = characterizes defocussing where D = O at tangential focus
    A = characterizes astigmatism. (A = 0 at the sagittal focus)
    C_1 = 1st type coma (classical)
    C_2 = 2nd type coma or mixed terms
    S_1 S_2 S_3 = characterize spherical aberration

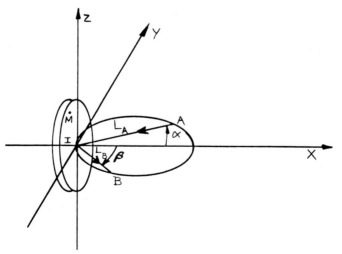

Figure 3. Parameters for the calculation of aberrations on the Rowland Circle

The locus of the spectrum is obtained when the T and D terms are equated to zero. Residual aberrations are given by the A, C_1, C_2 and S terms.

Analysis of these equations permits one to deduce that if the point A is on the Rowland Circle, then the locus of the spectrum is on the Rowland Circle.

In addition, one finds that the $\frac{Y^3}{2}$ term (1st type coma) vanishes but the other terms remain. Hence, in a Rowland Circle configuration, there is no 1st type coma; but astigmatism, 2nd type coma and spherical aberration remain.

4.   Production of an aberration corrected grating

The definition of a classically ruled grating presumes grooves that are parallel. It is possible, however, using holographic techniques, to modify the grooves' distribution in such a way that the locus of the spectrum can be relocated, resulting in the reduction or elimination of aberrations. It is also possible to maintain the Rowland Circle geometry whilst either reducing or eliminating aberrations. (an example is given later in this paper).

To accomplish such aberration correction, it is necessary to lay down grooves that instead of being parallel are a series of grooves of highly complex characteristics. This can be readily appreciated when one considers that the grooves are at the intersection of families of revolution hyperboloids with a spherical blank. Figure 4 shows two coherent point sources C and D which illustrate the production of an aberration corrected grating. If it is to work on the Rowland Circle, it is a Type II grating.

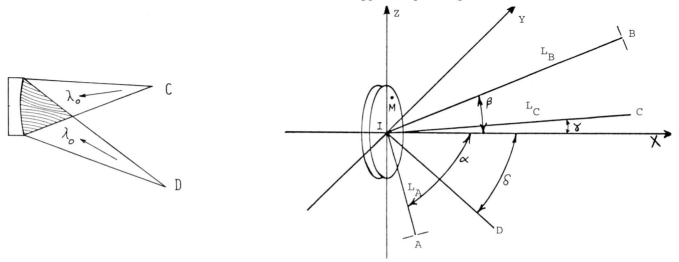

Figure 4. Recording of a Type II grating

Figure 5. Parameters for the calculation of aberration correction of a holographic grating

Under these circumstances, from Figure 5 the aberrant optical path would be defined by the following expression:

$$\Delta (M) = MA + MB - \frac{k\lambda}{\lambda o} (MC - MD) - \left[IA + IB - \frac{k\lambda}{\lambda o} (IC - ID) \right] \qquad (3)$$

$\Delta (M)$ can be expressed as a function of Y, Z, l_A l_B l_C l_D , β, γ, and δ. Total or partial correction of the aberrations without changing the working parameters could be obtained by properly choosing the values of γ, δ, l_C and l_D. For example, in this way the $\frac{Z^2}{2}$ term (astigmatism) can be reduced or even eliminated. If the height of the tangential focus is near zero at the center of the spectral range, astigmatism over the whole spectral range can be reduced overall by up to a factor of 10.

5. Classification

For the purposes of simplicity, concave holographic gratings can be classified as follows
Type I = ruled grating equivalent - no aberration correction
Type II = Rowland Circle configuration in which astigmatism is eliminated at one wavelength with some reduction in 2nd type coma.
Type III = stigmatic at three wavelengths and considerably reduced aberrations over the whole spectrum. This type of grating is the basis of many flat-field spectrograph gratings.
Type IV = monochromator gratings that are astigmatism free at least at one wavelength, with considerable reduction of other aberrations over whole spectrum. This type of grating is also the basis of new "scanning spectrographs" or "monograph gratings".

6. Ray Tracing

In this section of the paper, examples are given of Type I, Type II, and Type IV gratings.

A ray trace is conducted following the appropriate selection of recording parameters as described in Section 4.

A geometric ray trace is then carried out to provide the following analyses:
i. Spot diagrams: An examination of the imaging capability of the system starting from diffraction limited spot sources located at the entrance slit at the center of the optical axis, halfway up the slit and also at the top of the slit. (measured from the optical axis) A total of 62 rays are taken for analysis. This provides a good first approach to evaluating a grating's viability.
ii. The impulse response: The determination of the energy distribution in the direction of dispersion of the entire entrance slit. A total of 4000 rays are taken, equally distributed over the entrance slit and fall with equal density over the whole surface of the grating.

6.1 Spot Diagrams

12 rays per spot are programmed to fall onto the periphery of the grating, plus another 12 rays fall onto the periphery of an area defined by the outside grating diameter divided by $\sqrt{2}$. In this way two areas of equivalent energy are examined.

At the center of the entrance slit energy is symetrically distributed; hence, only the points on half the grating are traced. See Figure 6.

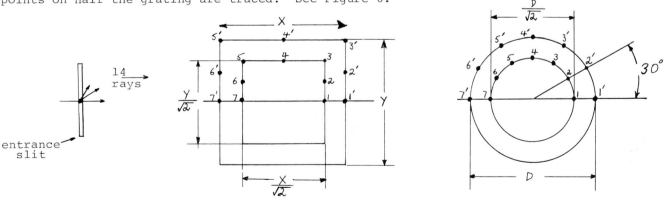

Figure 6.

The points are distributed at 30° intervals around a circular grating and equally distributed as shown in the case of a rectangular grating.

For the two points distributed in the upper half of the entrance slit, the whole surface of the grating is used. See Figure 7.

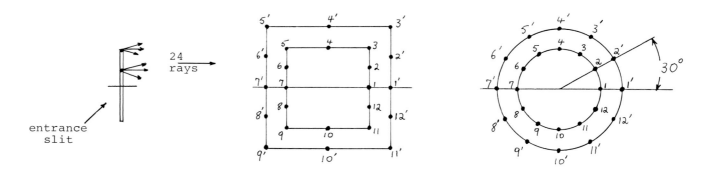

Figure 7.

6.2 The examples that follow, illustrate typical Spot Diagrams defining the image of 2 spots located on the entrance slit as imaged at the exit slit.

The X,Y axes show the vertical and horizontal components of the exit slit in millimeters and how the exit image at a particular diffracted wavelength would appear at this exit slit. This is the location of the tangential focus.

Generally speaking, enlargement of the image in a vertical direction is usually due to astigmatism and defocussing. Enlargement horizontally is due to a combination of all other possible aberrations.

6.3   Spot at center of entrance slit

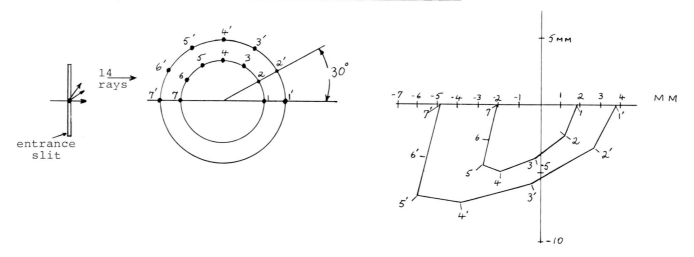

Figure 8.

It can be seen in Figure 8 that the image is inverted and below the optical axis, from this we know that the sagittal focus is between the grating and the exit slit. The exact location of the sagittal focus is, therefore, easy to calculate using simple geometry by connecting the points in the image back to the same numbered points on the grating. The location of the sagittal focus is at the point where the lines thus drawn intersect.

It can be seen that the image shows astigmatism of approximately 7mm.

6.4

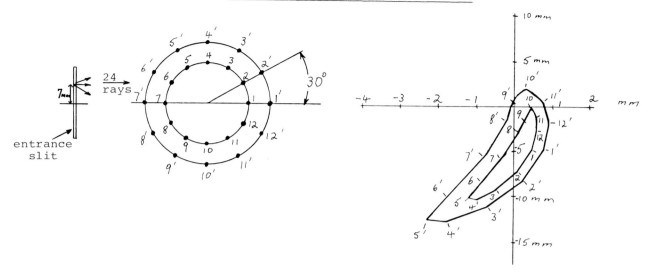

Figure 9.

Figure 9 shows the image characteristics of a spot located halfway up the entrance slit. Again the size of the image both vertically and horizontally can be readily calculated.

6.5

Impulse response

The impulse response of the system can be traced, assuming either infinitely narrow slits of finite height or slits of both finite height and width. In general, the impulse response, assuming infinitely narrow slits, defines the aberration limit of the system and, as a good rule of thumb, can be added to the resolution given by the dispersion of the grating multiplied by the entrance slit width. The convolution of impulse response with entrance and exit slit would give the true spectral response of the system.

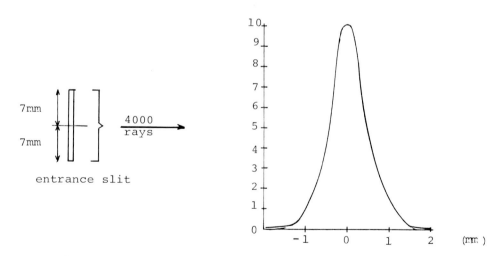

Figure 10. Impulse response

In the above diagram, the X axis shows the energy distribution in mm along the dispersion axis. The dispersion is given in units of wavelength per mm; for example Å/mm. The Y axis shows energy in arbitrary units.

From this profile, line symmetry, width at half height and width at the base can be determined. Given that a certain detector of finite size will be used, calculations can be made to determine geometric throughput when the above plot is used in conjunction with the spot diagrams identifying vertical image height.

6.6

Rowland Circle vs. Type II holographic gratings

The two ray traces shown below in Figure 11 and 12 plot the spot diagrams and impulse responses for both a conventional Rowland Circle spectrograph and also a Type II aberration corrected configuration.

In both studies the wavelength chosen was 3130Å, this being one of the wavelengths of the Berylium doublet and also close to the 3130Å Mercury doublet. The three "spots" on the entrance slit are always diffraction limited, but in both cases the impulse response is calculated for a finite slit in both height and width.

Wavelength under study 3130Å
Radius of curvature = 1000mm, ∝ = 35°, Ruled area 30 x 50mm, grating density 1800g/mm

entrance slit height 5mm
entrance slit width 0.02mm

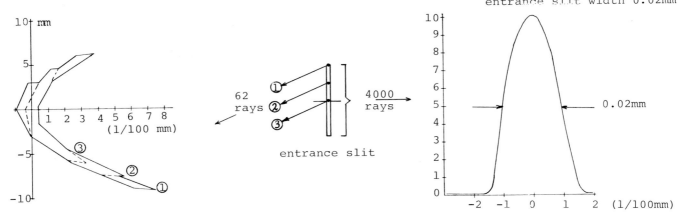

Figure 11. Rowland Circle

Wavelength under study 3130Å
Radius of curvature = 1000mm, ∝ = 35°, Ruled area 30 x 50mm, grating density 1800g/mm

entrance slit height 5mm
entrance slit width 0.02mm

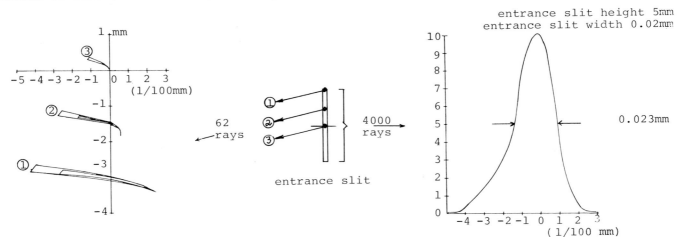

Figure 12. Type II aberration corrected

As can be readily seen in Figure 11, the height of a spot located anywhere on the entrance slit is imaged at the exit with a considerable degree of astigmatism, and some image curvature. In fact, a point at the top (or bottom) of the entrance slit (in this case 2.5mm above [or below] the optical axis) is imaged as a slightly curved line about 7.5mm in height.

The impulse response, however, shows a symmetric peak with a full width at half height of about 0.02mm (the same as the entrance slit width), demonstrating the ability of the Rowland Circle configuration to produce images free from coma. For this reason, the Rowland Circle mount has for many years been an excellent workhorse spectrograph.

A disadvantage, however, is that the exit slit should ideally be curved and also be 15mm higher than the entrance slit in order to prevent light being lost to the detector. In addition, the flux distribution is usually too low to permit the use of compact solid state detectors. However, as much of the energy is located on axis with a height of about 6mm, straight slits are generally used in analytical spectroscopic applications.

Figure 12 shows a Type II aberration corrected holographic grating used in an identical configuration.

A spot located anywhere on the entrance slit, is imaged at a height of between 0.4mm and 0.75mm. When compared to a conventional Rowland Circle mount in which a similarly located spot on the entrance slit has been seen to be up to 7.5mm in height, the reduction in astigmatism with this Type II grating is dramatic. The height of the exit slit can, therefore, be 6mm and capture virtually all available radiation.

The resolution, however, is not as good as the Rowland Circle configuration, the width at half height being 0.023mm, representing a 15% reduction in resolution due mainly to coma. It should be remembered that this study shows the results at just one wavelength and that aberrations may be more or less at other wavelengths.

Nevertheless, for many applications where throughput is more important than high resolution, the Type II grating can be very useful whilst making few compromises.

6.7   New instrument design - the "monograph grating"

The development of solid state detectors, such as 1/4", 1/2" and 1" diode arrays, have made the prospect of conducting spectroscopic experiments at a range of wavelengths simultaneously very attractive. This permits the study of reactions which take place very rapidly, saving the time normally spent in scanning to cover the same spectral range. In this case, the signal to noise ratio obtained in unit time is usually much improved over scanning techniques.

The monograph grating described below has been created to present a fixed exit focal field of 1/4" to match a commercially available diode array.

There is virtually no field curvature, and imaging is of very high quality overall. Over 1/4" a 280nm segment of a spectrum can be analyzed at any one time anywhere in the range from 600nm to 1200nm.

The aperture is F/2.95 permitting the efficient use of fibre optics. The whole system requires an area of only 2" x 4". The resolution as given by the full width at half the peak height is always better than 2.5nm anywhere in the spectrum, when operated with narrow slits.

Any wavelength in the range can be brought to the center of the spectrum simply by rotating the grating. No optic other than the grating itself is required.

Figure 13 shows the layout of the optical system with the following characteristics:
Grating size - 30 x 30mm   Holographic Type IV
Focal length - 100mm
Aperture - F/2.95
Groove density 213g/mm
Dispersion - 46nm/mm
Total spectral range - 600-1200nm
Flat field coverage - 1/4"
Spectral coverage over 1/4" - 280nm
Entrance slit height - 6nm
Entrance slit width - 0.05mm
Maximum resolution - better than 2.5nm
Minimum resolution with 0.05mm entrance slit - better than 5nm
Astigmatism never exceeds 0.007 H (where H = height of the grating)

For simplicity, let λ_1, and λ_2 be peripheral wavelengths on the flat spectral field. The position of the spectrum will not change, only the wavelengths present, as the grating is rotated.

The following ray traces provide examples of two wavelengths each located at an extremity of the spectral field after rotation of the grating.

Figure 14 - spot diagram and impulse response for 600nm at λ_1 (880nm would be located at λ_2)
Figure 15 - spot diagram and impulse response for 1200nm at λ_2 (920nm would be located at λ_1)

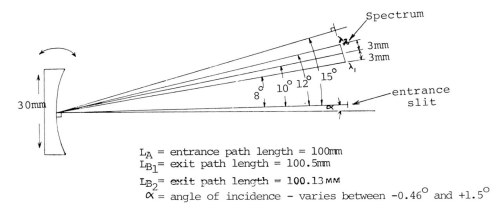

L_A = entrance path length = 100mm
L_{B_1} = exit path length = 100.5mm
L_{B_2} = exit path length = 100.13мм
α = angle of incidence - varies between -0.46^O and $+1.5^O$

Figure 13.   "Monograph" grating - scanning spectrograph

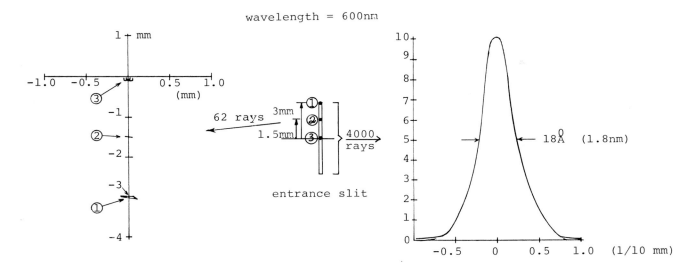

Figure 14.   Spot diagram and impulse response at 600nm

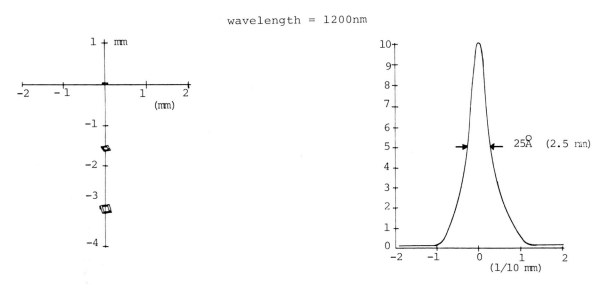

Figure 15.   Spot diagram and impulse response at 1200nm

The ray traces shown indicate that if an entrance slit 6mm high and very narrow is re-imaged onto a flat focal plane with detectors 6mm high by 0.05mm wide, virtually all diffracted radiation would be collected. In fact, computations show that geometric losses never exceed 1.6%. If the entrance slit is 0.05mm in width, the final image would always be less than 5nm at full width at half the peak height.

6.8   Conclusion

From the results obtained and presented in this paper, it is reasonable to conclude that the use of holographically generated diffraction gratings designed by modern computer ray trace techniques provide a viable mechanism in the design of new instruments.

It can be seen that imaging can be corrected for astigmatism even over wide wavelength ranges and that resolution can be maintained at very high levels.

References

1. Diffraction Gratings Ruled and Holographic - Handbook, published by Instruments SA, Inc., 173 Essex Avenue, Metuchen, N.J., USA, and ISA-Jobin-Yovon, 16-18 Rue du Canal, 91160 Longjumeau, France.
2. Hayat, G.S., et al, "Designing a New Generation of Analytical Instruments Around the New Types of Holographic Diffraction Gratings", Optical Engineering, Vol. 14, No. 5., pp 420-425. 1975.
3. Hayat, G.S., et al, "Holographic Grating Update", E.O.S.D.,April 1978.
4. G.W. Stroke, "Diffraction Gratings", Handbuch der Physik, Vol. XXIX.

Diffraction gratings ruled and holographic—a review

J. M. Lerner
Instruments SA, Inc., 173 Essex Avenue, Metuchen, New Jersey 08840
J. Flamand, J. P. Laude, G. Passereau, A. Thevenon
Jobin-Yvon, 16-18 Rue du Canal, 91160 Longjumeau, France

1.   Abstract

Diffraction gratings have been evolving steadily over the last ten years. Modern ruling engines have permitted classically ruled gratings to be more efficient with flatter diffracted wavefront than ever before. Examples include sophisticated laser gratings with up to 97.5% efficiency and flatness of better than $\lambda/3$ in 17th order 6328Å.

Holographically recorded diffraction gratings continue to evolve with the development of blazed plane gratings and with groove densities of up to 6000g/mm.

Concave holographic gratings now range from toroidal aberration corrected gratings, for use in the soft x-ray, to flat field spectrographs and scanning spectrographs operating out into the infra-red.

2.   Introduction

Over the last few years the spectroscopist has seen many new developments in diffraction gratings, ranging from new plane, ruled master gratings for waveguide lasers, to holographically recorded gratings, for use from soft x-ray to the infra-red.

This paper will attempt to briefly review and put into perspective some of these developments.

3.   Plane gratings

All diffraction gratings, whether they be classically ruled or holographic, obey the same fundamental diffraction grating equation which, when stated in its general form, can be written:[1,4]

$$\mathrm{Sin}\alpha + \mathrm{Sin}\beta = k\,n\,\lambda \tag{1}$$

where α = angle of incidence
β = angle of diffraction
k = order
n = groove density
λ = wavelength of interest

The dispersion of a grating can be deduced from equation (1) and can be written:

$$\frac{d\lambda}{d\lambda}\ \mathrm{nm/mm} = \frac{10^6\ \mathrm{Cos}\ \beta}{k\,n\,F} \tag{2}$$

where F = distance from the center of the grating to the locus of the spectrum.

The resolution of a grating can be generally stated by the equation:

$$R = kN = \frac{(\mathrm{Sin}\alpha + \mathrm{Sin}\beta)\,W}{\lambda} \tag{3}$$

where $N = nW$ $\tag{4}$
where R = resolution as defined by the Rayleigh criterion
N = total number of grooves on the surface of the grating
W = width of the grating

Modern gratings achieve between 80-90% of their theoretical resolution. Classically ruled gratings are not commonly ruled in excess of 2400g/mm, but holographic gratings can be recorded with up to 6000g/mm.

However, if maximum resolution is to be obtained in an instrument using plane gratings, it must be assumed that
(a) the collimating and focussing optics do not degrade the resolution.
(b) the optical design will produce aberration free images at any wavelength in the range of interest.

In practice, (a) and especially (b) may not permit theoretical resolution if only because the diffraction limit of the system may be inferior to that of the grating. The equations given all presume that the diffracted wavefront of the grating will not depart significantly from λ/4. Most high quality commercially available plane gratings typically meet this requirement and are rarely worse than λ/3.

The flatness of a grating is of special importance when the grating forms part of a resonator cavity. A laser grating, for example, must be very flat when used as an end-reflector in a laser cavity.

In this case, the ability to tune a wavelength at high resolution can be determined by the diffracted wavefront of the grating.

Figure 1 and 2 show the interferograms of a stainless steel master grating with a gold overcoat blazed at 10.6 micron with 150g/mm.

Figure 1 shows the flatness in zero order and Figure 2 shows the flatness at 6328Å in 17th order. (17th order of 6328Å is at 10.75 micron and was chosen because of its proximity to 10.6 micron).

The flatness is very close to λ/4 and can safely be used in a laser cavity to permit operation at maximum resolution.

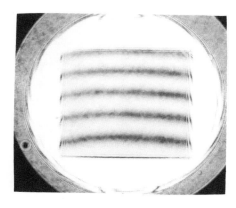

Figure 1 - zero order 6328Å   Figure 2 - 17th order 6328Å

3.1   Ghosts

In general, the term "ghost" is used to describe spurious lines caused by periodic imperfections in the grating groove spacing.[1,4]

There are two principal types of ghost:
(a) Rowland Ghosts that are associated with periodic errors in the lead screw of the ruling engine. These ghosts are usually symetrically located with respect to their parent line.
(b) Lyman Ghosts that are associated with external periodic vibrations. These ghosts are usually found large distances from the parent line.

The engines used at Instruments SA, Inc. are double interferometrically controlled and reduce Rowland ghosts to very low levels. Adequate isolation from outside influences reduces the occurrence of Lyman ghosts.

There are other ghosts associated with random vibrations that are non-periodic which, like Lyman ghosts, can only be eliminated by careful isolation of the ruling engine.

Practically, the intensity of a ghost is given by the following equation:
$$I_g = k^2 \Pi^2 Eo^2 n^2 Ip \qquad\qquad (5)$$
where
I_g = ghost intensity
k = order
Eo = periodic error
n = groove density
Ip = intensity of parent = 1.

It is of importance to note that ghost intensity is proportional to the square of groove density, the square of the order that the grating is used in, and the square of the magnitude of the periodic error.

Ghosts have plagued spectroscopists for many years, especially when using either high density ruled gratings or low density gratings in high orders. Some echelle gratings, for example, can show very severe ghosts. (an echelle grating is a low groove density grating used in high orders, typically from 10-1000th order). The magnitude of the problem can be seen if one reviews published lists of elemental emission lines. On almost every page certain emission lines are deleted as each edition of the list goes to the printer. Some lines were probably instrument artifacts due in significant part to the effect of grating ghosts. The early spectrographs employed were typically used in 1st, 2nd, 3rd, and even 4th order, using gratings ruled on non-interferometrically controlled engines.

As improved gratings became available, especially holographically recorded gratings, emission spectra became cleaner and artifacts were either eliminated altogether or at least considerably reduced. At the same time, elements that were said to "interfere" with other elements in an emission spectrum dramatically decreased.

(we deal with holographically recorded diffraction gratings in some depth later in this paper).

In any event, if is for the user to beware when purchasing a diffraction grating and to obtain from the supplier the magnitude of ghosts that may be present in their gratings.

3.2 <u>Stray light</u> in a classically ruled grating can be clearly shown if the grating is illuminated with monochromatic light, for example, from a laser. Very often under these circumstances it will be found that a continuous line or area of "stray light" is centered down the dispersion axis ('Y').

This stray light can be explained together with ghosts in the same Fourier analysis of the light distribution and intensity. In such an analysis, Rowland ghosts and Lyman ghosts are identified as the result of periodic errors as discussed in section 3.1, and "stray-light" or grass as a function of non-periodic errors (for example, non-periodic vibrations, roughness of grooves, etc.)

4.   Efficiency of classically ruled diffraction gratings

The efficiency of a diffraction grating is a measurement used to determine the magnitude of diffraction losses at a given wavelength and resolution. Efficiency is either quoted in relative terms; i.e., compared to the reflectivity of a mirror at the same wavelength with the same coating as the grating, or in absolute terms in which the exit radiation at a given wavelength is compared to incident radiation at the same wavelength. Efficiency curves that are published by manufacturers tend to show relative efficiency.

It should be noted, however, that concave gratings may show severe throughput losses due to aberrations; therefore, extrapolating efficiency curve data to calculate the throughput of a concave grating system requires great caution. [1,2,5]

The maximum efficiency of a classically ruled grating can be as high as 80% in unpolarized light and significantly higher in polarized light. Figure 4 shows a typical efficiency curve for blazed holographic gratings which tends to be similar to that of blazed ruled gratings.

5.   Holographic gratings

The term "holographically recorded diffraction grating" refers to the technique used to produce the grating.[1]

Two intersecting beams from a single longitudinal mode laser are used to produce interference fringes that alternately expose a photosensitive material deposited on an optically flat glass, which, when processed, produces grooves in relief.

The spacing between grooves is controlled by adjusting the angle α between the two laser beams. See Figure 3.

This technique produces grooves that are almost perfectly equidistant, thus completely eliminating all ghosts due to periodic errors. The overall quality of the surface is such that imperfections, roughness, etc., are very much less than those found in classically ruled gratings, thus reducing stray-light.

For the same reasons discussed in Section 3.2, stray-light is NEVER focussed down the optical axis as produced in classically ruled gratings. Stray-light that may be present is evenly distributed overall. This has the effect in an instrument of reducing background stray-light typically from 5-100 times that of a classically ruled grating, depending upon the spectral domain of use, the Ultra-Violet (UV) or Vacuum Ultra-Violet (VUV) benefitting the most. In the IR above 1 micron, the stray-light reduction of a holographic grating becomes increasingly less significant due to the fact that classically ruled gratings working in this spectral range are of low groove density.

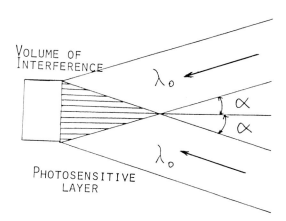

Figure 3. Recording a plane holographic grating

6. ## Efficiency of holographic gratings

Figure 4 shows a typical efficiency curve of a blazed holographic grating. Maximum efficiency averages around 60%, but is still somewhat lower than some of the better classically ruled gratings.

Non-blazed holographic gratings can often extend the spectral range over which the grating can be used by reducing the efficiency at one wavelength, but increasing efficiency at peripheral wavelengths. Non-blazed holographic gratings range in maximum efficiency from 30-55% and very often the efficiency profile is relatively flat.

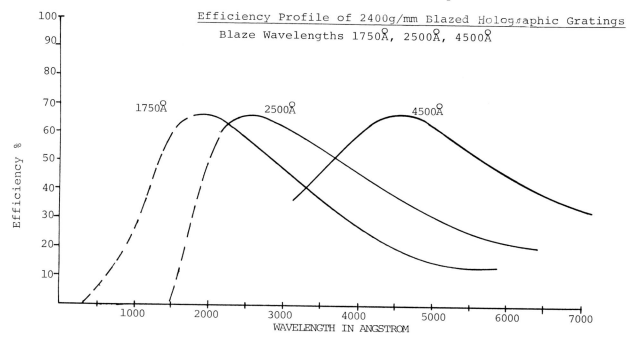

Figure 4.

7. <u>Production of aberration corrected gratings</u> is very important in an instrument design where a single grating is to form the entire optical system.[1,2,5]

A concave classically ruled grating operating on the Rowland Circle[5] shows astigmatism that reduces throughput coupled with field curvature that restricts the use of flat detectors such as diode arrays, etc.

Astigmatism and field curvature may be either reduced or eliminated by laying down grooves that, instead of being parallel, are a series of grooves of highly complex characteristics.[4]

Figure 5 shows an example in which the photoresist on a spherical blank is illuminated by laser beams emitted from two point sources, C & D. The grooves are located at the intersection of families of revolution hyperboloids and the spherical surface. Chemical processing of the exposed photoresist then leaves the grooves in relief, which are then vacuum deposited with aluminum or other appropriate coating.

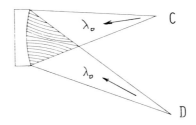

Figure 5.

Aberration corrected holographic gratings (ACHG) produced in this way have permitted new generations of instruments to be designed. The dominant feature of all instruments utilizing an ACHG is that the grating forms the sole optical component in the system. The elimination of other optics permits instruments to be designed that can be simple, inexpensive, robust, and still offer high resolution, throughput and stray-light rejection.

In general, holographically recorded concave gratings can be classified as follows:
Type I = ruled grating equivalent - no aberration correction[4]
Type II = Rowland Circle configuration in which astigmatism is eliminated at one wavelength with some reduction in 2nd Type Coma.
Type III = stigmatic at three wavelengths and considerably reduced aberrations over the whole spectrum. This type of grating is the basis of many flat-field spectrograph gratings. The grating remains in a fixed position with respect to the entrance slit and spectrum and can be used with flat-field detectors.
Type IV = monochromator gratings that are astigmatism free at least at one wavelength, with considerable reduction of other aberrations over the whole spectrum. This type of grating is also the basis of new "scanning spectrographs" of "monograph gratings". The wavelength or spectral range is changed by simple rotation of the grating.

The following examples show typical configurations of ACHG systems:
7.1-1. Type IV grating monochromator - In Figure 6 is shown an example in which the ACHG is the only optical element in an entire monochromator system. It was designed to operate at F3 and permits new fibre optic devices to be utilized. The wavelength is selected by rotation of the grating.

Grating size 30 x 30mm Type IV
Groove density 1200 g/mm
Focal length 100mm
Spectral range 200-800nm (can be extended to 1.2μ)
Resolution 0.5nm or as determined by the slit width
Aperture F3
γ=Fixed included angle

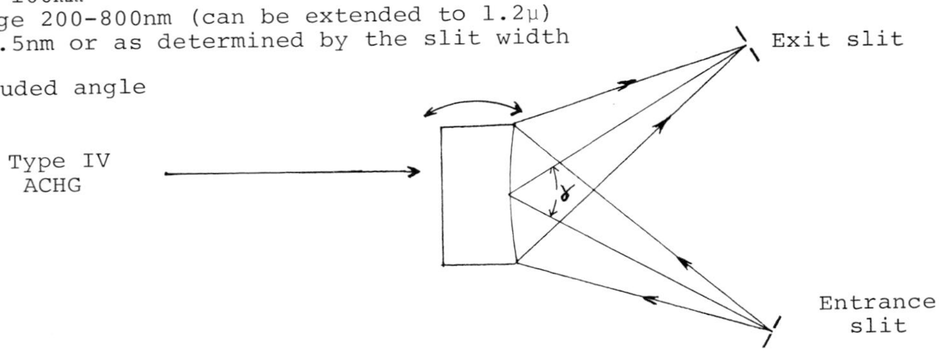

Type IV
ACHG

Exit slit

Entrance
slit

Figure 6. Concave grating spectrometer

7.2-2. <u>Type III flat-field spectrograph</u> - An example of such a configuration is shown in Figure 7 where the grating accepts radiation from the entrance slit and focusses the diffracted light across a 25mm flat field. This design permits the use of commercially available diode arrays and other solid state detectors.

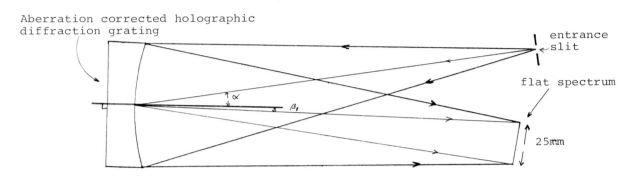

Figure 7. Type III flat field grating spectrograph

The resolution is usually determined by the width of the photosensitive device used in the detector.

Established flat field grating configurations

Spectral range	Groove density	l_a	l_{b1}	l_{b2}	$\propto$	β_1	Flat field
200nm-1.2μ	120 g/mm	210mm	206.2mm	208.2mm	9.19°	-7.8°	25mm
200-800nm	200 g/mm	210mm	206.2mm	208.2mm	9.65°	-7.33°	25mm
400-800nm	300 g/mm	210mm	206.2mm	208.2mm	11.98°	-5.01°	25mm
200-400nm	600 g/mm	210mm	206.2mm	208.2mm	11.98°	-5.01°	25mm

7.3-3. <u>Type IV scanning flat field or "monograph" gratings</u> - In this configuration, the grating is designed to produce a flat spectral field, just as in example 2. However, the wavelength range submitted can be changed by simple rotation of the grating.[5]

This means that the grating can be used with either solid state detectors or an exit slit or both. The example shown in Figure 8 has been described in detail in reference 4.

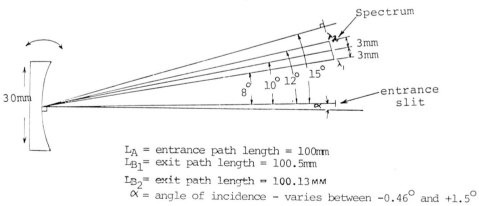

L_A = entrance path length = 100mm
L_{B1} = exit path length = 100.5mm
L_{B2} = exit path length = 100.13mm
$\propto$ = angle of incidence - varies between -0.46° and +1.5°

Figure 8. "monograph" grating - scanning spectrograph

8.   Ion etched gratings

Once a grating has been holographically generated, the grooves can be ion-etched into the substrate itself. This is of particular value where the presence of organic material is undesirable due to the destructive potential of high energy incident radiation, or if the grating will be used at high temperature.

In the soft X-ray and VUV there are two major advantages to this type of grating:
i. There is no organic material whatsoever present.
ii. Even orders; e.g., 2nd and 4th, are strongly attenuated and odd orders reduced in intensity.

It can be anticipated that many future advances in the state-of-the art of diffraction grating technology may be associated with ion-etching techniques.

9.-4. Toroidal grating monochromators and spectrographs for use in the VUV - The spectroscopist working in the VUV has traditionally used classically ruled concave gratings operating on the Rowland Circle. The problem with this configuration is that light losses due to astigmatism are very severe.[6]

One of the ways of reducing such aberrations is to produce a holographic aberration corrected grating on a toroidal blank. In this case, astigmatism is eliminated at one wavelength and is very low elsewhere in the spectrum. Toroidal grating monochromators (TGM) and toroidal grating spectrographs (TGS) may be used from 6Å to 1500Å. It is usually appropriate to ion-etch these gratings in order that organics be eliminated and the even orders attenuated.

TGM and TGS instruments are utilized in both synchrotron radiation and Tokamak fusion centers throughout the world. It is expected that the use of toroidal gratings will increase as new VUV facilities come on line during the coming years.

Figure 9 shows the layout of a TGM in which the wavelength is changed by simple rotation of grating, while the entrance and exit path lengths remain fixed.

Figure 10 shows a TGS in which the spectrum is flat and, depending on the configuration, could operate from 16Å to 1500Å with interchangeable gratings.

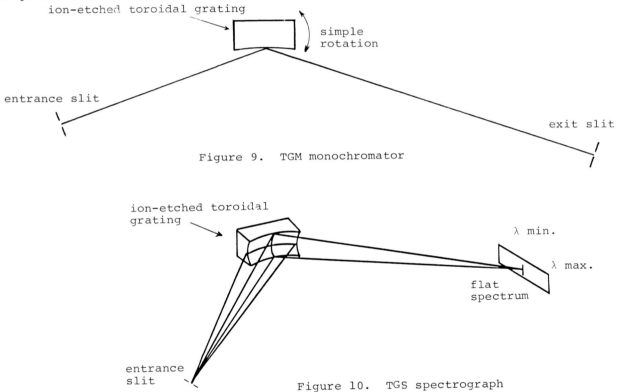

Figure 9. TGM monochromator

Figure 10. TGS spectrograph

References

1. Diffraction Gratings Ruled and Holographic - Handbook, published by Instruments SA, Inc., 173 Essex Avenue, Metuchen, N.J., USA, and ISA-Jobin-Yvon, 16-18 Rue du Canal, 91160 Longjumeau, France.
2. Hayat, G.S., et al, "designing a New Generation of Analytical Instruments Around the New Types of Holographic Diffraction Gratings", Optical Engineering, Vol.14, No.5., pp 420-425. 1975.
3. Hayat, G.S., et al, "Holographic Grating Update", E.O.S.D.,April 1978.
4. G.W. Stroke, "Diffraction Gratings", Handbuch der Physik, Vol. XXIX.
5. Lerner, et al, Proceedings SPIE, Vol. 240, 1980.
6. Flamand, et al, VI International Conference on Vacuum Ultraviolet Radiation Physics, Extended Abstracts, Vol. III, University of Virginia, 1980.

SESSION 3

DIFFRACTION THEORY I

Session Chairman
Allen Mann
Hughes Aircraft Company

Diffraction of light by gratings studied with the differential method

Michel Nevière

Laboratoire d'Optique Electromagnétique, E.R.A. du C.N.R.S. n° 597
Faculté des Sciences et Techniques, Centre de St-Jérôme, 13397 Marseille Cedex 4, France

Abstract

The differential method developped in the seventies reduces the problem of diffraction of light by a periodic structure to the numerical integration of a set of coupled differential equations. It works both for perfectly conducting gratings and for finite conductivity gratings, including dielectric ones. It can handle very easily dielectric overcoatings deposited on top of gratings and gives very low computation times. The principle of the method is explained and a review of its extensions and possibilities is given.

Introduction

Although many diffraction problems are studied with the aim of the integral equation technique, an alternative method has been developped which allows calculating the field diffracted by a periodic structure like a grating. The electromagnetic field, expressed in terms of its Fourier series, shows itself to be the solution of a set of coupled differential equations, which has to be integrated in the groove region, and whose solution is matched with plane waves expansions at the boundaries. We wish here to present the principle of the method, to outline its capacities and to discuss its field of applications.

The principle of the differential method

The differential method is applicable to any diffracting structure which is periodic with respect to one coordinate. In the case of a grating illustrated on fig.1, the structure has period d along the Ox axis. Its surface, whose equation is y = f(x), divides space into two regions. Region ① is vacuum, with refractive index $\nu_1 = 1$. Region ② is a metal or a dielectric. An incident plane wave with wavelength λ, falls on the grating under incidence θ, and we define $\alpha_0 = 2\pi/\lambda \sin \theta$. Since we are dealing with an electromagnetic wave, the problem is a vectorial one. In order to simplify it, it is classical to consider two fundamental cases of polarization : the TE (or P) polarization, for which vector $\vec{E}$ is parallel to the grating grooves ($\vec{E} = E(x,y)\hat{z}$) ; the TM (or S) polarization, for which vector $\vec{H}$ is parallel to the grooves ($\vec{H} = H(x,y)\hat{z}$). In both cases, the problem becomes scalar. It means that in order to get the diffracted field, we only have to determine the scalar function $E(x,y)$ or $H(x,y)$. Then the problem takes different forms if we assume region ② to be a real metal or to be perfectly reflecting. Let us begin with a real metal characterized by its complex refractive index ν_2 , and first describe the method for TE polarization.

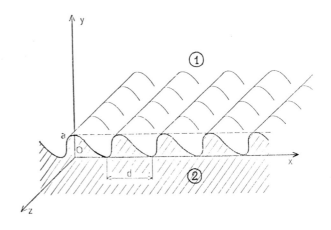

Figure 1. Schematic representation of a grating

Finite conductivity gratings used in TE polarization

In each homogeneous region, Maxwell equations lead to Helmholtz equations :

$$\Delta E + k_j^2 E = 0 \qquad\qquad j = 1, 2 \qquad\qquad\qquad (1)$$

with $k_j = \dfrac{2\pi}{\lambda} \nu_j$

Thus the electric field can be expressed in terms of a superposition of plane waves called Rayleigh expansion. Defining :

$$\beta_n^{(j)} = (k_j^2 - \alpha_n^2)^{1/2} \tag{2}$$

$$\text{with} \quad \alpha_n = \alpha_0 + n \frac{2\pi}{d} \tag{3}$$

$$\text{and} \quad \text{Re}\{\beta_n\} + \text{Im}\{\beta_n\} > 0 \quad ^1 \tag{4}$$

we get :

when $y \geqslant a$:

$$E(x,y) = E^i(x,y) + \sum_{n=-\infty}^{+\infty} B_n \exp\left[i(\alpha_n x + \beta_n^{(1)} y)\right] , \tag{5}$$

where E^i is the incident field equal to :

$$\exp\left[i(\alpha_0 x - \beta_0^{(1)} y)\right] ;$$

when $y \leqslant 0$:

$$E(x,y) = \sum_{n=-\infty}^{+\infty} T_n \exp\left[i(\alpha_n x - \beta_n^{(2)} y)\right] . \tag{6}$$

In the modulated region $0 < y < a$, instead of dealing with two Helmholtz equations, we can say that :

$$\Delta E + k^2(x,y)E = 0 , \tag{7}$$

$$\text{with} \quad k^2(x,y) = \begin{cases} k_1^2 & \text{if} \quad y > f(x) \\ k_2^2 & \text{if} \quad y < f(x) . \end{cases}$$

It is worth noting that eq.(7) holds in the sense of distributions,[2] because both the jumps of E and its normal derivative being zero, the Laplacien in the sense of distributions is actually equal to the Laplacian without precaution. Thus, it automatically takes into account the boundary conditions on the grating surface.[2] The only thing which remains is to match the solution in the modulated region with the Rayleigh expansion in the homogeneous regions.

In order to find the solution in the modulated region, we first use the pseudo-periodicity of the field to represent it by a series development :

$$E(x,y) = \sum_{n=-\infty}^{+\infty} E_n(y) \exp(i \alpha_n x) . \tag{8}$$

If we remark that, due to the periodicity of the profile, the function $k^2(x,y)$ is periodic with respect to Ox with period d, we can express it in terms of its Fourier series :

$$k^2(x,y) = \sum_{n=-\infty}^{+\infty} C_n(y) \exp(i n \frac{2\pi}{d} x) . \tag{9}$$

Thus, if we project eq.(7) on the $\exp(i \alpha_n x)$ basis, we get a set of ordinary coupled differential equations :

$$\frac{d^2 E_n(y)}{dy^2} = \alpha_n^2 E_n(y) - \sum_{m=-\infty}^{+\infty} C_{n-m}(y) E_m(y) . \tag{10}$$

If we introduce a column matrix $[E]$ whose elements are the functions $E_n(y)$ and a square matrix $[V]$, whose elements V_{nm} are defined by :

$$V_{nm} = \alpha_n^2 \delta_{nm} - c_{n-m} ,$$

eq.(10) can be written in matrix form :

$$\frac{d^2[E]}{dy^2} = [V] [E] . \tag{10'}$$

The numerical solution of this differential system has to match the analytical ones at the boundaries $y = 0$ and $y = a$ of the modulated region in such a way to satisfy the continuity of $E_n(y)$ and $dE_n(y)/dy$. In order to do that, we first replace the infinite set of equations by a set of $2N+1$ equations, from $n = -N$ to $n = +N$. We then perform $2N+1$ numerical integrations of (10') with particular starting values $[E]_p$, $p \in (-N, +N)$ given by :

$$E_{np}(-h) = \exp\left[i \; \beta_n^{(2)} \; h\right] \delta_{np} \qquad (h : \text{integration step})$$

$E_{np}(0) = \delta_{np}$, which are chosen in order to verify eq.(6).

Thus the numerical solutions $[E(y)]_p$ verify (6) and (10'), but not (5). Since the exact solution we are looking for is a superposition of $2N+1$ particular solutions of the form :

$$[E(y)] = \sum_{p=-N}^{+N} T_p [E(y)]_p \; ,$$

we determine the coefficients T_p by enforcing condition (5) at points $y = a$ and $y = a+h$:

$$\sum_{p=-N}^{+N} T_p E_{np}(a) = \exp\left[-i \; \beta_0^{(1)} \; a\right]\delta_{n,0} + B_n \exp\left[i \; \beta_n^{(1)} \; a\right] , \qquad (11)$$

$$\sum_{p=-N}^{+N} T_p E_{np}(a+h) = \exp\left[-i \; \beta_0^{(1)} \; (a+h)\right]\delta_{n,0} + B_n \exp\left[i \; \beta_n^{(1)} \; (a+h)\right] . \qquad (12)$$

Introducing :

$$A_{np} = \frac{E_{np}(a+h) \; \exp\left[i \; \beta_n^{(1)} a\right] - E_{np}(a) \; \exp\left[i \; \beta_n^{(1)} \; (a+h)\right]}{\exp\left[-i \; \beta_0^{(1)} (a+h)\right] \exp\left[i \; \beta_n^{(1)} (a)\right] - \exp\left[-i \; \beta_0^{(1)} a\right] \exp\left[i \; \beta_n^{(1)} (a+h)\right]}$$

$$A'_{np} = \frac{E_{np}(a+h) \; \exp\left[-i \; \beta_0^{(1)} a\right] - E_{np}(a) \; \exp\left[-i \; \beta_0^{(1)} (a+h)\right]}{\exp\left[-i \; \beta_0^{(1)} a\right] \exp\left[i \; \beta_n^{(1)} (a+h)\right] - \exp\left[-i \; \beta_0^{(1)} (a+h)\right] \exp\left[i \; \beta_n^{(1)} a\right]} \; ,$$

eq.(11) and (12) become :

$$\sum_{p=-N}^{+N} T_p A_{np} = \delta_{n,0} \qquad (13)$$

$$\sum_{p=-N}^{+N} T_p A'_{np} = B_n . \qquad (14)$$

Thus the resolution of the linear system (13) gives the T_p Rayleigh coefficients, which describe the field under the grating grooves. A simple matrix product (14) gives the B_n Rayleigh coefficients, which give the field above the grooves. The reflected absolute efficiencies $\mathcal{E}_n$ immediately follow :

$$\mathcal{E}_n = B_n B_n^* \frac{\beta_n^{(1)}}{\beta_0^{(1)}} . \qquad (15)$$

In the case of a dielectric grating, the transmitted efficiency are simply given by :

$$T_n T_n^* \frac{\beta_n^{(2)}}{\beta_0^{(1)}} \qquad (16)$$

As explained in ref.3, the numerical integration of (10') using the values of the function at points $y = -h$ and $y = 0$ as starting values is straightforward if we use a suitable algorithm, like Numerov algorithm. But the process supposes the continuity of the functions $C_{n-m}(y)$ and their derivatives at $y = 0$ and $y = a$. This condition is always verified, except if the grating profile presents plateaux like in the case of a lamellar grating. In that case, one must start the integration from the values of $E_{np}(0)$ and $\frac{d E_{np}}{dy}\Big|_{y=0}$, which are obtained from eq.(6) :

$$E_{np}(0) = \delta_{np}$$

$$\frac{d E_{np}(0)}{dy} = -i \; \beta_n^{(2)} \; \delta_{np} = -i \; \beta_n^{(2)} \; E_{np}(0) . \qquad (17)$$

The starting of the numerical integration can then be done by a Runge-Kutta algorithm for the first step,[4] and then be continued by the Numerov technique.

In both cases the differential method gives accurate results with very short computation times.

Finite conductivity gratings used in T.M. polarization

The function $H(x,z)$ again verifies Helmholtz equations (1) in each homogeneous region. But this time the normal derivative of $H(x,z)$ is not continuous at the grating surface, since the permittivity is discontinuous. Thus the equation $\Delta H + k^2(x,y)H = 0$ is not valid in the sense of the distributions and has to be replaced by a different propagation equation. If we recall that $\vec{H} = H(x,z)\hat{z}$, the Maxwell equation $\text{curl } \vec{H} = -i\omega\varepsilon\vec{E}$ implies :

$$\overrightarrow{\text{grad}}(H) \wedge \hat{z} = -i\omega\varepsilon\vec{E} \qquad \text{and} \qquad \text{curl } \vec{E} = i\omega\mu\vec{H}\hat{z}$$

lead to :

$$\text{curl }\left[-\frac{1}{i\omega\varepsilon}\;\overrightarrow{\text{grad}}(H)\wedge\hat{z}\right] = i\omega\mu_0 H\hat{z}$$

or to :
$$\text{div}\left[\frac{1}{k^2(x,y)}\;\overrightarrow{\text{grad}}\,H\right] + H = 0\;. \tag{18}$$

One can check that the divergence operator acts on a continuous quantity and thus, that (18) is valid in the sense of distributions. Outside the modulated zone, $k^2(x,y)$ is a constant and (18) reduces to a Helmholtz equation. Thus Rayleigh expansions (5) and (6), as well as eq.(8) and (9), are valid for $H(x,y)$. The principle of the method remains the same, except that we have to use a different algorithm to perform the numerical integration. Indeed, when one puts the Fourier series of $1/k^2(x,y)$ and $H(x,y)$ into (18), one founds a set of second order differential equations with first order derivative, which precludes the use of Numerov algorithm. In order to integrate the TM set of differential equations, we introduce function $\tilde{E}(x,y)$ defined by :

$$\tilde{E}(x,y) = \frac{1}{k^2(x,y)}\;\frac{\partial H_z}{\partial y} = \frac{E_x}{i\omega\mu_0} \tag{19}$$

From (18), we get :

$$\frac{\partial\tilde{E}}{\partial y} = -\frac{\partial}{\partial x}\left(\frac{1}{k^2}\;\frac{\partial H}{\partial x}\right) - H\;. \tag{20}$$

Projecting on the $\exp(i\alpha_n x)$ basis, with $\tilde{E}_n$ and H_n the components for $\tilde{E}$ and H, and β_n the Fourier coefficients of $1/k^2(x,y)$, eqs.(19) and (20) lead to an infinite set of ordinary first order differential equations :

$$\frac{dH_n}{dy} = \sum_{m=-\infty}^{+\infty} c_{n-m}\,\tilde{E}_m \tag{21}$$

$$\frac{d\tilde{E}_n}{dy} = \alpha_n\sum_{m=-\infty}^{+\infty}\alpha_m\,\beta_{n-m}\,H_m - H_n\;. \tag{22}$$

Thus we have a double infinity of unknown functions $\tilde{E}_n$ and H_n. If H is accurately represented by $2N+1$ terms of its Fourier series, it implies that we have $2(2N+1)$ functions to determine. The integration is initiated by taking particular values of $H_n(0)$ and $dH_n(0)/dy$ which verify (6) :

$$H_{np}(0) = \delta_{np}\;.$$

$$\left.\frac{dH_{np}}{dy}\right|_{y=0} = -i\,\beta_n^{(2)}\,\delta_{n,p}\;.$$

Thus (19) gives : $\tilde{E}_{np}(0) = \sum_m \beta_{n-m}(0)\,\dfrac{dH_{mp}(0)}{dy} = \dfrac{1}{k_2^{\,2}}\,H'_{np}(0)$ and (22) gives $\dfrac{d\tilde{E}_{np}(0)}{dy}$.

From the values of the functions and their derivatives at $y = 0$, the classical Runge-Kutta algorithm computes the values of $H_{np}(y)$ and $\tilde{E}_{np}(y)$ $\forall y$ in the modulated region and, in particular, at $y = a$. The use of (21) gives $dH_{np}(a)/dy$. We then enforce condition (5) by matching the numerical solution and its normal derivative with Rayleigh expansions at $y = a$:

$$\sum_{p=-N}^{+N} T_p\,H_{np}(a) = \exp\left[-i\beta_n^{(1)}a\right]\delta_{n,0} + B_n\exp\left[i\beta_n^{(1)}a\right]\;. \tag{23}$$

$$\sum_{p=-N}^{+N} T_p\,H'_{np}(a) = -i\beta_n^{(1)}\exp\left[-i\beta_n^{(1)}a\right]\delta_{n,0} + i\beta_n^{(1)}\,B_n\exp\left[i\beta_n^{(1)}a\right]\;. \tag{24}$$

The T_p Rayleigh coefficients are found through the resolution of the linear system :

$$\sum_{p=-N}^{+N} T_p \left[H'_{np}(a) - i \, \beta_n^{(1)} \, H_{np}(a) \right] = -2i \, \beta_n^{(1)} \, \exp\left[-i \, \beta_n^{(1)} a\right] \, \delta_{n,0}$$

and the B_n coefficients are obtained through eq.(23). The reflected efficiencies are again given by (15), while the transmitted efficiencies for a dielectric grating are now given by

$$\frac{1}{\nu_2^2} \, T_p \, T_p^* \, \frac{\beta_n^{(2)}}{\beta_0^{(1)}} \; .$$

The computation times are still competitive with those given by the integral method,[5] although they are from 2 to 3 times higher than for TE polarization. But numerical instabilities remain for deep grooves, especially for good reflectors.

Perfectly reflecting gratings.

The method is no longer utilizable for perfectly conducting gratings since the function $k^2(x,y)$ would be infinite in the metal and could not be described by a truncated Fourier series. In order to avoid this difficulty, we first look for a conformal mapping, which maps the xy plane onto a XY plane in such a way that the grating profile is mapped onto the X axis. Defining the complex variables $w = x + i y$ and $W = X + i Y$, the analytic function $w(W)$ which describes the conformal mapping is chosen in such a way to preserve the periodicity of the problem and the outgoing wave condition at infinity. It $f(x,y)$ is equal to the z-component of $\vec{E}$ or $\vec{H}$, depending on the polarization, which is transformed in $F(X,Y)$ by the conformal mapping, one can easily verify that the Helmholtz equation is turned into

$$\Delta F(X,Y) + k^2 \left| \frac{dw}{dW} \right|^2 F(X,Y) = 0 \tag{26}$$

and that the boundary condition is :

$$F(X,0) = 0 \qquad \text{for TE polarization} \tag{27}$$

$$\left. \frac{\partial F(X,Y)}{\partial Y} \right|_{Y=0} = 0 \qquad \text{for TM polarization.} \tag{28}$$

The determination of the function $w(W)$ for a given profile is explained in ref.6 and 4. It is shown that $w(W)$ is periodic with period d with respect to X, and that $|dw/dW|^2 \to 1$ if $Y \to \infty$ in an exponential manner. Thus, above a matching ordinate A, which turns to be closely related to d, $k^2|dw/dW|^2$ can be replaced by k^2. It means that the grating is equivalent to a perfectly reflecting mirror (see eq.(27,28) coated with a graded dielectric layer with thickness A, whose refractive index is periodic with respect to X. The function $k^2|dw/dW|^2$ plays the same role as the function $k^2(x,y)$ of the preceding paragraph and the resolution of the problem[7] is similar to the one used for TE polarization and finite conductivity gratings. The main difference is that the two fundamental cases of polarization do not present different propagation equations, but different boundary condition at $Y = 0$. Thus, a numerical integration from $Y = A$ to $Y = 0$ allows resolving the two cases at the same time. Their resolutions only differ by a matrix inversion after the end of the integration.

Extensions of the theory

The applications of the theory are not limited to bare gratings. The Differential Method can handle very easily the dielectric overcoatings with are deposited on top of vacuum UV aluminum gratings in order to prevent them from oxidizing. Only the function $k^2(x,y)$ has to be modified, but the computation times for dielectric coated gratings and for bare gratings are the same. Also, it can handle structures in which a dielectric grating or metallic grating is superimposed on a stack of dielectric or metallic layers, like the grating coupler device [8-10] used in integrated Optics or the devices used in photolithography.[11] Only the matching conditions (5) and (6) are to be changed. The method turned to be able to predict and optimize the coupling coefficient of grating couplers [8-10].

If the theory is able to predict grating efficiency, it was also generalized in order to calculate the Scattering Operator (S-matrix) used in some theoretical studies.[12,13] The determination of its poles and zeros in the complex plane enabled us to study the guided waves on the corrugated structure.[1,14] We were thus able to predict in a quantitative way grating anomalies, coupling of a laser beam into an optical waveguide, and phenomena of total absorption of light by a grating.

A special attention must be devoted to soft X-rays and X-UV gratings. The Electromagnetic Theories of gratings are generally developed in the resonance region, i.e. in a spectral domain for which the λ/d ratio is not too far from unity : $0.1 < \lambda/d < 10$. In order to get accurate values for the diffracted efficiencies, one has to keep in his computations all the real orders plus a few evanescent ones. In X-ray and XUV regions, even with the finest groove pitches which can be produced at present time, the λ/d ratio is so small that several hundred spectral orders may propagate. It means core memory storage and computation times which prohibit electromagnetic computations in this region. However, we recently discovered[15] that the Differential Method was able to predict accurate results for a given order efficiency by keeping only a few spectral orders around it. Due to the low modulated gratings which are used in this spectral domain, a given Fourier coefficient is only coupled with a few neighbours in the set of differential equations. The results obtained were not only found to be rapidly convergent when one increases the size of the truncated matrices. They also verified reciprocity relations and showed a good agreement with measurements.[16] Thus the Differential Method now allows us to perform at very low cost accurate analysis of soft X-rays and XUV gratings for which, due to the poor reflectivities of the surface, no numerical instability occurs.

The formalism was also extended to more complicated structures like stacks of gratings or multilayers coated gratings,[17] which are used as wavelength selectors at the end of laser cavities.

When the incident beam propagates outside the cross-section plane of a grating, the diffracted orders lie on a cone whose axis is the direction of the rulings, and the phenomenon is thus called "conical diffraction". For a perfectly conducting grating, the efficiencies in conical diffraction can be simply derived from the TE and TM efficiencies.[2] On the other hand, finite conductivity gratings lead to a much more complicated analysis, and the problem remains of vectorial nature. But the theory has been developped[18] and the computer codes predicted an interesting behaviour of a special conical diffraction mounting called the Generalized Maréchal and Stroke (G.M.S.) mount.[19] A more complicated vectorial problem was also resolved for the case of crossed (i.e. two dimension) gratings.[20] One must however be aware that the numerical instabilities found for good reflectors and TM polarization are likely to be found when one deals with these more complicated structures. But recent improvement,[21] which performs numerical integration from the top and the bottom of the grooves and matches the numerical solutions in the middle of the modulated region has increased by a factor 2 the groove depth range of validity of the method.

Although the present paper is only concerned with gratings, the Differential Theory also applies to other structures, provided they present a periodicity with respect to one coordinate. This is the case for dielectric or metallic cylinders of arbitrary shape,[22] which present a periodicity with period 2π with respect to the polar angle θ. Not only the diffracted field was calculated, but the Singularity Expansion Method[23] was developed for cylinders with finite conductivity.[22] Present extensions of the Theory which are now in progress, but have not yet been published concern dielectric resonators and non linear gratings.

Validity and applications of the method

No particular problem arises when one deals with dielectric transmission gratings, with the groove depths which are usually used. The remark still applies if the dielectric present losses, or if one deals with a metallic grating in a spectral range for which it presents a poor reflectivity (XUV or X-ray doamin). The case of good reflectors is however quite different. Since they present a complex refractive index whose square modulus is much greater than unity, the function $k^2(x,y)$ can hardly be described by a truncated Fourier series. One has to increase the size of the truncated matrices, which include evanescent orders with very large propagation constants $\beta_n^{(2)}$. Thus, numerical instabilities occurs, especially for TM polarization, which first show up as a lack of convergence of the results when the parameter N is increased. If the groove depth is high enough, they may give diffracted efficiencies greater than unity. Thus, it is worth checking the convergence of the results now and then when one deals with metallic gratings.

For TE polarization, good results can be obtained for most commercial gratings. For TM polarization, only shallow gratings can be investigated in the visible or infrared region. High modulated gratings must be investigated with the Integral Method.[5] However, when the instabilities does not occur, the Differential Method gives much lower computation times, particularly for TE polarization and dielectric coated gratings. The numerical difficulties found for TM polarization are of essential nature and also appear in other methods. But they appear for higher groove depths in the Integral Method, so that most practical gratings can be handled.

Many comparisons have been made between the results given by the two alternative methods. In their common domain of validity, the discrepancy on the diffracted efficiencies generally appear on the 4th figure, sometimes on the third or second (for TM polarization and medium modulated gratings). Excellent agreement is obtained for shallow gratings, for which discrepancies only appear on the 5th figure. Also the results have been checked against reciprocity and energy balance criterion, in the case of dielectric or perfectly reflecting gratings. They are generally verified with an accuracy of 10^{-6} or 10^{-5}. The reader interested in comparison with experiments can refer to ref.24, where he will find many theoretical and experimental curves drawn in the various spectral domains from infrared to XUV. Also, many typical applications can be found. Let us cite, for example, the introduction of strong anomalies in TE polarization due to a thick dielectric overcoating;[14] the optimization of X-ray and X-UV gratings ;[15,25] the study of the GMS mounting in X-UV.[19] Some applications would be presented at the conference.

Conclusion

The Differential Method is now a competitive technique to investigate grating efficiencies. Although for TM polarization its validity is limited on the side of high reflectivity, it is a very powerful tool for dielectric gratings, or metallic gratings used in UV or X-ray region for which other techniques are inoperative.

References

1. M. Nevière : "The Homogeneous Problem", in "The Electromagnetic Theory of Gratings ; Methods and Applications", R. Petit, ed., Springer-Verlag, Berlin, Heidelberg, New-York, 1980.
2. R. Petit :"A Tutorial Introduction", in "The Electromagnetic Theory of Gratings ; Methods and Applications", R. Petit, ed., Springer-Verlag, Berlin, Heidelberg, New-York, 1980.
3. M. Nevière, P. Vincent, R. Petit : Nouv. Rev. Optique, Vol. 5, pp.65-77, 1974.
4. P. Vincent : "Differential Methods", in "The Electromagnetic Theory of Gratings ; Methods and Applications", R. Petit, ed., Springer-Verlag, Berlin, Heidelberg, New-York, 1980.
5. D. Maystre : J. Opt. Soc. Am., vol. 68, pp.490-495, 1978.
6. M. Nevière, M. Cadilhac : Opt. Commun., vol. 3, pp.379-383, 1971.
7. M. Nevière, M. Cadilhac, R. Petit : IEEE Trans. on Antennas and Propagation, vol. AP-21 pp.37-46, 1973.
8. M. Nevière, R. Petit, M. Cadilhac : Opt. Commun., vol. 8, pp.113-117, 1973.
9. M. Nevière, P. Vincent, R. Petit, M. Cadilhac : Opt. Commun., vol. 9, pp.48-53, 1973.
10. M. Nevière, P. Vincent, R. Petit, M. Cadilhac : Opt. Commun., vol. 9, pp.240-245, 1973.
11. M. Nevière, Thèse d'Etat n° A.O. 11556, University of Marseille, France.
12. P. Vincent, M. Nevière : Optica Acta, vol. 26, pp.889-898, 1979.
13. P. Vincent, M. Nevière : Appl. Phys., vol. 20, pp.345-351, 1979.
14. M. Nevière : Proceedings of the SPIE's 24th Annual Technical Symposium, July 28-August 1, 1980, San Diego, California, U.S.A.
15. M. Nevière, J. Flamand : Nuclear Instruments and Methods, to be published.
16. M. Nevière : Proceedings of the VI International Conference on Vacuum Ultraviolet Radiation Physics, June 2-6, 1980, University of Virginia, Charlottesville, Virginia, U.S.A.
17. P. Vincent, Thèse d'Etat, University of Marseille, France, 1978.
18. P. Vincent, M. Nevière, D. Maystre : Nuclear Instruments and Methods, vol. 152, pp.123-126, 1978.
19. P. Vincent, M. Nevière, D. Maystre : Appl. Opt., vol. 18, pp. 1780-1783, 1979.
20. P. Vincent : Opt. Commun., vol. 26, pp. 293-296, 1978.
21. P. Vincent : Proceedings of the SPIE's 24th Annual Technical Symposium, July 28-August 1, 1980, San Diego, California, U.S.A.
22. P. Vincent : Appl. Phys., vol. 17, pp. 239-248, 1978.
23. C. Baum : In Transient Electromagnetic Fields, ed. by L.B. Felsen. Topics in Appl. Phys., vol. 10, Springer, Berlin, Heidelberg, New-York, 1976, pp. 129-179.
24. D. Maystre, M. Nevière, R. Petit : "Experimental Verifications and Applications of the Theory", in "The Electromagnetic Theory of Gratings ; Methods and Applications", R. Petit, ed., Springer-Verlag, Berlin, Heidelberg, New-York, 1980.
25. M. Nevière : Proceedings of the VI International Conference on Vacuum Ultraviolet Radiation Physics, June 2-6, 1980, University of Virginia, Charlottesville, Virginia, U.S.A.

Grating profile reconstruction

André Roger

Laboratoire d'Optique Electromagnétique, E.R.A. du C.N.R.S. n° 597
Faculté des Sciences et Techniques, Centre de St-Jérôme, 13397 Marseille Cedex 4, France

Abstract

A simple low-cost non destructive method has been developed for reconstructing grating profiles. It is based on an inverse scattering, electromagnetic rigorous theory. A computer program has been written for metallic gratings. It makes use of an efficiency curve in a constant deviation mounting and in TE polarization, and reconstitutes the grating profile. Thus the implementation of the reconstruction requires only optical measurements which can be easily performed and do not damage the grating surface. The program has been tested both on numerically simulated data and on actual physical measures, and gives good results.

Introduction

Over the past fifteen years have been developed rigorous electromagnetic grating theories,[1] which start from the grating profile, the material properties and Maxwell equations and compute the efficiencies, i.e. the power diffracted in each order. These theories take into account the finite conductivity of the materials as well as dielectric coatings.[2] Many computer programs are now at our disposal, which have been thoroughly tested by numerical criterions and successful comparisons with physical measurements of efficiencies. Now it would be of the highest interest to follow the inverse path and to compute grating profiles from efficiency curves. It would give a quick non-destructive and precise way of controling manufactured gratings. For this purpose, we have constructed a theoretical and numerical method.

Physical parameters

The profile to be reconstructed is described by the periodic function y = F(x) of period d, d being the groove spacing of the grating. Usually, d is not larger than a few wavelengths of the incident wave, and so the reconstruction of F(x) is a problem of inverse scattering. Moreover, the profile is not described by a finite number of parameters, but by a function, and this situation makes the difference between inverse scattering and parameter optimization methods. Let an electromagnetic plane wave, with wavelength λ and angle of incidence θ, be incident upon the grating surface (Figure 1). For the sake of simplicity, the electric

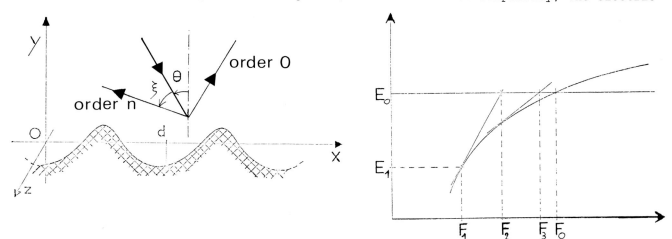

Figure 1 : The grating. The grooves are parallel to Oz, and the profile of the cross section is described by the periodic function y = F(x)

Figure 2 : The Newton algorithm

field is assumed to be parallel to the grooves (TE polarization). The grating is used in a constant deviation mounting in the nth order ; the angle of deviation is called ξ. With these conditions, λ and θ are related by :

$$\lambda = -\frac{2d}{n} \sin(\theta + \frac{\xi}{2}) \cos \frac{\xi}{2} . \tag{1}$$

The angle of diffraction θ_n of the nth order is given by :

$$\sin \theta_n = \sin \theta + n \frac{\lambda}{d} . \tag{}$$

We call $E(\theta)$ the efficiency in the nth order. The direct problem is to compute $E(\theta)$, $F(x)$ being known (and all other physical parameters being fixed). This problem has been thoroughly studied [1,2] for both the case of infinite and of finite conductivity materials. For a better understanding of the inverse problem, it is convenient to schematize the grating by an operator $\mathcal{O}$, which links the input function $F(x)$ and the output function $E(\theta)$. The direct problem is to compute the effect of the operator $\mathcal{O}$, the inverse problem is associated with the operator $\mathcal{O}^{-1}$.

The method

Description

The efficiency $E_0(\theta)$ ($\theta_1 < \theta < \theta_2$) being given, we want to solve the functional equation in F :

$$E_0(\theta) = \mathcal{O} . F(x) . \tag{2}$$

So as to avoid the substantial discontinuities which occur at the Rayleigh wavelengths, the interval (θ_1, θ_2) of variation of θ is chosen so that the number of diffracted orders remain constant in this interval. Of course, since equation (1) holds, the wavelength λ is simultaneously varying.

The operator $\mathcal{O}$ is very complicated, and cannot be explicitly written ; so of course it is not possible to solve equation (2) simply by analytical formulas. Our method is based on the use of the Newton Kantorovitch algorithm, which is a generalization of the well-known Newton algorithm for finding the zero of a real function. It is a step by step procedure, if converging, leads to an exact solution of equation (1). It starts with a first estimate $F_1(x)$ of the solution $F_0(x)$. This estimate has not to be precise ; for instance, a sine grating of same groove-depth-to-groove spacing ratio than $F_0(x)$ (with a relative accuracy of ± 30 %) is very often sufficient and yields convergence of the algorithm. The efficiency $E_1(\theta)$ corresponding to the grating profile $F_1(\theta)$ is computed by means of the program devoted to the direct problem. Then, instead of the non linear equation (2), we solve the linear equation (3) :

$$E_0(\theta) - E_1(\theta) = \mathcal{D} . (F_2(x) - F_1(x)) , \tag{3}$$

where $\mathcal{D}$ is the so-called Fréchet differential of $\mathcal{O}$, that is to say the linear operator tangent to $\mathcal{O}$, analogous to the tangent in the Newton method. Because equation (3) is only an approximation of equation (2), it does not give the exact solution $F_0(x)$ but a new approximation $F_2(x)$. Then the process is iterated and yields a sequence of functions $F_1(x)$, $F_2(x)$, ..., $F_n(x)$, which, under certain conditions, tends towards the exact solution $F_0(x)$.

Implementation

This method requires thus to compute the Fréchet differential $\mathcal{D}$, and then to invert equation (3). The computation of $\mathcal{D}$ has been achieved for perfectly conducting grating in a zero-deviation mounting in the -1 order [3,4] (Littrow mounting). Recently, it has been generalized to non zero deviation mountings in the nth order. The finite conductivity case is much more difficult to handle rigorously than the infinite case, but it has been possible to solve it by simple considerations. Let $E_i(\theta)$ and $E_c(\theta)$ be respectively the efficiencies of the perfectly conducting grating of profile $F(x)$, and of the same grating with finite conductivity. It is now well known [5] that, in TE polarization, E_i and E_c are simply related :

$$E_c(\theta) = R(\theta) \cdot E_i(\theta) , \tag{4}$$

$R(\theta)$ being a factor very close to the reflectance of the metal with finite conductivity, and thus nearly independent of the profile $F(x)$. Let $\mathcal{D}_c$ and $\mathcal{D}_i$ be respectively the Fréchet differentials for the conducting and infinitely conducting grating. By definition of the Fréchet differential, the following formulas hold at first order in δF :

$$\delta E_c(\theta) = \mathcal{D}_c \cdot \delta F(x) ,$$

$$\delta E_i(\theta) = \mathcal{D}_i \cdot \delta F(x) ,$$

where δF, δE_c, δE_i are a small variation of the profile $F(x)$ of the grating, and the corresponding variations of the efficiencies. Differentiation of equation (4) yields :

$$\delta E_c(\theta) = \delta R(\theta) \cdot E_i(\theta) + R(\theta) \delta E_i(\theta) ,$$

and the properties of $R(\theta)$ described above imply that $\delta R(\theta) = 0$. We obtain a simple expression for $\mathcal{D}_c$:

$$\delta E_c(\theta) = \mathcal{D}_c \cdot \delta F(x) = R(\theta) \delta E_i(\theta) = R(\theta) \mathcal{D}_i \cdot \delta F(x) .$$

Thus if $\mathcal{D}_i$ is known, [3,4] $\mathcal{D}_c$ can be easily computed. It is worth noting that even if $\mathcal{D}_c$ is only approximatively computed, the algorithm, when it converges, gives an exact solution of equation (2). Indeed, at the n-th step, we can write an equation similar to equation (3) :

$$E_0(\theta) - E_n(\theta) = \mathcal{D}_c \cdot (F_{n+1}(x) - F_n(x)) .$$

If the algorithm converges, it implies that $(F_{n+1}(x) - F_n(x))$ tends towards zero when n tends towards infinite, and thus $E_n(\theta)$ tends towards $E_0(\theta)$. An eventual lack of precision on $\mathcal{D}_c$ reduces only the domain of convergence of the algorithm and not the precision of the final solution.

Fundamental instability

On the other hand, equation (3) appears to be a Fredholm integral equation of the first kind, which is an ill-posed problem at Hadamard's sense.[6] It means that the solution $F_2(x) - F_1(x)$ of equation (3) does not depend continuously on the data $E_0(\theta) - E_1(\theta)$. In other words, an arbitrarily small variation of $E_0(\theta)$ can induce an arbitrarily large variation of $F_2(x)$, and one cannot get rid of this fundamental instability by increasing the experimental precision. A special mathematical technique, called Tikhonov-Miller regularization, must be employed in order to restore the stability.

Numerical examples

A computer program has been written on the basis of these considerations. First, theoretical reconstitutions have been achieved in the following manner : the grating profile $F_0(x)$ being given, one computes the corresponding efficiency $E_0(\theta)$ by using the "direct" program. Then a random error is added to $E_0(\theta)$ in order to simulate experimental errors, and one tries to reconstitute $F_0(x)$. Two examples are given in figures 3 and 4.

Since the grating profile $F(x)$ is periodic, it can be represented by a Fourier series. The computer reconstitutions have shown the following property : Fourier components of low order are easily detected, but high order Fourier components have very little influence on the efficiency and thus their detection requires high precision on the efficiency curve. This is the physical expression of the fundamental instability of this inverse scattering problem. In consequence, the profiles of blazed gratings are more difficult to reconstruct than those of nearly-sine gratings. If more precision on the profile is required, it is necessary to decrease the wavelength λ of the incident wave. For instance, in Littrow mounting, it has been numerically observed that the efficiency curve in the region where four orders are diffracted permits better reconstitutions than the efficiency curve corres-

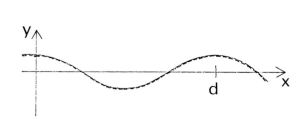

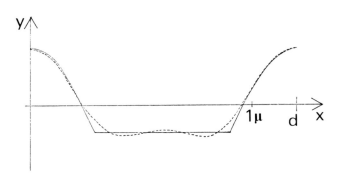

Figure 3. Reconstruction of a sine profile. The full line is the true profile, and the dotted line the reconstructed one. The grating is made of aluminum and is used in Littrow mounting in the -1 order with four diffracted orders ; groove spacing d = 1 μ ; deviation ξ = 0. in the -1 order ; angle of incidence 12° ≤ θ ≤ 19° ; wavelength 0.416 μ ≤ λ ≤ 0.651 μ ; a random error of 0.1 % has been added to the efficiency curve.

Figure 4. Reconstruction of the profile of a grating made with aluminum. The full line is the true profile, the dotted line is the reconstructed profile. Groove spacing d = 1,2 μ ; deviation ξ = 91° in the -1 order ; range of wavelength 0.589 μ ≤ λ ≤ 0.776 μ ; range of the angle of incidence 66° ≤ θ ≤ 73° ; random error of 0.1 %.

ponding to two diffracted orders. In terms of wavelength, one can hardly distinguish details smaller than $\lambda/4$. The computed profile can be easily checked : it is enough to apply the program to two different efficiency curves, for instance in two different orders, and to compare the two reconstructed profiles, which must be the same. This validity criterion has always been well satisfied by our computations.

Recently, the computer program has been tried on efficiency curves measured at the Royal Institute of Technology in Stockholm. Comparisons have been effected between computed profiles and electron microphotographs. The inverse scattering method appears to be efficient and precise ; these results will be published in a future paper.

Conclusion

Several generalizations of the method can be considered. The problem of infinitely conducting cylinders has already been studied.[7] On the other hand, it could be interesting to generalize the method to gratings used in TM polarization, though in this case the agreement between theoretical and experimental curves is less satisfying, probably because the electromagnetic field is much more sensitive to non periodic small defects of the profile in TM polarization than in TE polarization. Anyway, the inverse scattering method seems already to work well in TE polarization, and it could be a good non destructive way for testing grating profiles, since it requires only purely optical measurements.

References

1. Petit, R., "Electromagnetic grating theories : limitations and successes", Nouv. Rev. Optique, n° 6, pp. 129-135, 1975.
2. Maystre, D., "A new general integral theory for dielectric coated gratings", J.O.S.A., Vol. 68, pp. 490-495, 1978.
3. Roger, A., Maystre, D., "The perfectly conducting grating from the point of view of inverse diffraction", Opt. Acta, Vol. 26, pp. 447-460, 1979.
4. Roger, A., Maystre, D., "Inverse scattering method in electromagnetic optics : application to diffraction gratings", to be published in J.O.S.A.
5. Loewen, E.G., Nevière, M., Maystre, D., "Grating efficiency theory as it applies to blazed and holographic gratings", Applied Optics, vol. 16, pp. 2711-1720, 1977.
6. Roger, A., Maystre, D., "Determination of the index profile of a dielectric plate by optical methods", S.P.I.E., vol. 136, pp.26-28, 1977

7. Roger, A., "Computation of the shape of a perfectly conducting cylinder via knowledge of the bistatic or back scattering cross section", Proceedings of the International U.R.S.I. Symposium, Munich, 1980.

A simple method for determining the groove depth of holographic gratings

Sten Lindau
Department of Physics II and Institute of Optical Research
The Royal Institute of Technology, S-100 44 Stockholm, Sweden

Abstract

A method for determining the groove depth of a sinusoidal holographic grating is presented, which makes use of a laser to measure the efficiency in two simple geometries; at normal incidence, and in Littrow configuration. The efficiencies measured are compared to theoretical values to determine the groove depth. Measured efficiencies are plotted versus groove depth and appear to fit very well to corresponding theoretical curves. Some gratings were also studied by a scanning electron microscope. The groove depths measured from the SEM photographs are compared to those derived from efficiency measurements.

Introduction

The performance of a grating is usually displayed in graphical form, showing the efficiency as a function of wavelength for a given configuration. Such efficiency measurements are often very time-consuming, and careful adjustments are needed to achieve the right incidence angles.

If the groove density and coating material are specified, the efficiency is determined by the groove profile. Hence, if we can find the profile by some method, it is possible to calculate the efficiency of the grating in any configuration, for all wavelengths of interest.

The groove profile is, however, generally very difficult to measure, due to the small dimensions. The required resolution can be achieved by a scanning electron microscope (SEM). In order to get an edge-on view of the profile, the grating has to be destroyed, which limits the usefulness of this approach.

Another, direct method that has been used for determining groove profiles, makes use of the Talystep[1], which is an electromechanical device. A diamond stylus is drawn across the grating surface, its vertical movement being recorded.

The method to be presented here, was first suggested by Wilson[2]. Its simplicity lies in two facts:
1. A holographic grating, exposed in a symmetrical configuration, comes out with a more or less sinusoidal groove profile. Thus, it is not necessary to determine how the profile looks in detail; it is well described by one single parameter, the groove depth.
2. The groove depth is determined by measuring the efficiency for one wavelength, in two simple configurations. The measurements are done directly in a laser beam, and no collimating optics are needed.

The Wilson method

Measurement configuration

Suppose that we want to determine the groove depth of a holographic grating with a given groove frequency. We then have to decide, which configuration to use for the efficiency measurements, and the laser wavelength.

Natural choices for the setup are normal incidence, and Littrow configuration. For both cases, the correct incidence angle is easy to adjust accurately. In Littrow configuration, the -1 order is diffracted back along the incident beam, and the same is true for the 0 order in normal incidence mount. It is not so easy to measure the efficiency of the back diffracted beams, hence we decide for the following procedure:
1. For normal incidence, the efficiency of the -1 order is measured.
2. In Littrow configuration, the 0 order efficiency is determined.
The measurements are performed for booth TE and TM polarization.

Choice of laser

Which laser wavelength to choose depends on the groove frequency of the grating. In our laboratory we use 1800 grooves/mm as a standard for our holographic gratings, a frequency often used.

If we want to use the configurations mentioned above, we cannot choose too long a wavelength. If the wavelength is longer than the grating period, which in this case is 556 nm, only the 0 order is propagating in normal incidence configuration. A suitable laser is the He-Cd laser, which has a wavelength of 441.6 nm.

Theoretical efficiency values

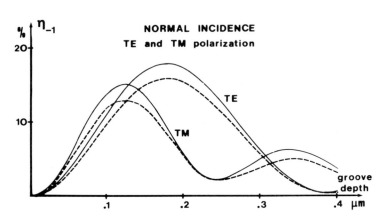

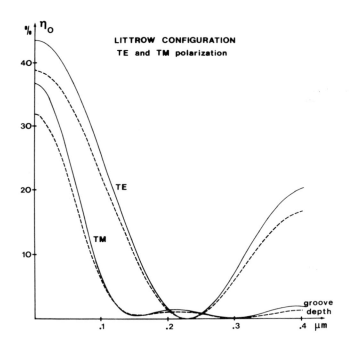

The measurements are compared to theoretical efficiency values, calculated using the rigorous finite conductivity theory of Maystre[3]. The theoretical values are preferably plotted versus groove depth, giving what will be referred to as Wilson Curves.

Figure 1 shows a set of Wilson curves for gold-coated gratings with 1800 grooves per millimeter. The wavelength is the He-Cd laser line at 442 nm.

The upper curves (full line) are computed using a complex refractive index, $N = 1.405 - i \cdot 1.90$, for evaporated gold. A different refractive index was used for the dashed curves (see Discussion).

Determining the groove depth

For each efficiency measurement, the corresponding groove depth, h, is obtained from the appropriate Wilson curve. In this way we get four values of h, which should not differ too much, and we take the mean as the groove depth.

From the oscillating nature of the curves, it is immediately clear, that it is not sufficient with only one measurement. Several different groove depths can, for example, give an efficiency of 5 % for the TM polarization in normal incidence. By studying the three other curves, one decides which one of the different possible depths to choose.

For gratings of depth greater than about 0.15 μm, the Littrow TM curve cannot be used for determining h, since large variations in h give only a slight change in efficiency.

Figure 1. Efficiency as a function of groove depth for a gold-coated grating with 1800 grooves/mm.

Experiments

In our laboratory, a special interferometer setup has been developed[4], which includes electronic exposure control. In this way, any desired exposure can be achieved with high accuracy.

However, the weak link of the holographic grating fabrication process appears to be development. The strength of the developer depends on a large number of factors, such as storage conditions, temperature, age etc.

For these experiments, we tried to keep the problems of the developer at a minimum, by exposing several different gratings on a single substrate. So, on the final plate, the only difference between any two gratings is the exposure; development and coating ought to be the same for all of them.

The photoresist used was the positive resist Shipley AZ-1350, which was spin coated onto the glass blank at 3000 rpm, giving a layer thickness of about 0.5 μm. After coating, the substrates were baked for 15 minutes at 70 ^{O}C.

A number of different exposures were tried, to obtain gratings of depth varying from about 0.05 μm to 0.40 μm. The groove frequency 1800 gr/mm was used throughout the experiments.

The exposed gratings were developed for 30 seconds in AZ-303, diluted 1:10. Finally, the gratings were coated with gold, by vacuum evaporation. The thickness of the gold layer was chosen so as to give "zero" transmission for plane coatings (which, however could not be achieved for deep gratings). Measurements afterwards showed the thickness to be about 0.24 microns.

Spectroscopic gratings are usually coated with aluminium. The reason for chosing gold for these experiments, was that we wanted to circumvent problems experienced with deep Al-coated gratings. The aluminium has a tendency to crystallize, giving a rather rough surface. Gold, which has a much higher atomic weight, does not show these problems.

The efficiencies were measured as described above, and several measurements were done in each configuration.

Some gratings were afterwards studied by SEM. The specimens were prepared by scoring the glass substrate perpendicular to the grooves, and then cracking, to get a cross section of the grating. To avoid plastic deformation of the resist, the cracking was done in liquid nitrogen[5].

Results

Efficiency measurements

The groove depth, h, of each grating was determined from the efficiency measurements. For deep gratings, as mentioned earlier, only three of the Wilson curves could be used. The measured efficiencies were then plotted versus h. The result is shown in fig.2, together with theoretical Wilson curves for comparison.

Groove depth determination by SEM

Those gratings that are numbered in fig.2 (of which no.1, 3 and 4 were made on the same substrate), were also studied by SEM. The resulting photographs are shown in fig.3, together with measured depth, h. The values in brackets are the h-values as derived from the efficiency measurements.

The upper profile on each photograph is the grating profile; the lower one is the interface between the gold coating and the resist.

Discussion

The experimental values agree fairly well with the theoretical efficiency curves, but there are some discrepancies:
1. For high efficiencies, the experiments give efficiencies that are found to be well below the theoretical values. This can be explained by assuming the reflectance to be lower than the value used for computation.
   When we, by trial and error, choose a different refractive index (N = 1.4 - i·1.7) the curves dashed in fig.1 were obtained, which give excellent agreement with experiment for small groove depths.
2. For groove depths greater than about 0.15 μm, the efficiencies measured for the TM polarization in Littrow mounting are substantially higher than the predicted values. This is another reason why these measurements cannot be used for determining the groove depth, apart from the arguments raised earlier.

If we compare the groove depths obtained from diffraction measurements, with those from the SEM micrographs, it is seen that the latter values are generally lower, the difference increasing with greater depth. Possible explanations may be:

The Wilson curves are computed for sinusoidal gratings, and with an infinitely thick coating. The profile of a holographic grating is affected by booth the development and the coating, and will in general not come out sinusoidal. The departure from sinusoidal form is most clearly seen on grating no.2, which shows peaks of rounded form and sharp valleys. The

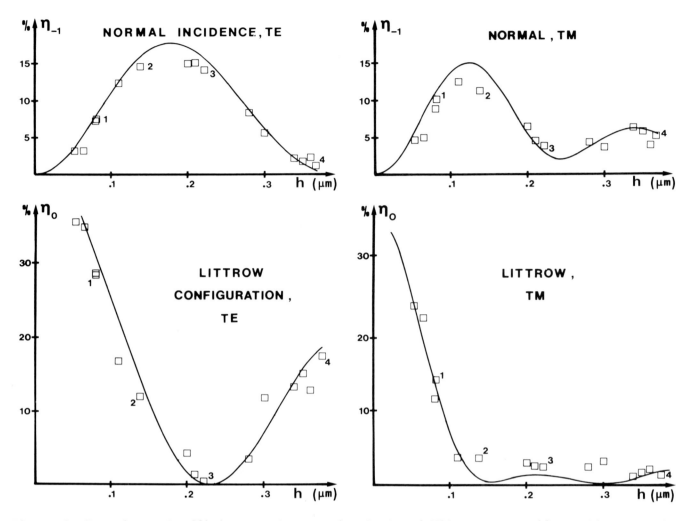

Figure 2. Experimental efficiency values, and calculated Wilson curves (for gold, N = 1.405-i·1.90) plotted versus groove depth h.

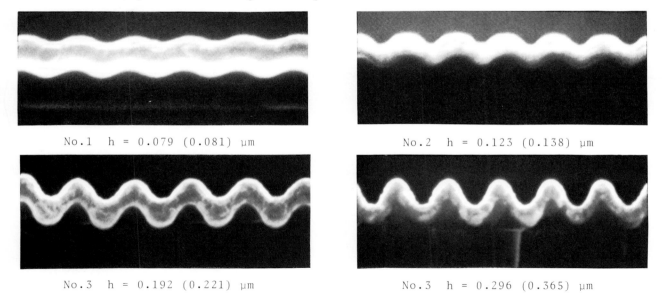

No.1 h = 0.079 (0.081) μm

No.2 h = 0.123 (0.138) μm

No.3 h = 0.192 (0.221) μm

No.3 h = 0.296 (0.365) μm

Figure 3. SEM micrographs of some grating profiles, together with measured groove depth, h.

same effect, but not so marked, can be observed on no.3 and 4, where also the difference between grating profile and resist profile is obvious.

When comparing an ideal, perfectly sinusoidal grating with a distorted one, apparently, one should compare the first Fourier components (as obtained from, e.g., SEM photographs) instead of the groove depths. This would give better agreement for the measurements reported here.

It is not known in detail how the finite thickness of the gold layer affects the efficiency. All of the deeper gratings made for these experiments have had a quite noticeable transmission.

Another, more trivial contribution to the discrepancy in groove depth determination, could be sought in the uncertainty about what happens to the profile, when the grating is prepared for SEM observation.

Conclusions

The method described has several attractive features. The measurements are easily done in a few minutes, and since the test is non-destructive, it could be used as a suitable criterion for checking production runs.

The efficiency values measured appeared to conform well with theoretical efficiencies for a wide range of different groove depths. SEM photographs of deep gratings indicate, however, that the values for the groove depth as determined from the efficiency measurements may be different from the real depth.

But the interesting property of a grating generally is the efficiency, and not the absolute value of the groove depth. Hence the groove depth as determined by the Wilson method is a relevant property of the grating, and provides information on the efficiency in other configurations.

One might well think of extending the Wilson method to other types of gratings than the sinusoidal ones. Provided that development and coating procedures are very well controlled and are kept fixed, the final groove profile is determined by the exposure. Diagrams, similar to the Wilson curves could be drawn, showing the efficiency as a function of exposure, rather than groove depth.

References

1. Verrill, J.F., "A study of blazed diffraction grating groove profiles using an improved Talystep stylus", Opt.Acta, Vol. 23, p.425 (1976)
2. Wilson, I.J., "Diffraction grating groove and mounting dependent properties", Thesis, University of Tasmania (1977)
3. Maystre, D., "A new integral theory for dielectric coated gratings", J.Opt.Soc.Am. Vol.68, p.490, (1978)
4. Johansson,S.,Nilsson,L-E., Biedermann,K. and Kleveby,K., "Holographic diffraction gratings with asymmetrical groove profiles", Applications of Holography and Optical Data Processing, ed. E.Marom, A.A.Friesem and E.Wiener-Avnear, p.521, Pergamon Press 1977
5. Brandes,R.G. and Curran,R.K. "Modulation transfer function of AZ111 photoresist", Applied Optics, vol.10, p.2101 (1977)

Numerical study of the coupling of two guided-waves by a grating with varying period

Patrick Vincent, Michel Cadilhac

Laboratoire d'Optique Electromagnétique, E.R.A. au C.N.R.S. n° 597 GRECO Microondes
Faculté des Sciences et Techniques, Centre de St-Jérôme, 13397 Marseille Cedex 4, France

Abstract

In Integrated Optics, the coupling of one mode propagating into a multimode slab waveguide to a single mode slab waveguide can be achieved by the use of a grating with varying period. A method to compute the coupling efficiency of the grating is proposed here. The fir t step is based on the differential formalism of diffraction and gives phenomenological coefficients characterizing the grating. Then an adiabatic approximation is used to take into account the variations of the grating period and the coupling coefficient is found by the numerical integration of a set of two coupled differential equations along the length of the device.

Introduction

The coupling of incident waves to guided modes through a dielectric grating has been studied extensively in the context of integrated optics applications[1-5]. The situation examined here exhibit some analogy with this problem, but is more complicated: it is the coupling between two modes propagating in two different waveguides with different propagation constants and coupled by a grating. The coupling phenomena occurs when the normalized spatial frequency of the grating is equal to the difference of the two propagation constants of the modes. However the incertainty upon the exact index profile of the guides leads us to the use of a grating having a slow varying period. We first introduce the phenomenological coefficients characterizing the grating. Then the power in each guide can be computed as a function of the length of the grating. This computation requires an adiabatic approximation valid if the coupling length is large compared to the grating period and if this period varies slowly.

Problem

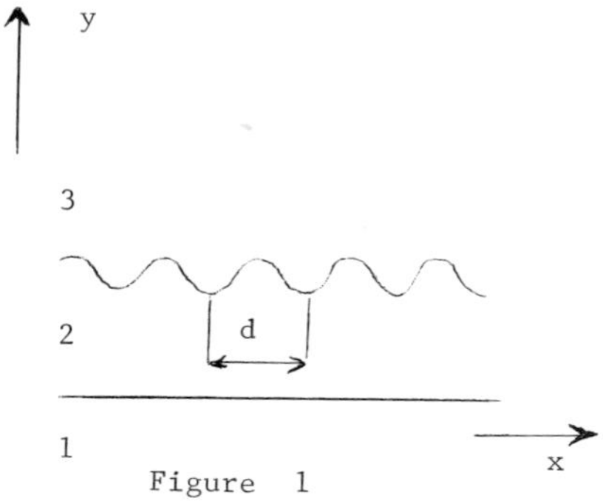

Figure 1

Let us consider a slab waveguide with a plane interface between medium 1 and medium 2 and a corrugated one between 2 and 3.(fig. 1) The grating has its grooves parallel to the z-axis and a given period d. We assume an $\exp(-i\omega t)$ time dependence. For the sake of simplicity we take the electric field $\underline{E}$ parallel to the z-axis, but the same considerations apply also to the other case of polarization. Thus the propagation problem reduces to a scalar one. According to Floquet theorem, let us write:

$$E(x,y) = f(x,y) \exp(i\alpha x), \qquad (1)$$

where f is a periodic function with respect to x (period d) and α is called the propagation constant of the field along the x-axis. The TE modes are the solutions of the propagation equations that verify an extended outgoing (or evanescent) wave condition when y tends towards infinity. Usually this condition implies that α must have particular values: the so-called propagation constnts of the modes.

When the grating couples the waves propagating inside the guide to non-evanescent waves, the mode is leaky and α is a complex number. Let us write (n being an integer and d the grating period) :

$$K = 2\pi/d \quad , \quad \alpha_n = \alpha + nK \quad , \quad \beta_{n,j}^2 = \omega^2 \varepsilon_j \mu_0 - \alpha_n^2 \text{ with } \text{Re}(\beta_{n,j}) + \text{Im}(\beta_{n,j}) \geqslant 0 . \quad (2)$$

Thus, outside the grating's grooves, the field in the j th layer can be expanded in plane waves:

$$E(x,y) = \sum_n \left[e_{n,j}^+ \exp(i\beta_{n,j} y) + e_{n,j}^- \exp(-i\beta_{n,j} y) \right] \exp(i\alpha_n x) \qquad (3)$$

The outgoing wave condition is satisfied if : $e_{n,1}^+ = 0 \quad \forall n$ and $e_{n,3}^- = 0 \quad \forall n$.

We examine now how these modes are related to the problem of the coupling between two parallel waveguides.

Coupling to a guided mode

We assume now that one mode of the corrugated waveguide (fig.1) is excited by a wave propagating above the grating in medium 3 with a propagation constant α :

$$E(x,y) = e^{-}_{0,3} \exp(i\alpha x) \exp(-i\beta_{0,3} y) \qquad (4)$$

This wave is reflected by the corrugated waveguide in different orders corresponding to the various values of the integer n . We assume that, in our case, only the specular reflection is important and that others waves are negligible. Thus the reflected field is represented by the coefficient $e^{+}_{0,3}$. Since we deal with a linear problem this coefficient is a linear function of the amplitude of the incident wave, i.e. the coefficient $e^{-}_{0,3}$; thus we write:

$$e^{+}_{0,3} = R \, e^{-}_{0,3} \qquad (5)$$

where R is the reflection coefficient of the corrugated surface.

Let us consider now $e^{+}_{0,3}$, R and $e^{-}_{0,3}$ as functions of the propagation constant of the field α . As defined in the preceding section, the modes are solutions of the propagation equations when the extended outgoing wave coindition is satisfied i.e. if $e^{-}_{0,3} = 0$. Thus when α is equal to α_g the propagation constant of the mode, we have :

$$e^{-}_{0,3}(\alpha_g) = 0 \quad \text{with} \quad e^{+}_{0,3}(\alpha_g) \neq 0 \ .$$

This shows that α_g is a singularity for the function $R(\alpha)$. Moreover, the equation (1) does not defined the propagation constant uniquely, and all the values $\alpha_g + nK$ are also singularities for R .

Let us take now an incident wave packet, defined by :

$$E^{i}(x,y) = \int e^{-}_{0,3}(\alpha) \exp(i\alpha x - i\beta_{0,3} y) \, d\alpha , \qquad (6)$$

with the modulus of $e^{-}_{0,3}(\alpha)$ negligible outside a neihbourhood (small compared to K) of $\alpha_A = \alpha_g - K$. We assume also that the singularity of $R(\alpha)$ is a simple pole and that, in this neighbourhood; $R(\alpha)$ is accurately represented by the first term of its Laurent expansion:

$$R(\alpha) = \frac{t}{\alpha - \alpha_g + K} \qquad (7)$$

From eq. (5) we get :

$$e^{+}_{0,3}(\alpha) = \frac{t}{\alpha - \alpha_g + K} \, e^{-}_{0,3}(\alpha) \qquad (8)$$

Thus the reflection on the corrugated surface is characterized by two complex phenomenological constants : α_g and the coupling coefficient t.

Coupling between two waveguides

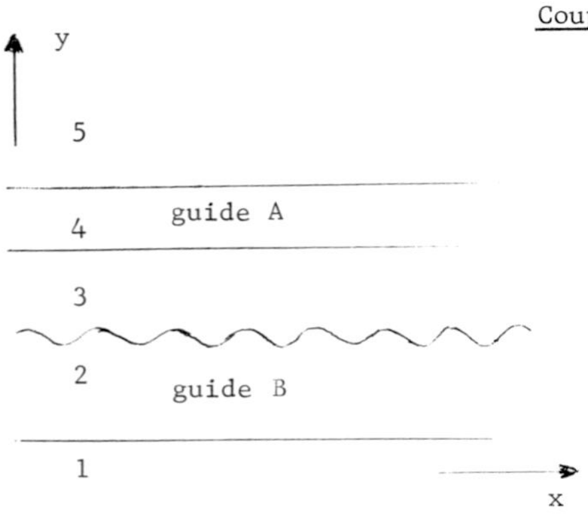

Figure 2.

In a previous paper[3] , we have proposed a formalism based on equation (8) to compute the coupling efficiency of a grating illuminated by a laser beam. The incident beam was described by the function $e^{-}_{0,3}$ which was a data of the problem. For the device describe in figure 2, the situation is somewhat different because the function $e^{-}_{0,3}$ is unknow. These waves are linked to the waves going upward i.e. the $e^{+}_{0,3}(\alpha)$ function by the reflection coefficient of the upper guide (guide A).

This coefficient which, of course, is also a function of the propagation constant α can be described by a formula similar to (7) except for the translation K of the propagation constant of the mode since this guide is not corrugated. Thus we write:

$$e^{-}_{0,3}(\alpha) = \frac{t_A}{\alpha - \alpha_A} \, e^{+}_{0,3}(\alpha). \qquad (9)$$

For clarity let us write:

$$\hat{\Psi}_A(\alpha) = e^{-}_{0,3}(\alpha) \quad , \quad \alpha_B = \alpha_g \quad , \quad t_B = t \quad ,$$

$$\hat{\Psi}_B(\alpha) = e^{+}_{0,3}(\alpha) .$$

Thus the field propagating between the two waveguide must satisfy the two equations:

$$\hat{\Psi}_A(x) = \frac{t_A}{\alpha - \alpha_A} \hat{\Psi}_B(\alpha) , \qquad (10)$$

$$\hat{\Psi}_B(\alpha) = \frac{t_B}{\alpha - \alpha_B + K} \hat{\Psi}_A(\alpha) . \qquad (11)$$

These equations give us the propagating constants of the modes related to the entire structure i.e. the two coupled guides: it is the values of α that make these equations compatible.

In fact, we don't need here these constants, but we look for the x-dependent solutions, i.e. the field as a function of the variable x when it is known by its initial value at abscissa x_0. To get these solutions let us return to equation (6) which defines a wave packet. For y=0, the relation between the function $e^{-}_{0,3}$ and the real field (i.e. the effective field) in the y=0 plane is a Fourier transform . Thus eq. (10) and eq. (11) can be interpreted as two relation between the Fourier transforms of the field $\hat{\Psi}_A$ and $\hat{\Psi}_B$.

Let us write:

$$\Psi_A(x) = \int \hat{\Psi}_A(\alpha) \exp(i\alpha x) \, d\alpha ,$$

$$\Psi_B(x) = \int \hat{\Psi}_B(\alpha) \exp(i\alpha x) \, d\alpha .$$

Then from equation (10) and equation (11), we get :

$$\frac{1}{i} \frac{d\Psi_A}{dx} = \alpha_A \Psi_A + t_A \Psi_B \qquad (12)$$

$$\frac{1}{i} \frac{d\Psi_B}{dx} = (\alpha_B - K) \Psi_B + t_B \Psi_A \qquad (13)$$

These equations are suitable to perform the adiabatic approximation needed to compute the field when the grating period and, consequently, K is a function of x: we just have to change K into the function K(x).

Computation of the field

Let us write:

$$\alpha_A = \alpha'_A + i\alpha''_A \quad , \quad \alpha_B = \alpha'_B + i\alpha''_B \quad (\alpha''_A , \alpha'_A , \alpha''_B \text{ and } \alpha'_B \text{ real })$$

If $\alpha'_A \alpha'_B > 0$, the two coupled modes propagate in the same direction, if this quantity is negative it is the oposite. Let us first examine the first case. We assume now that the grating has a finite extend and lies between x_0 and x_1 (fig. 3). and that the mode propagating in the guide A towards the positive values of x excites one mode of the guide B. Thus the suitable initial conditions to integrate eq. (12) and (13) are:

$$\Psi_A(x_0) = 1 \quad \text{and} \quad \Psi_B(x_0) = 0 .$$

The numerical integration from x_0 to x_1 gives Ψ_A and Ψ_B; from these two function it is easy to compute the field in each guide, then the coupling coefficient.

Let us deal now with the second case. We have a retro-coupling phenomena and the mode excited in the guide B propagates towards the negative values of x. The initial conditions are :

$$\Psi_A(x_1) = 1 \quad \text{and} \quad \Psi_B(x_1) = 0 .$$

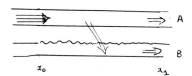

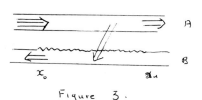

Figure 3.

and the numerical integration is performed from x_1 to x_0. The coupling coefficient is then deduced from the ratio of the energy flux at x_0.

Transformation of the equations

The adiabatic approximation is valid if $K(x)$ varies slowly and if $\alpha'_B - K \simeq \alpha'_A$. In order to improve the accuracy of the numerical integration we define:

$$K(x) = K_0 + K_1(x) \quad , \quad U_A(x) = \exp(-i\alpha'_A x)\, \Psi_A(x)$$

$$U_B(x) = \exp(-i\alpha'_A x)\, \Psi_B(x)$$

Thus equations (12) and (13) can be written:

$$
\begin{cases}
\dfrac{1}{i}\dfrac{dU_A}{dx} = i\alpha''_A\, U_A + t_A\, U_B \\[2em]
\dfrac{1}{i}\dfrac{dU_B}{dx} = \left(i\alpha''_B - K_1\right) U_B + t_B\, U_A
\end{cases}
$$

It is also useful to normalize the coupling coefficients with respect to the energy flux. Let us write ϕ_A and ϕ_B the energy flux in guide A and B; these flux are proportionnal to the square modulus of the amplitudes, thus we can define normalized amplitudes and normalized coupling coefficients by:

$$
\begin{cases}
\phi_A = A\,|U_A|^2 \\[1em]
\phi_B = B\,|U_B|^2
\end{cases}
\Rightarrow
\begin{cases}
U^*_A = \sqrt{A}\; U_A \\[1em]
U^*_B = \sqrt{B}\; U_B
\end{cases}
\text{ and }
\begin{cases}
t^*_A = \sqrt{\dfrac{A}{B}}\; t_B \\[1em]
t^*_B = \sqrt{\dfrac{B}{A}}\; t^*_A
\end{cases}
$$

Computation of the phenomenological coefficients

The computation of α_A, α_B, t_A, t_B is performed using the differential formalism of diffraction. The propagation constants are computed as the zeros of a complex function by an iterative method, previously described in our paper on the grating coupler theory[1,2]. A detailed discussion about the properties of these constants can be found in ref. 6. The coupling coefficients are computed using and arbitrary value for α' (e.g. the real parts of α_A and α_B)always with the differential formalism. Thus, these for constants are deduced from Maxwell equations with a rigorous method. Remarks that our programs gives also the others coefficients of the field, allowing to check the validity of our assumptions.

Conclusion

We propose a formalism to investigate the coupling between two waveguides based on two coupled differential equations. An important point is that the coefficients of these equations are found by solving rigorously Maxwell equations, that is not the case in usual perturbation theories. Thus, it is easy to check the validity of our formalism. The method is described with the corrugation on the receiving guide It can be easily extended to other cases i.e. when the corrugation is on the first guide or when more than two waves are involved in the coupling phenomena. Then the number of differential equations will be equal to the number of waves used, but the differential formalism is useful to compute all the coefficients needed. We hope to be able to give some numerical results in a near future .

<u>References</u>

1. M. Neviere, R. Petit, M. Cadilhac, Opt. Commun.,Vol. 8, pp113-117,1973.
2. M. Neviere, P. Vincent, R. Petit, M. Cadilhac, Opt. Commun., Vol. 9,pp48-53, 1973.
3. M. Neviere, P. Vincent, R. Petit, M. Cadhilhac, Opt. Commun., Vol. 9, pp240-245, 1973.
4. D. Marcuse, Bell. Syst. Techn. J., Vol 55,p1295, 1976.
5. T. Tamir, S.T. Peng, Appl. Phys., Vol.14, p235, 1977.
6. M. Neviere, " The Homogeneous problem", in " The Electromagnetic Theory of Gratings: Methods and Applications", R. Petit ed., Springer-Verlag, Heidelberg, New York, 1980.

Multiwave analysis of reflection gratings

M. G Moharam, T. K. Gaylord

School of Electrical Engineering, Georgia Institute of Technology, Atlanta, Georgia 30332

Abstract

A new exact coupled-wave analysis is applied to the diffraction of electromagnetic waves by longitudinally-periodic media (reflection gratings). The analysis is formulated in a simple state equation matrix form easily implemented on a digital computer. The relative intensities of the two waves, the diffracted (reflected) and transmitted waves, are calculated for a wide range of parameters. These exact results are then compared to results obtained with the approximate 1) two-wave modal analysis, 2) multiwave coupled-wave analysis, and 3) two-wave coupled-wave analyses. Exact calculations for even-order Bragg incidence (which can not be handled with approximate coupled-wave analyses) are included.

Introduction

Planar reflection gratings have applications in many different areas such as acousto-optics, integrated optics, holography, and spectroscopy. The diffraction of electromagnetic waves by these periodic structures has been extensively studied over many years.

Most of the previous work, whether utilizing the rigorous modal approach[1-9] or the simpler but approximate coupled-wave approach,[10-14] has been for transversely-periodic media. These are unslanted transmission gratings (the direction of the periodicity is parallel to the grating surface). However, recent applications such as grating couplers, grating filters, distributed Bragg reflector lasers, and distributed feedback lasers are either slanted or reflection type gratings. Kogelnik[12] and Magnusson and Gaylord[13] have analyzed general slanted gratings using the two-wave and the multiwave coupled-wave analyses respectively. Their analyses are not exact because of approximations introduced into the coupled-wave approach such as 1) neglecting higher-order waves,[12] 2) neglecting the second derivatives of the field amplitudes,[12,13] and 3) neglecting the boundary diffracted waves.[12,13] Bergstein and Kermisch[5] analyzed slanted gratings using the modal approach but their analysis is for very small modulations (two-wave approximation). Chu and Kong[9] have studied the general slanted grating problem using an exact modal approach but they did not present calculated results for reflection or slanted gratings because of the extremely difficult numerical problem resulting from the modal formulation.[15]

In this paper, reflection gratings bounded by two different media are analyzed rigorously using an exact coupled-wave approach that has recently been developed.[15] The analysis is formulated in simple state equation matrix form and numerical solutions are easily obtained. The intensities of the diffracted and the transmitted waves are calculated for a wide range of parameters. These exact results are compared with those from the approximate two-wave modal[5] and two-wave and multiwave coupled-wave analyses of Kogelnik[12] and of Magnusson and Gaylord.[13] The applicability of these approximate theories is discussed. The case of light incidence at other than first-order Bragg condition is also treated.

Theory

The grating under consideration is a lossless (pure phase) sinusoidal permittivity grating with an obliquely-incident wave polarized perpendicular to the plane of incidence (designated TE, H, or s polarization). The modulated region is bounded by two different homogeneous media with relative permittivities (dielectric constants) ε_1 and ε_3, as shown in Fig. 1. All three regions have the permeability of free space. The relative permittivity in the modulation region of thickness d is given by

$$\varepsilon(x,z) = \varepsilon_2 + \Delta\varepsilon \cos[K(\sin\phi \; x - \cos\phi \; z)] \tag{1}$$

where $K = 2\pi/\Lambda$ and Λ is the grating period. The quantities ε_2 and $\Delta\varepsilon$ are the average relative permittivity and amplitude of the modulation in the relative permittivity. The slant angle is ϕ and for the case of pure reflection gratings ϕ is zero. In the modal and coupled-wave analyses, ϕ approaches zero but is not equal to zero because that causes the grating not to be strictly periodic and the Floquet theorem which is the basis for the modal and the coupled-wave analyses, will not apply. As outlined in Ref. (15) the assumed solutions of the wave equation in the two unmodulated and the modulated regions for the normalized electric fields are:
In region 1 (z < 0),

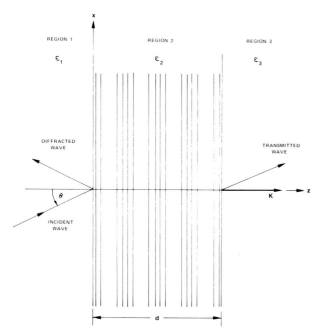

Figure 1. Geometry for diffraction of an electromagnetic wave by a planar grating.

$$E_1 = \exp[-j(\beta_0 x + \xi_{10}z)] + \sum_i R_i \exp[-j(\beta_i x - \xi_{1i}z)] \quad (2)$$

In region 3 ($z > d$),

$$E_3 = \sum_i T_i \exp[-j(\beta_i x + \xi_{3i}(z-d))] \ . \quad (3)$$

In region 2 ($0 < z < d$),

$$E_2 = \sum_i S_i^{'}(z) \exp[-j(\beta_i x + \xi_{2i}z)] \quad (4)$$

where $\beta_i = k_1 \sin\theta - iK \sin\phi$, $\xi_{\ell i} = (k_\ell^2 - \beta_i^2)^{\frac{1}{2}}$ for $\ell = 1,3$, $\xi_{2i} = k_2 \cos\theta' - iK \cos\phi$, $k_\ell = 2\pi(\varepsilon_\ell)^{\frac{1}{2}}/\lambda$ for $\ell = 1,2,3$, where λ is the free-space wavelength of the light, θ is the angle of incidence in region 1, and θ' is the angle of refraction in region 2. R_i and T_i are the amplitudes of i-th reflected and i-th transmitted waves and are to be determined. $S_i^{'}(z)$ is the amplitude of the i-th wave anywhere in the modulated region, and is to be determined by solving the scalar wave equation

$$\nabla^2 E_2 + (2\pi/\lambda)^2 \varepsilon(x,z)E_2 = 0 \quad (5)$$

in the modulated region. It is important to note that by taking ϕ approaching, but not equal to, zero all of the individually calculated B_i's converge to the same value. That is, all of the reflected waves (R_i's) and all of the transmitted waves (T_i's) have virtually the same propagation constant but are still individually distinguishable. With an understanding of this limiting process, ϕ can be taken to be equal to zero in the calculations.

To find $S_i^{'}(z)$, Eqs. (1) and (4) are substituted into Eq. (5) with ($\phi = 0$) resulting in the system of coupled-wave equations:

$$\frac{1}{\rho B^2} \frac{d^2 S_i(v)}{dv^2} = [1 - (2i/B)] \frac{dS_i(v)}{dv} - \rho i(i-B) S_i(v) + S_{i+1}(v) + S_{i-1}(v) \quad (6)$$

where $\rho = 2\lambda^2/\Lambda^2\Delta\varepsilon$, $B = 2\Lambda(\varepsilon_2)^{\frac{1}{2}} \cos\theta'/\lambda$, and $v = j\pi\Delta\varepsilon z/2\lambda(\varepsilon_2)^{\frac{1}{2}} \cos\theta' = j\gamma(z/d)$. Here B is the Bragg condition parameter, it is unity for first Bragg incidence. The quantity ρ is the regime parameter which determines the boundary for the two-wave Bragg diffraction regime ($\rho > 10$) for reflection gratings. Thus, this parameter has a significance for reflection gratings similar to its significance in transmission gratings.[16]

Equation (6) can be written in matrix state equation form[15] as

$$\begin{bmatrix} \underline{S}' \\ \underline{S}'' \end{bmatrix} = \begin{bmatrix} b_{rs} \end{bmatrix} \begin{bmatrix} \underline{S} \\ \underline{S}' \end{bmatrix} \quad (7)$$

where $\underline{S}$, $\underline{S}'$, and $\underline{S}''$ are the column vectors $S_i(v)$, $dS_i(v)/dv$, and $d^2 S_i(v)/dv^2$ respectively, and [b] is the coefficient matrix determined from Eq. (6). The solution of Eq. (7) gives the field amplitude $S_i(v)$ as

$$S_i(v) = \sum_m C_m w_{im} \exp(q_m v) \quad (8)$$

where q_m are the eigenvalues of the matrix [w] composed by the eigenvectors of the coefficient matrix [b]. Note that w_{im} is the m-th element of the row of the matrix [w] that corresponds to the i-th wave (not the i-th row). C_m together with T_i and R_i are determined by matching the tangential electric and magnetic fields at the two boundaries. The boundary conditions give[15]

$$2 \delta_{io} \xi_{1i} d = \sum_m C_m w_{im}(\xi_{2i} d + \xi_{1i} d - q_m \gamma), \quad (9)$$

$$0 = \sum_m C_m \, w_{im} (\xi_{2i} d - \xi_{3i} d - q_m \gamma) \exp(j q_m \gamma), \tag{10}$$

where $\xi_{\ell i} d = \rho \gamma B^2 (\varepsilon_\ell / \varepsilon_2 - \sin^2 \theta')^{\frac{1}{2}} / 2 \cos \theta'$ for $\ell = 1, 3$, and $\xi_{2i} d = \rho \gamma B (B - 2i)/2$. The values of C_m are found by solving this system of linear equations. Then R_i and T_i are calculated from[15]

$$\delta_{i0} + R_i = \sum_m C_m \, w_{im} \tag{11}$$

and

$$T_i = \sum_m C_m \, w_{im} \exp[j(q_m \gamma - \xi_{2i} d)], \tag{12}$$

where δ_{i0} is the Kronecker delta function. As ϕ approaches zero all the reflected waves collapse into one wave of intensity $\sum_i R_i R_i^*$ and all the transmitted waves collapse into one wave of intensity $\sum_i T_i T_i^*$. Clearly the sum of these two individual sums should be equal to the input intensity [normalized to unity in Eq. (2)]. Therefore, there may be many waves in the modulated region, but there is only one reflected and one transmitted wave in the two unmodulated regions as ϕ approaches zero.

Comparison to Previous Analyses

Although many authors state that their analyses apply to reflection gratings, very few results have been presented for these gratings. Authors[9] using the modal approach to analyze reflection gratings did not present any calculated results because solutions of the modal formulation are extremely difficult to obtain for this case. In most coupled-wave analyses, the second derivatives of the field amplitudes and the boundary diffraction are neglected. That is, coupled-wave analyses calculate the wave amplitudes inside the grating, not those of the diffracted waves leaving the grating. In the present analysis both the wave amplitudes inside and outside the grating are calculated. Kogelnik[12] analyzed reflection gratings using the coupled-wave approach where he assumed a two-wave regime in addition to the other approximations mentioned above. Our present analysis reduces to his analysis in the limit when $1/\rho B^2$ approaches zero and retaining only the zero-order and first-order waves in Eq. (6). Kogelnik[12] obtained closed-form expressions for the field amplitudes in the modulated region, which may be written in terms of our parameters as

$$S_0(v) = \cosh(\gamma + jv)/\cosh \gamma, \tag{13}$$

$$S_1(v) = -j \sinh(\gamma + jv)/\cosh \gamma, \tag{14}$$

$$S_i(v) = 0, \qquad i \neq 0, 1. \tag{15}$$

In this case, the reflected intensity

$$I_r = R_1 R_1^* = S_1(0) S_1^*(0) = \tanh^2 \gamma \tag{16}$$

The transmitted intensity is the difference between the incident intensity and the reflected intensity. Note, that in obtaining S_0 and S_1 Kogelnik assumed that $S_0(0) = 1$ and $S_1(j\gamma) = 0$ (thus, neglecting boundary reflections). Magnusson and Gaylord[13] extended Kogelnik's analysis to allow for all the higher-order waves in the modulated region. They neglected boundary diffraction and the second derivatives of the field amplitudes. The present analysis reduces to their analysis if $1/\rho B^2 = 0$ in Eq. (6). They also did not present calculated results for reflection gratings. But, if one calculates all the S_i field amplitudes in the modulated region, then the intensity of the reflected wave leaving the grating is

$$I_r = \sum_{i>0} (2i - 1) S_i(0) S_i^*(0) \; . \tag{17}$$

Note, that to obtain S_i, Magnusson and Gaylord[13] assumed $S_i(j\gamma) = 0$ for $i > 0$, $S_i(0) = 0$ for $i < 0$, and $S_0(0) = 1$, thus, neglecting boundary diffraction. It is important to note that in these approximate coupled-wave analyses that neglect in the second derivatives of the field amplitudes, solutions are not possible for incidence at even-order Bragg conditions (i.e., $B = 2, 4, 6, \ldots$) because of even-order Bragg conditions, the coefficient of the first derivative of i-th wave also vanishes $[1 - 2i/B = 0$ in Eq. (6)]. The results of the Bergstein and Kermisch[5] two-wave modal analysis can be obtained from the present formulation by keeping only the zero-order and first-order waves in Eqs. (2) through (4) and neglecting all higher-order waves.

In the next section the intensities of the transmitted and diffracted waves are calculated using the present exact coupled-wave analysis and these intensities are then compared to the approximate results of Kogelnik,[12] of Magnusson and Gaylord,[13] and of Bergstein and Kermisch.[5]

Results and Discussion

Following the procedure outlined in section II, the intensities of the reflected and transmitted waves are calculated for the first Bragg incidence B = 1, over a range of grating strengths γ from zero to 3.0, and for several values of the parameter ρ. The average permittivities of all three regions are taken to be equal so that comparisons can be made with the previous approximate coupled-wave analyses.[12,13] In Fig. 2, the calculated intensities of the reflected wave are plotted versus the grating strength γ for two different values of the parameter ρ. The transmitted and reflected intensities sum to unity. The intensity of the reflected wave increases initially in a parbolic manner and finally reaches saturation at a maximum value of unity as γ approaches infinity. This corresponds to 100% conversion of the incident wave into the reflected wave. Nearly 100% conversion is possible with moderate grating strengths, for example more than 97% conversion occurs at γ = 2.5. The errors in the calculated diffracted intensities due to the approximation in the Kogelnik[12] (two-wave), in the Magnusson and Gaylord[13] (multiwave) coupled-wave analyses, and in the Bergstein and Kermisch[5] modal analysis are plotted in Fig. 3. The absolute error is defined as the difference between the exact reflected wave intensity and the intensity calculated using each of the three approximate methods. Figure 3 shows that both the multiwave coupled-wave analysis of Magnusson and Gaylord,[13] and the two-wave modal analysis of Bergstein and Kermisch[5] produce roughly the same error and the Kogelnik[11] two-wave analysis produces slightly more error. This indicates that the second derivatives of the field amplitudes are as important as the higher-order waves for reflection gratings.

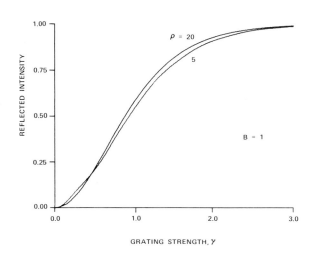

Figure 2. Calculated intensity of the diffracted (reflected) wave for incidence at the first Bragg condition (B = 1) and $\varepsilon_1 = \varepsilon_2 = \varepsilon_3$.

In Fig. 4, the calculated diffracted intensities of the reflected wave are plotted for the case of the average relative permittivity inside the grating being twice that of the surrounding regions. This corresponds, for example, to a hologram grating in air. Direct comparison in this case is only possible with the two-wave modal analysis. This is because the approximate coupled-wave theories are developed for the same average permittivity inside and outside of the grating region. Figure 5 shows the absolute error introduced by using the two-wave modal analysis relative to the present exact analysis.

The reflected intensity for incidence at the second Bragg condition is plotted versus γ in Fig. 6 for several values of ρ. Figure 6 shows that the reflected intensity is very small for large ρ and that smaller values of ρ are needed to produce significant second-order diffraction. This indicates that ρ which has been shown to delineate the two-wave diffraction regime in transmission gratings, plays the same role for reflection gratings.

Conclusion

The diffraction of a light wave by planar reflection gratings bounded by two different media has been analyzed using the recently developed exact rigorous coupled-wave approach. The solution is cast in a simple matrix form which is suitable for straightforward solution using standard digital computer program packages. The present analysis is compared with previous approximate coupled-wave and modal analyses of reflection gratings. It is shown that the approximations introduced in previous work cause significant errors in the calculated results and cause the formulation to breakdown for light incident at even-order Bragg conditions. Calculated results for incidence at higher-order Bragg conditions shows that the reflected intensity is very small unless the regime parameter ρ is small. The present analysis may be applied to grating couplers, grating filters, distributed feedback lasers, distributed Bragg reflector lasers, and other devices that use reflection gratings.

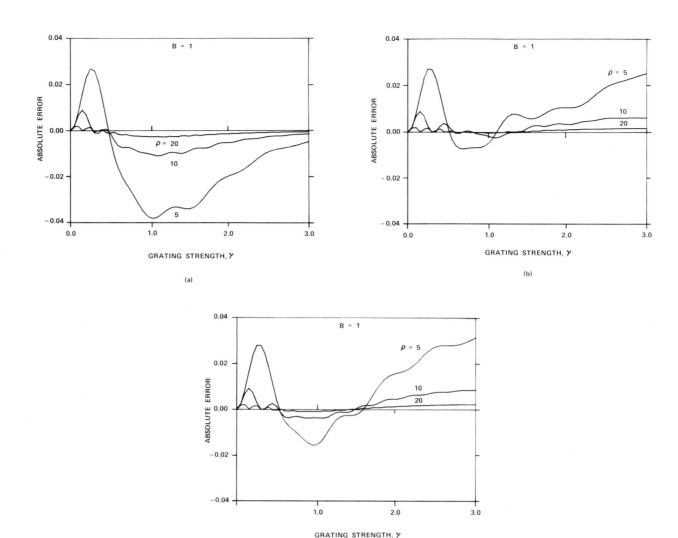

Figure 3.   Absolute error in the diffraction efficiency of the diffracted (reflected)
            wave defined as the difference between the diffraction efficiency calculated
            using the present exact analysis and the diffraction efficiency calculated
            using various approximate theories:
            (a)   Kogelnik's two-wave coupled-wave analysis;
            (b)   Magnusson and Gaylord's multiwave coupled-wave analysis; and
            (c)   Bergstein and Kermisch's two-wave modal analysis.

Acknowledgments

This work was sponsored by the National Science Foundation under Grant No. ECS-7919592
and by the Joint Services Electronics Program under Grant No. DAAG29-78-C-0005.

References

1.   T. Tamir, H. C. Wang, and A. A. Oliner, IEEE Trans. Microwave Theory Tech. MTT-12,
323 (1964).
2.   T. Tamir and H. C. Wang, Can. J. Phys. 44, 2073 (1966).
3.   T. Tamir, Can. J. Phys. 44, 2461 (1966).
4.   C. B. Burckhardt, J. Opt. Soc. Am. 56, 1502 (1966).
5.   L. Bergstein and D. Kermisch, Proc. Symp. Modern Opt. 17, 655 (1967).
6.   R. S. Chu and T. Tamir, IEEE Trans. Microwave Theory Tech. MTT-18, 486 (1970).
7.   R. S. Chu and T. Tamir, Proc. IEE 119, 797 (1972).
8.   F. G. Kaspar, J. Opt. Soc. Am. 63, 37 (1973).
9.   R. S. Chu and J. A. Kong, IEEE Trans. Microwave Theory Tech. MTT-25, 18 (1977).
10.   P. Phariseau, Proc. Indian Acad. Sci. 44A, 165 (1965).

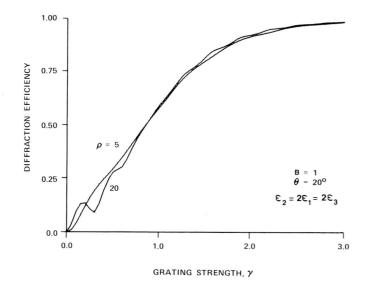

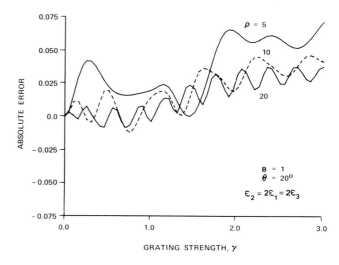

Figure 4. Calculated intensity of the diffracted (reflected) wave for incidence at the first Bragg condition (B = 1) and $\varepsilon_1 = \varepsilon_2/2 = \varepsilon_3$.

Figure 5. Absolute error in the diffraction efficiency of the diffracted (reflected) wave introduced by Bergstein and Kermisch's two-wave modal analysis in comparison to the present exact analysis.

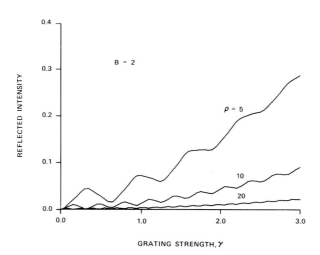

Figure 6. Calculated diffraction efficiency of the diffracted (reflected) wave for incidence at the second Bragg condition (B = 2).

11. W. R. Klein and B. D. Cook, IEEE Trans. Sonic Ultrasonics SU-14, 123 (1967).
12. H. Kogelnik, Bell Sys. Tech. J. 48, 2909 (1969).
13. R. Magnusson and T. K. Gaylord, J. Opt. Soc. Am. 67, 1165 (1977).
14. J. A. Kong, J. Opt. Soc. Am 67, 825 (1977).
15. M. G. Moharam and T. K. Gaylord, J. Opt. Soc. Am. (submitted).
16. M. G. Moharam, R. Magnusson, and T. K. Gaylord, Opt. Comm. 32, 14 (1980).

Surface grating on dielectric slab waveguide between parallel perfectly conducting plates: numerical solution

Jesus Alfonso Castaneda, Rodolfo F. Cordero-Iannarella
University of California, Los Angeles

A Numerical Solution for the interacting TE mode fields of a surface grated dielectric slab waveguide between parallel perfectly conducting plates is presented. The structure is assumed to operate in the reflection mode (where the dominant coupling is between counter-running film guided waves). The structure between plates provides an approximation to the open structure problem, where the plates are absent and the substrate and superstrate are both infinite in extent. With this approximation to the open structure it is possible to handle some cases of operation of the open structure near cutoff.

Introduction

The structure considered is the surface perturbed asymmetric dielectric slab waveguide between parallel perfectly conducting plates (Figure 1). The perturbation is uniform to a depth h from the film-superstrate interface and sinusoidal along the length. The interest is in the application of this closed waveguide structure as an approximation to open grated waveguide structures (Figure 2). The continuous radiation mode spectrum, which is characteristic of open structures, poses special problems. This is especially true when the open structure is operating near cutoff. Under these circumstances a substantial interaction of the guided modes with the radiation field spectrum can be expected. For an accurate determination of the coupling properties of such a device these effects must be included in the analysis. In contrast to the open waveguide structure, the closed structure has only a discrete mode spectrum. The asymmetric dielectric slab waveguide between parallel plates, however, is still a three region problem. Moreover, it's mode spectrum is also comprised of three mode types, film modes, substrate modes, and superstrate modes.

The device is assumed to have been designed to operate in the reflection mode, that is, the strongest although not necessarily the majority interaction is between a forward traveling and a backward traveling near-phase-matched pair of film modes. Moreover, the structure parameters are chosen such that the phase mismatch between the incident mode and the super-

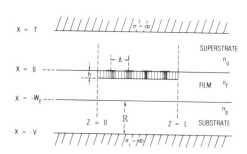

Figure 1 - Dielectric slab waveguide with sinusoidal perturbation between parallel plates, i.e. the closed structure.

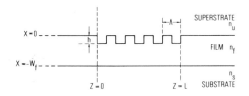

Figure 2 - Open structure with rectangular corrugation perturbation.

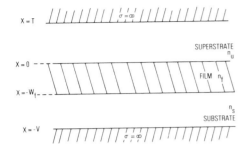

Figure 3 - Unperturbed waveguiding sturcture: the asymmetric dielectric slab waveguide sandwiched between parallel perfectly conducting plates.

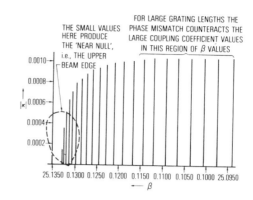

Figure 4 - Line plot of the substrate mode spectrum and coupling coefficients' magnitudes.

strate modes preclude any significant coupling between them. The incident mode is TE to z and in view of the structure and perturbation symmetries the balance of the excited modes are likewise TE modes. The problem treated is rather specific. Still it serves to illustrate the procedure adopted for its solution. From that solution it is possible to characterize not only the extant film guided mode fields but also the substrate modes' fields which when the structure is used as an approximation to the open structure correspond to the radiation leakage through the substrate.

The analysis consists of a direct application of coupled mode theory as in the case of the simple two wave interaction grating structure. However, in our case, there are a multitude and variety of modes, not just two waves. Since this is strictly an application of the coupled mode theory as popularized by Yariv and others we are encumbered by the same restrictions as in those applications. Chief among these restrictions is the requirement that the interaction be weak locally, albeit the cummulative interaction level can be large.

To effect the solution the steps are, in order: 1) The derivation of the mode functions for the unperturbed structure of Figure 3, 2) the substitution of these into the scalar wave equation for the perturbed structure and the subsequent reduction of it to a set of first order coupled linear differential equations in the mode weighting amplitude functions, 3) the evaluation of the mode coupling coefficients. In connection with the numerical solution of the set of equations for the mode weighting amplitude functions we must consider, 4) the conversion of the boundary value problem to an initial value problem, 5) the numerical integration, and lastly 6) the superposition of the modes weighted by the determined mode weighting amplitudes' values at the grating ends in order to construct the scattered fields outside the grating regions.

Analysis

If for a grating structure the permitivity differs from that of the unperturbed waveguide by $\Delta \epsilon (x,z)$ then the equation governing the TE fields can be written as:

$$\left\{ \frac{\partial^2}{\partial x^2} + \frac{\partial^2}{\partial z^2} + \omega^2 \mu \, \epsilon(x) \right\} E_y = \omega^2 \mu \, \Delta\epsilon(x,z) \, E_y \qquad (1)$$

$$\epsilon(x) = \epsilon_o n^2(x),$$
$$n(x) = n_u, \quad T \geq x \geq 0$$
$$= n_f, \quad 0 > x > -W$$
$$= n_s, \quad -W > x \geq -V$$

For the structure of figure 1

$$\Delta\epsilon(x,y) = 0, \qquad x > 0$$

$$= \epsilon_0(n_f^2 - n_u^2) \sin(\frac{2\pi}{\Lambda} z), \quad 0 \geq x \geq h$$

$$= 0, \qquad h > x$$

The form of equation (1) and in particular the appearance of the perturbation term on the right side, is suggestive of the physical interpretation of the coupled mode analysis as applied to perturbed structures. As it appears the perturbation term plays the role of a distributed source function. The perturbed structure is euphemistically represented as an unperturbed structure with embedded distributed sources. It is expected that those sources, and the perturbation they model, can naturally excite only the modes of the unperturbed structure. Taking the course suggested by this, the field E_y is expressed as a super-position of the eigenmodes of the structure of figure 3.

$$E_y = \sum_m A_m^{\pm}(z) E_m(x) \ e^{\pm j\beta mz} \qquad (2)$$

The $\left|A_m^{\pm}(z)\right|$ are the mode weighting functions, the $\left|Em(x)\right|$ the mode functions of the un-perturbed structure, and the $\left|\beta m\right|$ the corresponding propagation constants or eigenvalues.

Before proceeding to the substitution of this expansion into equation (1) we digress to specify the mode functions $\left|Em(x)\right|$ and the propagation constants $\left|\beta m\right|$.

Field Solutions to the boundary value problem, referring to the structure of Figure 3, can be characterized as a complete set composed of the three solution groups of film modes, substrate modes and superstrate modes. The assumption, that there is leakage only into the substrate region, allows us to limit the solution to that of the film modes and substrate modes. The differential equation for the mode functions is

$$\frac{d^2 Ey(x)}{dx^2} + (k^2 n^2(x) - \beta^2) \ Ey(x) = 0 \qquad (3)$$

where n(x) is defined as for equation (1). The two mode groups of interest, i.e. the film modes and the substrate modes, are distinguished as follows:

i) for the film modes $\beta > n_s k$, and the mode function is a standing wave in the film region and has hyperbolic function forms in the substrate and the superstrate regions.

ii) for the substrate modes $n_s k > \beta > n_u k$, and the mode functions have standing wave forms in both the film and substrate regions with a hyperbolic form in the superstrate region.

For film modes we use the relations:

$$u^2 = \beta^2 - k^2 n_u^2 ,$$

$$f^2 = k^2 n_f^2 - \beta^2 , \qquad\qquad (4)$$

$$s^2 = s_f^2 = \beta^2 - k^2 n_s^2 > 0.$$

Applying the boundary conditions at the plates and the dielectric interfaces the exact forms of the film mode functions is found to be:

$$E(x) = -B \frac{\sin(\theta)}{\sinh(uT)} \cdot \sinh[u(x-T)] \quad , \quad T \geq x \geq 0$$

$$= B \sin[fx + \theta] , \qquad\qquad 0 \geq x \geq -W \qquad (5)$$

$$= B \frac{\sin(\theta - fW)}{\sinh(sR)} \cdot \sinh[s(x + V)] , \quad -W \geq x \geq V,$$

where θ is the principal value as given by

$$\tan(\theta) = -\frac{f}{u} \text{Tanh}(uT) \ . \tag{6}$$

The eigenvalue equation is

$$m\pi = fW - \arctan\left[\frac{u}{f} \text{cotanh}(uT)\right] - \arctan\left[\frac{s_F}{f} \text{cotanh}(s_F R)\right] \ ,$$
$$m = 0,1,\ldots,M, \tag{7}$$

which taken together with equations (4) is used to determine the $|\beta_m|$, the mode propagation constants. The expression for the coefficient B derives its form from a forced normalization, i.e., in the case of undisturbed propagation we require the power carried in the mth mode be unity when the mode weighting is unity. The expression is:

$$B^2 = \frac{4\omega\mu}{\beta} \cdot \frac{1}{W}_{\text{effective}} \tag{8}$$

where $W_{\text{effective}}$ is defined as

$$W_{\text{eff}} = W + \frac{1}{u} \cdot \left[\frac{\text{Tanh}(uT)\left(u^2 + f^2 - f^2 \frac{2uT}{\text{Sinh}(2uT)}\right)}{f^2 \text{Tanh}^2(uT) + u^2} \right]$$
$$+ \frac{1}{s} \cdot \left[\frac{\text{Tanh}(sR)\left(s^2 + f^2 - f^2 \frac{2sR}{\text{Sinh}(2sR)}\right)}{f^2 \text{Tanh}^2(sR) + s^2} \right] \tag{9}$$

For the substrate modes, where

$$s^2 = s_s^2 = \beta^2 - k^2 n_s^2 < 0, \tag{10}$$

the mode function expressions are identical to those quoted for the film modes. The eigenvalue equation, however, is different in form. In terms of the real variable r, defined as

$$r = js_s ,$$

the eigenvalue equation is:

$$p\pi = rR + \arctan\left\{ \frac{r}{f} \tan\left[fW + \arctan\left(\frac{f}{u} \tanh(uT)\right)\right]\right\} \tag{11}$$
$$p = 1,2,\ldots,P.$$

Returning to the consideration of the perturbed structure, the substitution of (2) into (1) is made. The resulting relation in the mode weighting amplitude functions $\{A_m(z)\}$ can be greatly simplified. First the surviving second derivatives are deleted, this in virtue of the assumption that for each weighting function the following is true:

$$\left|\frac{dA}{dz}\right| >> \frac{2\pi}{\beta} \left|\frac{d}{dz}\left(\frac{dA}{dz}\right)\right| \ . \tag{12}$$

Beyond this the orthogonality of the mode functions themselves can be exploited. The coupled differential equations take their final forms as:

$$\frac{dA_n^{(+)}(z)}{dz} = \sum_m \kappa_{mn}\, e^{-j\phi_{mn}\cdot z}\, A_m^{(-)}(z)$$

$$\tag{13}$$

$$\frac{dA_n^{(-)}(z)}{dz} = \sum_m \kappa_{mn}\, e^{j\phi_{mn}\cdot z}\, A_m^{(+)}(z)$$

with the phase mismatches ϕ_{mn} given by

$$\phi_{mn} = \beta_m - \beta_n - \frac{2\pi}{\Lambda} \tag{14}$$

and the coupling coefficients κ_{mn} as

$$\kappa_{mn} = \frac{\omega\epsilon_0(n_f^2 - n_u^2)}{8} \int_{-h}^{0} \mathscr{E}_m(x)\,\mathscr{E}_n(x)\,dx \;, \tag{15}$$

or explicitly:

$$\kappa_{mn} = \frac{k^2(n_f^2 - n_u^2)}{2[\beta_m\,\beta_n\, W_{eff}^{(m)}\, W_{eff}^{(n)}]^{1/2}} \cdot$$

$$\frac{f_m\,\mathrm{Tanh}(u_m T)}{[f_m^2\mathrm{Tanh}^2(u_m T) + u_m^2]^{1/2}} \cdot \frac{f_n\,\mathrm{Tanh}(u_n T)}{[f_n^2\mathrm{Tanh}^2(u_n T) + u_n^2]^{1/2}} \cdot$$

$$\left\{ \frac{\mathrm{Sin}(f_m - f_n)X}{2(f_m - f_n)}\left(1 + \frac{u_m u_n}{f_m f_n}\,\mathrm{Coth}(u_m T)\,\mathrm{Coth}(u_n T)\right) \right.$$

$$+ \frac{\mathrm{Sin}(f_m + f_n)X}{2(f_m + f_n)}\left(1 - \frac{u_m u_n}{f_m f_n}\,\mathrm{Coth}(u_m T)\,\mathrm{Coth}(u_n T)\right)$$

$$+ \frac{\mathrm{Cos}(f_m + f_n)X}{2(f_m + f_n)}\left(\frac{u_n}{f_n}\,\mathrm{Coth}(u_n T) + \frac{u_n}{f_m}\,\mathrm{Coth}(u_m T)\right)$$

$$\left. + \frac{\mathrm{Cos}(f_m - f_n)X}{2(f_m - f_n)}\left(\frac{u_n}{f_n}\,\mathrm{Coth}(u_n T) - \frac{u_m}{f_m}\,\mathrm{Coth}(u_m T)\right) \right\} \begin{array}{l} X = 0 \\ \\ X = -h \end{array} \tag{16}$$

The numerical integration applied to the set of equations (13) is a simple state space forward walk. The basic relation for this procedure is supplemented and the computational efficiency is improved by approximating:

$$A_n^{(\pm)}(z_2) = A_n^{(\pm)}(z_1) + \int_{z_1}^{z_2} \frac{dA_n^{(\pm)}}{dz}\, dz$$

$$\approx A_n^{(\pm)}(z_1) + \sum_m \kappa_{mn}\, A_m^{(\mp)}(z_1) \left(\frac{e^{(\mp)j\phi_{mn}z_2} - e^{(\mp)j\phi_{mn}z_1}}{(\pm)j\phi_{mn}} \right) \quad (17)$$

The task remaining is that of recasting the boundary value problem as an initial value problem. For the cases considered we assume an incident wave amplitude of unity at z = 0. All other forward traveling wave amplitudes are zero at this point. At the end of the grating region (i.e., z = L) all backward traveling waves vanish.

Much as is done in the analytical procedures, in this numerical procedure the superposition principle can be used to form a solution that matches the boundary conditions. The first step is to generate a complete set of linearly independent solutions.

Any specific set of initial conditions can be considered as an element in the space of all possible sets of initial conditions. We can identify certain special sets of initial conditions of this space as linearly independent and spanning the space. Moreover, the differential equations system is linear and the integration from the initial values (at z = 0) to the final values (at z = L) can be characterized as a linear transformation by the matrix $\underline{\underline{T}}$ (an M x M matrix for a system of M order differential equations). Any solution we may be interested in is necessarily associated with an element of the initial conditions space. If $\underline{A}$ is the initial conditions vector and $\underline{B}$ is the final values vector, then these are related by $\underline{\underline{T}}$ according to

$$\underline{B} = \underline{\underline{T}} \cdot \underline{A} \,. \quad (18)$$

Vectors $\underline{A}$ and $\underline{B}$ are M-dimensional since there are M wave amplitudes to be specified at z = 0. It follows that the set of solutions derived from the complete and linearly independent set of initial conditions vectors will also be linearly independent and complete. The set of orthogonal initial conditions vectors are:

$$\underline{a}_1 = (1,0,0,\ldots,0)$$
$$\underline{a}_2 = (0,1,0,\ldots,0)$$
$$\underline{a}_3 = (0,0,1,\ldots,0)$$
$$\cdot$$
$$\cdot$$
$$\cdot$$
$$\underline{a}_M = (0,0,0,\ldots,1)$$

These are in fact orthonormal. They are easily produced and the corresponding solutions can routinely be obtained by way of equation (19). The elements of the matrix $\underline{\underline{T}}$ can be found by considering the M equations of the form:

$$\underline{\underline{T}} \cdot \underline{a}_n = (t_{n1},\ t_{n2},\ldots,t_{nM}) \quad (19)$$

This equation corresponds to an integration of the coupled equations starting with all wave amplitudes set to zero except the nth which is set equal to one. With M separate integrations we can completely specify the matrix $\underline{\underline{T}}$. We now address the boundary value problem.

For the general initial conditions vector $\underline{A}$ we have

$$\underline{A} = (\alpha_1,\alpha_2,\alpha_3,\ldots,\alpha_M)$$

and thus

$$\underline{\underline{T}} \cdot \underline{A} = \left(\sum_k \alpha_k \, t_{1k}, \ \sum_k \alpha_k \, t_{2k}, \cdots, \ \sum_k \alpha_k t_{Mk} \right).$$

If $\alpha_1 = 1$ (corresponding to an incident wave amplitude of unity at $z = 0$) and $\alpha_j = 0$ for $2 \leq j \leq N$ (which for the case $N = M/2$ corresponds to the zero amplitude at $z = 0$ of all other forward traveling waves) then the results are:

$$\underline{B} = \underline{\underline{T}} \cdot \underline{A} = \left\{ (t_{11} + \sum_{N+1}^{M} \alpha_k \, t_{1k}), (t_{21} + \sum_{N+1}^{M} \alpha_k t_{2k}), \cdots \right.$$

$$\left. \cdots, \ (t_{M1} + \sum_{N+1}^{M} \alpha_k t_{Mk}) \right\}$$

We now require that the last $M - N$ components of $\underline{B}$ vanish, which for the case $N = M/2$ corresponds to setting the value of all backward traveling waves to zero amplitude at the grating end $z = L$). We finish with the resulting equations:

$$\sum_{N+1}^{M} \alpha_k t_{jk} = -t_{j1}, \ N+1 \leq j \leq M \tag{20}$$

From this system of $M - N$ equations we can determine the $M - N$ unknowns α_k. Note that it is not necessary to evaluate all the elements of $\underline{\underline{T}}$, in fact only $M - N + 1$ integrations, of the equation set (19), are required.

To complete the solution, the integration is carried out once more using the completely specified initial condition vector $\underline{A}$ to obtain the final values vector $\underline{B}$:

$$\underline{A} = \left\{ A_m^{(\mp)}(0) \right\} \ , \ \underline{B} = \left\{ A_m^{(\mp)}(L) \right\} . \tag{21}$$

Since outside the perturbation region (i.e., outside the grating length) there is no coupling, the superposition with the end values represents the complete solution to the backscattering and forward scattering problems:

$$E_y(x,z,t) = \sum_m A_m^{(-)}(0) \ \mathscr{E}_m(x) \ e^{j(\omega t - \beta_m z)} \ , \ \text{for } z < 0, \tag{22a}$$

$$E_y(x,z,t) = \sum_m A_m^{(+)}(L) \ \mathscr{E}_m(x) \ e^{j(\omega t + \beta_m z)}, \ \text{for } z > L. \tag{22b}$$

Numerical Examples

We consider two cases. For both, the structure is defined by the following material constants and dimensions (refer to Figure 1):

$$n_u = 1.99, \qquad n_f = 2.01, \qquad n_s = 2.00$$

$$W_f = .38 \mu m, \qquad h = .15 \mu m, \qquad = .5 \ m$$

$$T = \infty, \qquad R = V - W_f = 50 \mu m.$$

The two cases differ only in that the considered grating lengths are different. The results are in the form of plots of the backscattered intensity in the substrate region (Figures 5 and 6).

The propagation constants for the single film mode and the substrate modes are found from the solution of equations (7) and (11) respectively. There is no computed difference between the film mode eigenvalue of the waveguide structure detailed above and the true open waveguide structure (where $T = \infty$ and $R = \infty$). The value is, to machine accuracy:

$$\beta_1 = 25.13555891251522 \qquad .$$

For R = 50μ the structure supports a total of 39 substrate modes (or 78 substrate waves). Only the 20 lower order modes are important. The solution procedure is carried out keeping 23 of these in addition to the single film mode, for a total of 24 modes or 48 waves. The mathematical problem is that of a system of 48 first order coupled differential equations.

The perturbation period is selected as

$$\Lambda = \frac{\pi}{\beta_1} = .124986\mu\mathrm{m}.$$

The coupling coefficients are found from equation (16). For κ_{11} which directly couples the counter-running film mode wieghting functions, we find

$$\kappa_{11} = -.0048\mu\mathrm{m}^{-1}$$

The balance of the coupling coefficients coupling directly to the incident mode are the substrate mode related ones:

$$\kappa_{n1}, \quad n=2,3,\ldots,24.$$

These and the eigenvalue spectrum of interest are graphically represented in the line plot of Figure 4.

Case 1, Figure 5, is for a perturbation length L = 250μm. Case 2, Figure 6, is for L = 125μm. In both figures the curves are normalized to the respective peak intensities computed at the plane z = -250μm. The beam shape characteristics are interesting. A length change not only affects the level of the substrate radiation but actually modifies the pattern. If we identify the rays through the points of maximum intensity the shift of the beam direction between the two cases becomes evident. The grating length change not only broadens the radiation beam but scans it downward away from the film-substrate interface. Although the fundamental phase mismatches per grating period are equal for both cases, the cumulative effect of these phase mismatches over the respective lengths will be different (since the lengths are different). We expect that the spectrum of the radiation leaked to the substrate depend on the length of the grating. A multimode interaction contrasts with the simple two wave interaction in that length changes not only affect the level of interaction (e.g., transmitted or reflected intensity) but also modify the field pattern (such as the broadening and scanning of the beam between cases 1 and 2).

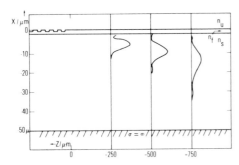

Figure 5 - Case 1. L = 250μm. Plots representing the backscattered substrate beam. The curves represent intensity profiles of the beam as computed at the planes against which the respective curves are plotted. The curves at the -500μm and -750μm planes are normalized to the peak value computed at the -250μm plane.

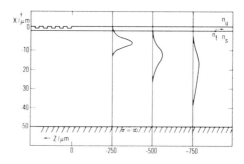

Figure 6 - Case 2. L = 125μm. Plots representing the backscattered substrate beam. The curves represent intensity profiles of the beam as computed at the planes against which the respective curves are plotted. The curves at the -500μm and -750μm planes are normalized to the peak value computed at the -250μm plane.

References

1. Yariv, A. "Coupled-mode Theory for Guilded-wave Optics," IEEE Journal of Quantum Electronics, Vol. QE-9, No. 9, pp. 919-933, September 1973.

2. Cordero, R. F. "Thin-film Periodic-waveguide Electro-optic Switches and Amplitude Modulators: A Theoretical Examination," Wave Electronics, Vol. 3, pp. 137-143, 1977-1978.

3. Kerner, S., Alexopoulos, N. G., Cordero-Iannarella, R. F., "On the Theory of Corrugated Optical Disk Waveguides," IEEE Trans. on Microwave Theory and Techniques, Vol. MTT-28, No.1, January 1980.

4. Alexopoulos, N. G. and Kerner, S.,"Coupled power theorem and orthogonality relations for optical disk waveguides," Journal Optical Society of America, Vol 67, No. 12, December 1977.

PERIODIC STRUCTURES, GRATINGS, MOIRÉ PATTERNS AND DIFFRACTION PHENOMENA

Volume 240

SESSION 4

DIFFRACTION THEORY II

Session Chairman
Erwin G. Loewen
Bausch and Lomb, Incorporated

High-efficiency diffraction grating theory

J. M. Elson

Michelson Laboratory, Physics Division, Naval Weapons Center, China Lake, California 93555

Theory

Outlined here is a derivation of the Rayleigh-Fano equations and how they may be applied to light scattering from periodic and random surface irregularity. Consider a two-dimensional surface with deviation about the mean surface level described by $\zeta(x,y)$. A vector normal to the surface at the point $\vec{\rho} = (x,y)$ is given by $\vec{n} = \hat{z} - \zeta_x \hat{x} - \zeta_y \hat{y}$, where $\zeta_x = \partial\zeta/\partial x$ and $\zeta_y = \partial\zeta/\partial y$. As a starting point, we specify that the tangential components of the electric and magnetic fields be continuous and that the normal components of the displacement and magnetic fields be continuous across the surface. In the case of the continuity of the tangential components, we have

$$\frac{\vec{n} \times [\vec{E}_+(\vec{\rho},\zeta(\vec{\rho})) - \vec{E}_-(\vec{\rho},\zeta(\vec{\rho}))]}{(1 + \zeta_x^2 + \zeta_y^2)^{1/2}} = 0 \quad , \tag{1}$$

where $\vec{E}_+(\vec{\rho},z)$ and $\vec{E}_-(\vec{\rho},z)$ are the electric field vectors above (+) and below (-) the rough surface where the fields have been evaluated at the surface boundary $z = \zeta(\vec{\rho})$. Since Eq. (1) is equated to zero, we may delete the denominator. Considering the case of p-polarized fields, the electric and magnetic fields may be written

$$\vec{E}_\pm(\vec{\rho},z) = \pm \int d^2k \ N_\pm(k) \left[\frac{\hat{k}q_\pm \mp \hat{z}k}{(\omega/c) \ \varepsilon_\pm}\right] e^{i\vec{k}\cdot\vec{\rho}} \ e^{\pm iq_\pm z} \tag{2}$$

and

$$\vec{H}_\pm(\vec{\rho},z) = \int d^2k \ N_\pm(k) \ (\hat{k} \times \hat{z}) \ e^{i\vec{k}\cdot\vec{\rho}} \ e^{\pm iq_\pm z} \quad , \tag{3}$$

respectively, where the $e^{-i\omega t}$ time dependence has been suppressed. Equations (2) and (3) refer to scattered light having wave vector $(\vec{k},q_\pm)$, where $(\omega/c)^2\varepsilon_\pm = k^2 + q_\pm^2$ and $\varepsilon_\pm$ are the dielectric constants for the upper (+) and lower (-) medium. The time-averaged Poynting vector for the upper medium may be calculated from Eqs. (2) and (3) as $\vec{P} = \frac{c}{8\pi} (\vec{E}_+ \times \vec{H}_+^*)$.

To calculate the total power scattered into the upper medium, we evaluate $\int \vec{P}\cdot\hat{z} \ d^2\rho$, where the integration is over the surface area. This calculation yields the scattered power per unit solid angle as

$$\frac{dP}{d\Omega} = \frac{\pi}{2} \frac{\omega^2}{c} \sqrt{\varepsilon_+} \cos^2\theta \ |N_+(k)|^2 \quad . \tag{4}$$

The angle θ is the polar scattering angle as measured from the $\hat{z}$ direction and $d\Omega = \sin\theta d\theta d\phi$, where ϕ is the azimuth angle measured from the $\hat{x}$ direction.

We may append an incoming beam to Eqs. (2) and (3). The specular and diffracted (or diffuse) scattered fields are described by the $N_+(k)$ coefficient. The transmitted fields are similarly described by $N_-(k)$. The incident beam is assumed to be in the (x,z) plane at incident angle θ_0 from the $\hat{z}$ direction. With the incoming beam included, Eqs. (2) and (3) may be used in the boundary condition equations, such as Eq. (1), and this yields integral equations for $N_\pm(k)$, i.e.,

$$\frac{(\varepsilon_+-\varepsilon_-)}{\varepsilon_-} \int d^2k \ N_-(k)\psi(k-k',q_+'-q_-) \left[\frac{q_+'q_-\cos(\phi-\phi') + kk'}{q_+'-q_-}\right]$$

$$= -2\sqrt{\varepsilon_+}L^2q_0 \ \cos\sigma \ \delta_{k',k_0} \tag{5a}$$

and

$$\int d^2k \ N_+(k)\psi(k-k',q_+-q'_-) \left[\frac{q_+q'_-\cos(\phi-\phi') + kk'}{q_+-q'_-}\right]$$

$$= \varepsilon_+ \left[\frac{(q_0 q'_-\cos\phi'-k_0 k')\cos\sigma}{\sqrt{\varepsilon_+}\ (q_0+q'_-)} + \frac{(\omega/c)q'_-\sin\phi'\sin\sigma}{q_0+q'_-}\right] \psi(\vec{k}_0-\vec{k}', -q_0-q'_-) \quad . \tag{5b}$$

Equations (5) refer to diffracted or scattered fields which are polarized parallel to the plane of diffraction or scattering (p-polarized). The angle σ refers to the incident polarization where σ is measured relative to the incident plane. When $\sigma = 0$ or $\pi/2$, the incident polarization is parallel (p-polarized) or perpendicular (s-polarized) to the incident plane, respectively. The $\psi(\vec{K},Q)$ function is an integral over the surface area defined by

$$\psi(\vec{K},Q) = \int d^2\rho \ e^{i\vec{K}\cdot\vec{\rho}} \ e^{iQ\zeta(\vec{\rho})} \tag{6}$$

and is important in that it contains the input regarding the surface irregularity $\zeta(\vec{\rho})$. Note that Eq. (5a) is homogeneous except when $k_0 = k'$ by virtue of the Kronecker δ-function. L^2 is the surface area, and other parameters are defined by $\vec{k}_0 = k_0\hat{x}$, $k_0 = (\omega/c)\sqrt{\varepsilon_+}\sin\theta_0$, $q_0 = (\omega/c)\sqrt{\varepsilon_+}\cos\theta_0$, $q_\pm = [\varepsilon_\pm(\omega/c)^2-k^2]^{1/2}$, $q'_\pm = [\varepsilon_\pm(\omega/c)^2-k'^2]^{1/2}$. The corresponding equations for s-polarized diffracted or scattered fields are

$$(\frac{\omega}{c})(\varepsilon_+-\varepsilon_-) \int d^2k \ \frac{M_-(k)\psi(k-k',q'_+-q_-)\cos\phi}{q'_+-q_-} = 2L^2\sqrt{\varepsilon_+}\sin\sigma\cos\theta_0\delta_{k_0,k'} \tag{7a}$$

and

$$\int d^2k \ \frac{M_+(k)\psi(k-k',q_+-q'_-)\cos(\phi-\phi')}{q_+-q'_-}$$

$$= \frac{\psi(k_0-k', -q_0-q'_-)(\cos\phi'\sin\sigma-\cos\theta_0\sin\phi'\cos\sigma)}{q_0+q'_+} \quad . \tag{7b}$$

Equations (5) and (7) are discussed by Toigo et al.[1] and references therein. We now consider the application of Eq. (5b) to specific cases of interest. Equations (5a) and (7) may be applied in a manner very similar to Eq. (5b).

Periodic Roughness

Consider the case of a sinusoidal profile grating described by $\zeta(\vec{\rho}) = (H/2)\sin\alpha x$. Using this expression for $\zeta(\vec{\rho})$ in Eq. (6) yields

$$\psi(\vec{K},Q) = (2\pi)^2\delta(K_y) \sum_{n=-\infty}^{\infty} J_n(QH/2)\delta(K_x+ \alpha) \quad , \tag{8}$$

where $\delta(y)$ is the Dirac δ-function and J_n is the Bessel function of order n. In Eq. (8) the expression

$$\exp[i(\frac{QH}{2} \sin\alpha x)] = \sum_{n=-\infty}^{\infty} J_n(\frac{QH}{2})\exp(in\alpha x)$$

has been used.

Use of Eq. (8) in Eq. (5b) allows the d^2k integration to be performed easily. The net result is an infinite series on the left side of Eq. (5b), each term of which contains a coefficient $N_+(k'_x-n\alpha)$ $(-\infty \leqslant n \leqslant \infty)$ and an infinite series on the right-hand side, each term of which contains a Dirac δ-function of the form $\delta(k_0-k'_x+m\alpha)$ $(-\infty \leqslant m \leqslant \infty)$. The infinite sums may be limited to $(-M \leqslant n \leqslant M)$. By choosing $k'_x = k_0+m\alpha$ $(-M \leqslant m \leqslant M)$, the result is $2M+1$ equations and unknowns $N_+(k_0-(n-m)\alpha)$. Such a linear equation system is readily solvable by existing computer algorithms. It is straightforward to compute the $\psi(\vec{K},Q)$ integral for rectangular and triangular profiles. In this case, similar linear equation systems are obtained.

Random Roughness

In the case of random roughness, the surface irregularity is not known and thus must be treated in a statistical manner. To illustrate, we rewrite Eq. (5b) in an abbreviated form,

$$\int d^2k \, N_+(k) f(k,k') \psi(\vec{k}-\vec{k}',q_+-q'_-) = g(k_o,k') \psi(\vec{k}_o-\vec{k}',-q_o-q'_-) \quad . \tag{9}$$

We now multiply both sides of Eq. (9) by its complex conjugate and perform an ensemble average, denoted by $< >$:

$$\int d^2k \, d^2k'' \, N_+(k) N_+^*(k'') f(k,k') f^*(k'',k') <\psi(\vec{k}-\vec{k}',q_+-q'_-) \psi^*(\vec{k}''-\vec{k}',q''_+-q'_-)>$$

$$= |g(\vec{k}_o,\vec{k}')|^2 \, <|\psi(\vec{k}_o-\vec{k}',-q_o-q'_-)|^2> \quad . \tag{10}$$

If we now assume a Gaussian distribution of the random roughness variable $\zeta(\vec{\rho})$ about the mean, then we may use a result of the Central Limit Theorem to evaluate the ensemble averages in Eq. (10).[2] This yields

$$<\psi(\vec{k}-\vec{k}',q_+-q'_-) \psi^*(\vec{k}''-\vec{k}',q''_+-q'_-)> = (2\pi)^2 \delta^2(\vec{k}''-\vec{k}) \int d^2\tau \, e^{i(\vec{k}-\vec{k}')\cdot\vec{\tau}}$$

$$\times \exp[-\delta^2 Re\{(q_+-q'_-)^2 - G(\vec{\tau})|q_+-q'_-|^2\}] \quad , \tag{11}$$

where $G(\vec{\tau})$ is the autocorrelation function ($G(0) = 1$) and δ^2 is the mean square roughness. The Dirac δ-function $\delta^2(\vec{k}''-\vec{k})$ simplifies Eq. (10) to

$$\int d^2k |N_+(k)|^2 \, |f(k,k')|^2 \, <|\psi(\vec{k}-\vec{k}',q_+-q'_-)|^2>$$

$$= |g(\vec{k}_o,\vec{k}')|^2 \, <|\psi(k_o-k',-q_o-q'_-)|^2> \quad . \tag{12}$$

Equation (12) is an integral equation for $|N_+(k)|^2$. Using $|N_+(k)|^2$ in Eq. (4) yields the angle resolved scattering $dP/d\Omega$. The crux of Eq. (11) involves the integral

$$I = \int d^2\tau \, e^{i(\vec{k}-\vec{k}')\cdot\vec{\tau}} \exp\{\delta^2 G(\vec{\tau})|q_+-q'_-|^2\} \quad . \tag{13}$$

As an example, a Gaussian autocorrelation function $G(\vec{\tau}) = \exp(-\tau^2/\sigma^2)$ may be assumed, where σ is a correlation length. Using the Gaussian $G(\vec{\tau})$ in Eq. (13) and expanding $\exp\{\delta^2 G(\vec{\tau})|q_+-q'_-|^2\}$ yields

$$I = \int d^2\tau \, e^{i(\vec{k}-\vec{k}')\cdot\vec{\tau}} \{1 + \delta^2|q_+-q'_-|^2\exp(-\tau^2/\sigma^2) + \frac{\delta^4|q_+-q'_-|^4}{2} \exp(-2\tau^2/\sigma^2)...\} \quad , \tag{14}$$

which integrates to

$$I = (2\pi)^2 \delta(\vec{k}-\vec{k}') + \pi\delta^2\sigma^2|q_+-q'_-|^2\exp(-|\vec{k}-\vec{k}'|^2\sigma^2/4)$$

$$+ \frac{\pi\delta^4\sigma^2|q_+-q'_-|^4}{4} \exp(-|\vec{k}-\vec{k}'|^2\sigma^2/8) + ... \quad . \tag{15}$$

One interesting result of this expansion is that it introduces multiple correlation lengths σ, $\sigma/\sqrt{2}$, etc. This is a direct consequence of retaining the higher order terms of the theory.

Conclusion

It is seen that the Rayleigh-Fano equations have great potential to investigate the effects of both periodic and random surface irregularity on the reflection of light.

References

1. Toigo, F., Marvin, A., Celli, V., and Hill, N. R., "Optical properties of rough surfaces: General theory and the small roughness limit," _Phys. Rev._, Vol. B15, pp. 5618-5626. 1977.

2. Bendat, J. S., and Piersol, A. G., _Random Data: Analysis and Measurement Procedures_, Wiley, New York, 1971, pp. 64-65.

Planar grating diffraction: multiwave coupled-wave theory and modal theory

M. G. Moharam, T. K. Gaylord, R. Magnusson

School of Electrical Engineering, Georgia Institute of Technology, Atlanta, Georgia 30332

Abstract

General slanted planar gratings bounded by two different media, have been analyzed using a new rigorous coupled-wave approach. The analysis is unifying and is formulated in a simple matrix form easily implemented on a digital computer. Sample calculations for transmission, slanted, and reflection gratings are presented. This new approach is compared to the widely-used modal approach and to the approximate coupled-wave approaches. It is shown that, for the present coupled-wave approach, solutions are easily calculated for general slanted gratings whereas solutions using the modal approach are available only for unslanted (pure transmission) gratings.

Introduction

The diffraction of electromagnetic waves by planar gratings has been extensively studied in recent years. These periodic structures have applications in several diverse areas such as acousto-optics, integrated optics, holography, and spectroscopy. Several different techniques have been used to analyze the diffraction of electromagnetic waves by spatially-modulated media. The most common of these methods are the coupled-wave approach[1-7] and the modal approach.[8-17] Both of these approaches are capable of producing rigorous formulations without approximations. In this form they are exact and equivalent.[18] However, the coupled-wave approach offers superior physical insight into wave diffraction phenomena and frequently yields simple analytical results. In this approach several approximations typically are made in order to obtain solutions. These assumptions and approximations are 1) neglecting boundary diffraction,[1-5,7] 2) neglecting the second derivatives of the field amplitudes,[1-5,7] and 3) retaining only one diffracted wave,[1,2,4,6] (in addition to the transmitted wave). These assumptions and approximations are generally valid for gratings with very small modulations. On the other hand, the modal approach is an exact analysis, but solutions are available only for the unslanted pure transmission case where the grating vector is parallel to the grating surface.

Several authors have attempted to analyze general slanted gratings. Kogelnik[4] and Magnusson and Gaylord[7] applied the coupled-wave approach, with the various assumptions discussed above, to this problem. Clearly these analyses are not exact and in some cases they give clearly incorrect results due to the use of these approximations. Bergstein and Kermisch[12] applied the modal approach to slanted gratings but they had to assume very weak modulation, and thus the two-wave regime approximation, in order to obtain solutions. Chu and Kong[17] have also presented a generalized modal approach formulation for the general slanted gratings case. However, they did not present any calculated results for the general slanted gratings. This is because the formulation for slanted gratings in the modal approach results in an extremely complicated transcendental relationship which is not directly solvable in general. This extreme numerical difficulty makes the modal approach essentially unusable in analyzing the general slanted grating problem.

In this paper, an exact rigorous coupled-wave approach is presented to analyze the diffraction of an electromagnetic plane wave incident obliquely at a planar grating bounded by two different media. This analysis is thus applicable 1) to holographic gratings in air or other media ($\varepsilon_1 = \varepsilon_3 \neq \varepsilon_2$), 2) to acousto-optic gratings within a medium ($\varepsilon_1 = \varepsilon_3 = \varepsilon_2$), and 3) to grating couplers such as are used in integrated optics ($\varepsilon_1 \neq \varepsilon_2 \neq \varepsilon_3 \neq \varepsilon_1$). The planar grating is, in general, slanted (i.e. the grating vector is at an angle with respect to the grating surface). The problem is formulated in a simple matrix form that is easily solved numerically. Sample calculations for transmission, reflection, and general slanted gratings are presented. The present rigorous coupled-wave approach is then compared in detail to the modal approach.

Theory

As in most of the previously cited work, the problem under consideration is the diffraction of an obliquely incident plane wave on a lossless (pure phase) sinusoidal grating with the incident wave polarized perpendicular to the plane of incidence (H-mode). Therefore the electric field will have only one component (in the y-direction of Fig. 1). The relative permittivity in the modulated region ($0 < z < d$) is

$$\varepsilon(x,z) = \varepsilon_2 + \Delta\varepsilon \cos[K(x \sin\phi + z \cos\phi)] \tag{1}$$

where ε_2 is the average dielectric constant, $\Delta\varepsilon$ is the amplitude of the sinusoidal relative permittivity, and $K = 2\pi/\Lambda$ where Λ is the grating period. The dielectric constant in the unmodulated regions ($z < 0$ and $z > d$) is ε_1 and ε_3 respectively. It is assumed that the three regions each have the permeability of free space.

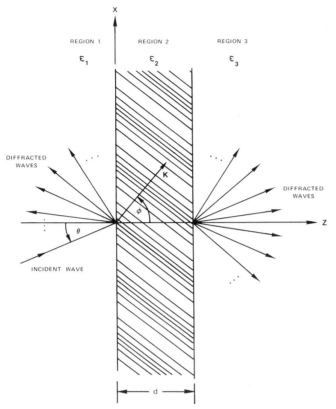

Figure 1. Geometry for planar grating diffraction.

The general approach to the planar grating problem involves finding solutions of the wave equation in each of the three regions and then matching the tangential electric and magnetic fields at the two interfaces ($z = 0$ and $z = d$) to determine the unknown constants (resulting from solving the differential wave equation). In region 1, the periodic boundary at $z = 0$ diffracts the incident wave into a spectrum of plane waves traveling back into region 1 ($z < 0$). The normalized electric field wave amplitudes in region 1 may be expressed as[8-10]

$$E_1 = \exp[-j(\beta_0 x + \xi_{10} z)] + \sum_i R_i \exp[-j(\beta_i x - \xi_{1i} z)] \tag{2}$$

where $\beta_i = k_1 \sin\theta - i K \cos\phi$ for any integer i (the wave index), $\xi_{\ell i} = (k_\ell^2 - \beta_i^2)^{\frac{1}{2}}$ for $\ell = 1, 3$ (the region index), $k_\ell = 2\pi(\varepsilon_\ell)^{\frac{1}{2}}/\lambda$ for $\ell = 1,2,3$, $j = (-1)^{\frac{1}{2}}$, λ is the free-space wavelength, θ is the angle of incidence in region 1, and R_i is the amplitude of the i-th reflected wave and is to be determined from the matching of the electric and magnetic fields. In region 3 ($z > d$) the spectrum of transmitted plane waves may be expressed as[16,17]

$$E_3 = \sum_i T_i \exp\{-j[\beta_i x + \xi_{3i}(z - d)]\} \tag{3}$$

where T_i is the amplitude of the i-th transmitted wave to be determined from the field matching at $z = d$. In region 2, the modulated region, ($0 < z < d$) the electric field may be expressed as

$$E_2 = \sum_i S_i'(z) \exp[-j(\beta_i x + \xi_{2i} z)] \tag{4}$$

where $\xi_{2i} = k_2 \cos\theta' - i K \cos\phi$, θ' is the angle of refraction inside the modulated region, and $S_i'(z)$ is the amplitude of the i-th wave at any point within the modulated region. These amplitudes are to be determined from solving the modulated-region wave equation

$$\nabla^2 E_2 + (2\pi/\lambda)^2 \varepsilon(x,z) E_2 = 0. \tag{5}$$

Note that the three electric fields E_1, E_2, and E_3 in the three regions are phase matched along the two interfaces. It is important to point out that the above analysis is valid for all slant angles ϕ except when the slant angle is identically zero (pure reflection grating). In this case, the modulation is no longer periodic and Eqs. (2) through (4) are not valid because they are derived from the Floquet theorem which is correct only for periodic structures. However, the above analysis can be used to analyze pure reflection gratings ($\phi = 0$) by considering $\phi = \delta$ and allowing δ to become an arbitrarily small quantity. All of the R_i and T_i amplitudes for the reflected and transmitted waves may then be calculated. As ϕ approaches zero, all of the reflected waves collapse together into one wave of intensity $\sum_i |R_i|^2$ and all of the transmitted waves collapse into one wave of intensity $\sum_i |T_i|^2$.

To obtain the diffracted amplitudes $S_i'(z)$, Eqs. (1) and (4) are substituted into Eq. (5) resulting in the infinite set of coupled-wave equations

$$(\mu^2/\rho) \frac{d^2 S_i(v)}{dv^2} = (1 - 2i\mu \cos\phi) \frac{dS_i(v)}{dv} - \rho i(i - B) S_i(v) + S_{i+1}(v) + S_{i-1}(v) \tag{6}$$

with $\rho = 2\lambda^2/\Lambda^2\Delta\varepsilon$, $B = 2\Lambda(\varepsilon_2)^{\frac{1}{2}} \cos(\phi - \theta')/\lambda$, $\mu = \cos(\phi - \theta')/B \cos\theta'$, and $v = j\pi\Delta\varepsilon z/2\lambda\cos\theta'$

$(\varepsilon_2)^{\frac{1}{2}} = j\gamma(z/d)$. This system of coupled-wave equations has been derived <u>without</u> the common assumptions and approximation associated with previous coupled-wave analyses[1-7] such as neglecting the second derivatives of the amplitudes. Therefore, the present analysis is as rigorous and as exact as the modal approach. There are four composite parameters in Eq. (6). The regime parameter ρ determines the boundary between the Bragg regime ($\rho > 10$) and the intermediate diffraction regime.[19] The parameter B represents the Bragg condition, i.e. for incidence at the p-th Bragg condition B = ρ, (note that B is not an integer for off-Bragg incidence). The last parameter γ is the widely used grating modulation parameter. Equation (6) may be written in matrix form as

$$
\begin{bmatrix} \underline{S}' \\ \underline{S}'' \end{bmatrix} = \begin{bmatrix} b_{rs} \end{bmatrix} \begin{bmatrix} \underline{S} \\ \underline{S}' \end{bmatrix}
\tag{7}
$$

where $\underline{S}$, $\underline{S}'$, and $\underline{S}''$ indicate the column vectors of S_i, dS_i/dv, and d^2S_i/dv^2 respectively. The quantity [b] is the coefficient matrix specified from Eq. (6). Equation (7) corresponds to a "state equation" in the state space description of linear systems. The system of differential equations given by Eq. (7) has a relatively simple and straightforward solution, obtainable in terms of the eigenvalues and the eigenvectors of the coefficient matrix [b]. It is

$$
S_i(v) = \sum_m C_m w_{im} \exp(q_m v)
\tag{8}
$$

where q_m is the m-th eigenvalue and w_{im} is the m-th element of the row in the matrix [w] composed of the eigenvectors, corresponding to the i-th wave (not the i-th row of the matrix [w]). The coefficients C_m are unknown constants to be determined together with R_i and T_i by matching the tangential electric and magnetic fields at the two boundaries (z = 0 and z = d). The four quantities to be matched and the resulting boundary conditions are tangential E at z = 0:

$$
R_i + \delta_{i0} = \sum_m C_m w_{im},
\tag{9}
$$

tangential H at z = 0:

$$
\xi_{1i}d(R_i - \delta_{i0}) = \sum_m C_m w_{im}(q_m\gamma - \xi_{2i}d),
\tag{10}
$$

tangential E at z = d:

$$
T_i = \sum_m C_m w_{im} \exp[j(q_m\gamma - \xi_{2i}d)],
\tag{11}
$$

tangential H at z = d:

$$
- \xi_{3i}d\, T_i = \sum_m C_m w_{im}(q_m\gamma - \xi_{2i}d)\, \exp[j(q_m\gamma - \xi_{2i}d)],
\tag{12}
$$

where δ_{i0} is the Kronecker delta function. Eliminating T_i and R_i from these equations gives

$$
2\,\delta_{i0}\,(\xi_{1i}d) = \sum_m C_m w_{im}(\xi_{2i}d - q_m\gamma + \xi_{1i}d),
\tag{13}
$$

$$
0 = \sum_m C_m w_{im}(\xi_{2i}d - q_m\gamma - \xi_{3i}d)\exp(jq_m\gamma)
\tag{14}
$$

Note that

$$
\xi_{2i}d = (\rho\gamma/2u^2)(1 - 2i\,\mu\,\cos\phi),
\tag{15}
$$

$$
\xi_{\ell i}d = (\rho\gamma/2u^2)[(\varepsilon_\ell/\varepsilon_2\,\cos^2\theta') - (\tan\theta' - 2i\mu\,\sin\phi)^2]^{\frac{1}{2}}, \quad \ell = 1,\, 3
\tag{16}
$$

The system of linear equations given by Eqs. (13) and (14) can be solved for C_m and then R_i, T_i can be calculated from Eqs. (9) and (11). Note that the number of equations available is exactly equal to the number of unknowns. For example, if n waves are retained in the analysis then there will be n unknown values each of R_i and of T_i and 2n unknown values of C_m. This is because the coefficient matrix [b] in Eq. (7) is a 2n x 2n matrix and therefore, has 2n eigenvalues and thus there are 2n unknown values of C_m. Alternatively, this may be viewed as being due to the n coupled-wave equations each being a second-order differential equation, and thus there are 2n roots or eigenvalues and 2n unknown constants C_m to be determined from the boundary conditions. Therefore, the total number of unknowns is 4n and Eqs. (9) through (12) provide 4n linear equations in these unknowns.

To summarize, the algorithm used to solve this problem proceeds as follows: First the coefficient matrix [b] is constructed and then eigenvalues and eigenvectors are calculated (typically using a computer library program). The system of linear equations, Eqs. (13) and (14) is then constructed and solved for C_m (using a technique such as Gauss elimination). Equations (9) and (11) are then used to calculate the diffracted amplitudes R_i and T_i. Power conservation requires that the sum of the diffraction efficiencies for all of the propagating waves be unity. That is

$$\sum_i (DE_{1i} + DE_{3i}) = 1 \tag{17}$$

where DE_{1i} and DE_{3i} are the diffraction efficiencies in regions 1 and 3 respectively. These diffraction efficiencies are given by

$$DE_{1i} = Re(\xi_{1i}/\xi_{10}) R_i R_i^* \tag{18}$$

and

$$DE_{3i} = Re(\xi_{3i}/\xi_{30}) T_i T_i^* . \tag{19}$$

The real part of the ratio of the propagation constants occurs when the time average power flow density is obtained by taking the real part of the complex Poynting vector. The quantity $Re(\xi_{\ell i}/\xi_{\ell 0})$ is just the usual ratio of the cosine of the diffraction angle for the i-th wave to the cosine of the angle of the zeroth-order wave for the ℓ-th medium. The results of sample calculations for the diffraction efficiencies for a pure transmission ($\phi = \pi/2$), pure reflection ($\phi = 0$), and for a general slanted grating case ($\phi = \pi/3$) are shown in Figs. 2 through 4 for $\varepsilon_1 = \varepsilon_2 = \varepsilon_3$ and in Figs. 5 through 7 for $\varepsilon_2 = 2\varepsilon_1 = 2\varepsilon_3$.

Comparison to Previous Analyses

A. Modal Approach

Basically, the rigorous coupled-wave approach presented in this paper and the widely-used modal approach analyze the planar grating diffraction problem by solving the wave equation in the three regions (Fig. 1) and then matching the tangential electric and the magnetic fields at the two boundaries to determine all of the unknowns. The main difference between the two approaches is in the technique used to find solutions of the wave equation in the modulated region. In the present coupled-wave approach, the resulting system of coupled-wave equations is formulated into a simple matrix form where the solution is readily obtained by calculating the eigenvalues and the eigenvectors of the coefficient matrix constructed from the coupled-wave equations. The eigenvalue problem, although not simple, has been very extensively studied and numerous efficient and straightforward computer programs, to calculate the eigenvalues and the eigenvectors, are available in typical computer program libraries. By contrast, the modal approach requires that a very complicated (for the general slanted grating case) transcendental relationship in the form of a continued fraction expansion be solved to find the wavenumbers and their corresponding coefficients that are needed to solve the wave equation in the modulated region. This is the primary difficulty when applying the modal approach to the analysis of slanted gratings. This is due to the fact that there is no systematic technique for solving this transcendental continued fraction relation.[8] For the modal approach the transcendental relationship can also be formulated as a matrix. This matrix is always n x n (as opposed to 2n x 2n for the present coupled-wave analysis). Only for unslanted, physically symmetric case is the resulting matrix for the modal approach in standard eigenvalue form. The corresponding wavenumbers needed in the solution are the positive and negative square roots of the eigenvalues. The vector of coefficients, which is the same for both the positive and negative wavenumbers, is the corresponding eigenvector of the problem. For slanted gratings the resulting modal approach n x n matrix is not in the form of a standard eigenvalue problem. Further, the wavenumbers occurring in the matrix cannot be systematically determined. In this case the wavenumbers must be determined through some trial-and-error procedure. Likewise, the vector of coefficients must be similarly independently determined. This is why the modal approach has been used extensively and successfully to analyze unslanted transmission gratings, while it has not been used to calculate results for general slanted gratings.

B. Approximate Coupled-Wave Analyses

Previous coupled-wave analyses have typically been approximate. The present coupled-wave analysis is exact and rigorous and it reduces to each of the previous analyses with the appropriate approximations and simplifications. For example, it reduces to Kong's analysis[6],

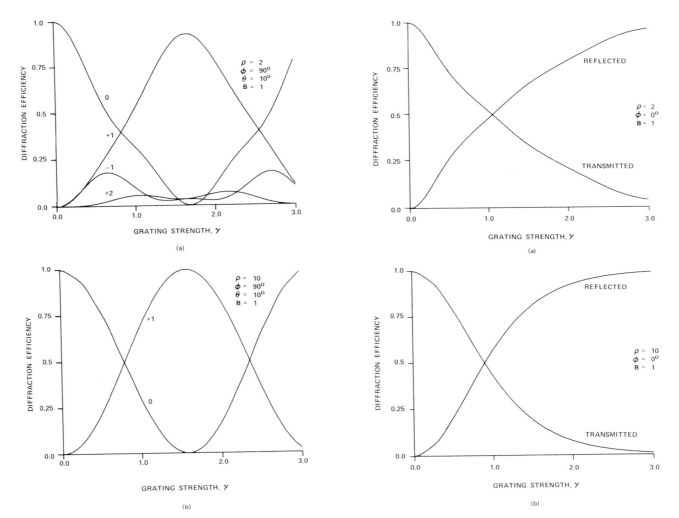

Figure 2. The diffraction efficiencies of the transmitted waves for pure transmission grating ($\phi = 90°$) with (a) $\rho = 2$ and (b) $\rho = 10$. The diffraction efficiencies of all reflected and transmitted waves not shown in the figure are less than 0.01. Here and in Figs. 3 and 4, $\varepsilon_1 = \varepsilon_2 = \varepsilon_3$.

Figure 3. The diffraction efficiencies of the transmitted and reflected waves for a pure reflection gratings with (a) $\rho = 2$ and (b) $\rho = 10$. These two waves are the only waves that propagate in the unmodulated regions.

if a two-wave regime is assumed (that is, retaining only the first-order and zero-order waves and neglecting all higher order waves). Kogelnik's[4] analysis is obtained by assuming two-wave regime, neglecting boundary diffraction, and neglecting second derivatives of the field amplitudes in Eq. (6). This last approximation which is very common in almost all previous coupled-wave analyses[1-5,7] implies that the parameter μ^2/ρ was assumed to be very small in these analyses. Magnusson and Gaylord's[7] analysis is obtained by neglecting boundary diffraction and the second derivatives of the field amplitudes. It is important to note that in previous coupled-wave analyses,[1-5,7] the amplitudes and intensities of the diffracted waves are calculated inside the modulated region. These analyses are based on solving the half-space grating problem. They do not (and cannot, because of the various approximations) solve the problem of general planar slab grating bounded by two different media.

Conclusion

The diffraction of a plane electromagnetic wave incident obliquely on a planar grating bounded by two different media has been analyzed using an exact and rigorous coupled-wave approach. The direction of the periodicity of the grating may have any arbitrary angle with respect to the boundaries. The solution has been formulated in a simple matrix form

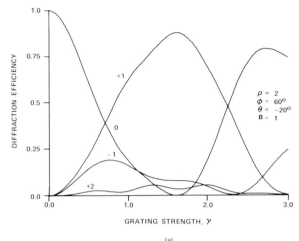

(a)

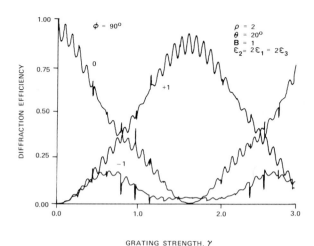

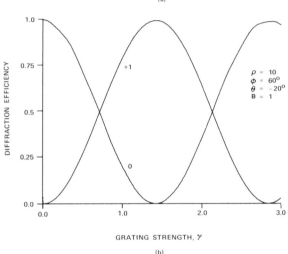

(b)

Figure 4. The diffraction efficiencies of the transmitted waves for a general slanted grating φ = 60° with (a) ρ = 2 and (b) ρ = 10. The diffraction efficiencies of all reflected and transmitted waves not shown in the figure are less than 0.01.

Figure 5. The diffraction efficiencies of the transmitted waves with ρ = 2 and $\varepsilon_2 = 2\varepsilon_1 = 2\varepsilon_3$ for transmission grating (φ = 90°).

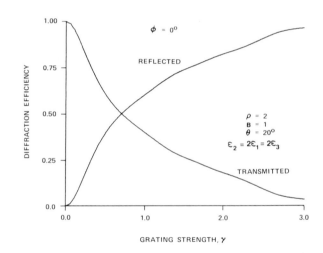

Figure 6. The diffraction efficiencies of the transmitted and diffracted waves with ρ = 2 and $\varepsilon_2 = 2\varepsilon_1 = 2\varepsilon_3$ for reflection gratings (φ = 0°).

that is easily implemented on a digital computer. Sample calculations have been presented for transmission, reflection, and general slanted gratings. The present coupled-wave approach has been compared to the modal approach and to approximate coupled-wave approaches. It is shown that the present approach is useful in analyzing any planar slanted or unslanted grating.

The modal approach, by comparison, leads to systematic solutions only for unslanted gratings. The present method of analysis is useful in all applications where planar gratings are utilized. However, it is especially valuable in applications where slanted gratings and reflection gratings are used such as in grating couplers and distributed feedback lasers since it is extremely difficult to use the modal approach in these cases. This analysis may also be straightforwardly extended to absorption gratings and to mixed phase and absorption if desired.

Acknowledgments

This work was sponsored by the National Science Foundation under Grant No. ECS-7919592 and by the Joint Services Electronics Program under Grant No. DAAG29-78-C-0005.

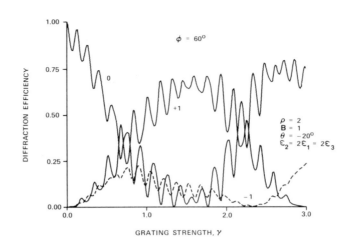

Figure 7. The diffraction efficiencies of the transmitted waves with $\rho = 2$ and $\varepsilon_2 = 2\varepsilon_1 = 2\varepsilon_3$ for general slanted gratings ($\phi = 60°$).

References

1. R. R. Aggrawal, Proc. Indian Acad. Sci. 31, 417-426 (1950).
2. P. Phariseau, Proc. Indian Acad. Sci. 44A, 165-170 (1965).
3. W. R. Klein and B. D. Cook, IEEE Trans. Sonic Ultrasonics SU-14, 123-134 (1967).
4. H. Kogelnik, Bell Sys. Tech. J. 48, 2909-2947 (1969).
5. G. L. Fillmore and R. F. Tynan, J. Opt. Soc. Am. 61, 199-203 (1971).
6. J. A. Kong, J. Opt. Soc. Am. 67, 825-829 (1977).
7. R. Magnusson and T. K. Gaylord, J. Opt. Soc. Am. 67, 1165-1170 (1977).
8. T. Tamir, H. C. Wang, and A. A. Oliner, IEEE Trans. Microwave Theory Tech. MTT-12, 323-335 (1964).
9. T. Tamir and H. C. Wang, Can. J. Phys. 44, 2073-2094 (1966).
10. T. Tamir, Can. J. Phys. 44, 2461-2494 (1966).
11. C. B. Burckhardt, J. Opt. Soc. Am. 56, 1502-1509 (1966).
12. L. Bergstein and D. Kermisch, Proc. Symp. Modern Opt. 17, 655-680 (1967).
13. R. S. Chu and T. Tamir, IEEE Trans. Microwave Theory Tech. MTT-18, 486-504 (1970).
14. R. S. Chu and T. Tamir, Proc. IEE 119, 797-806 (1972).
15. F. G. Kaspar, J. Opt. Soc. Am. 63, 37-45 (1973).
16. S. T. Peng, T. Tamir, and H. L. Bertoni, IEEE Trans. Microwave Theory Tech. MTT-23, 123-133 (1975).
17. R. S. Chu and J. A. Kong, IEEE Trans. Microwave Theory Tech. MTT-25, 18-24 (1977).
18. R. Magnusson and T. K. Gaylord, J. Opt. Soc. Am. 68, 1777-1779 (1978).
19. M. G. Moharam, T. K. Gaylord, and R. Magnusson, Opt. Comm. 32, 14-18 (1980).

Guided waves on corrugated surfaces and their link with grating anomalies and coupling phenomenon

Michel Nevière

Laboratoire d'Optique Electromagnétique, E.R.A. du C.N.R.S. n° 597
Faculté des Sciences et Techniques, Centre de St-Jérôme, 13397 Marseille Cedex 4, France

Abstract

The existence of a complex pole of the reflection coefficient of a metallic plane, or of the zero order efficiency of a metallic grating, implies the existence of a homogeneous solution of Maxwell equations. Such a solution, which exists in the absence of any incident wave, is a guided wave on the plane or corrugated surface and is responsible for several strange phenomena such as grating anomalies, total absorption of light by a grating, or the coupling of a laser beam into an optical waveguide by means of a photoresist grating. For TM polarization, the guided wave is usually due to plasmon resonance in the metal. But similar phenomena can be observed for TE polarization provided one superimposes a dielectric overcoating with convenient thickness on top of the metallic grating.

Introduction

What we intend to deal with consists of several curious phenomena, well known to experimentalists, or recently discovered. They are gratings anomalies, total absorption of a plane wave by a grating, as well as the coupling of a laser beam into an optical waveguide by means of a holographic thin film coupler. All these phenomena have the same origin in common : they are all connected with the excitation of surface waves along the periodic structure. Such a surface wave carries energy parallel to the mean plane of the surface, but is also slightly attenuated in that direction. Thus, it is referred to by many authors as a "leaky wave". From a mathematical point of vue, it is a solution of Maxwell equations and the associated boundary conditions on the grating surface, without any wave impinging on the structure, i.e. a "homogeneous" solution. We wish to show how the determination of the homogeneous solution can enlighten the study of the response of the periodic structure to a given excitation.

Reflection of a plane wave on a plane interface

Brewster absorption

The figure describes the notations which will be used hereafter. Region ① is vacuum, region ② is a lossless dielectric with refractive index ν_2. An incident plane wave, with TM polarization ($\vec{H} = H(x,y)\hat{z}$) falls on the plane dielectric interface under incidence θ. Defining $\alpha = k_0 \sin\theta$, with $k_0 = 2\pi/\lambda_0$, it gives rise to a reflected and a transmitted waves with amplitudes $r(\alpha)$ and $t(\alpha)$. The field in the 2 half spaces can be written in the form :

region ① :

$$H(x,y) = \exp\left[i(\alpha x - \beta^{(1)}y)\right] + r(\alpha)\exp\left[i(\alpha x + \beta^{(1)}y)\right] \tag{1}$$

$$\text{with} \quad \beta^{(1)} = k_0\cos\theta = \sqrt{k_0^2 - \alpha^2} \tag{2}$$

region ② :

$$H(x,y) = t(\alpha)\exp\left[i(\alpha x - \beta^{(2)}y)\right] \tag{3}$$

$$\text{with} \quad \beta^{(2)} = \sqrt{k_0^2\nu_2^2 - \alpha^2} \tag{4}$$

The continuity of the tangential components of the electromagnetic field at the interface leads to the Fresnel formulae :

$$r(\alpha) = \frac{\nu_2^2\beta^{(1)} - \beta^{(2)}}{\nu_2^2\beta^{(1)} + \beta^{(2)}} \tag{5}$$

$$t(\alpha) = \frac{2\nu_2{}^2\beta^{(1)}}{\nu_2{}^2\beta^{(1)} + \beta^{(2)}} \qquad (6)$$

Eq.(5) shows the existence of a <u>zero</u> of $r(\alpha)$, obtained when $\nu_2{}^2\beta^{(1)} = \beta^{(2)}$. By the use of (2) and (4), the value $\hat{\alpha}$ of α corresponding to the zero is found to be given by :

$$\hat{\alpha} = k_0\nu_2/\sqrt{1 + \nu_2{}^2} \qquad (7)$$

In order to get rid of the wavevector k_0, let us introduce the normalized parameter $\delta = \alpha/k_0 = \sin\theta$. The value $\hat{\delta}$ corresponding to $\hat{\alpha}$ is thus :

$$\hat{\delta} = \frac{\hat{\alpha}}{k_0} = \frac{\nu_2}{\sqrt{1 + \nu_2{}^2}} \qquad (7')$$

Since ν_2 is real, $\hat{\delta}$ is a real number. Moreover it is less than unity. Thus, it is possible to find a real angle of incidence θ_B in such a way that $\sin\theta_B = \hat{\delta} = \nu_2/\sqrt{1 + \nu_2{}^2}$, which is equivalent to $\mathrm{tg}\,\theta_B = \nu_2$. When a plane wave falls on the interface under incidence θ_B, the reflection coefficient is null and θ_B is nothing else than the Brewster incidence.

Resonance of a lossy interface

Now what happens for an interface between vacuum and a <u>lossy</u> material ? Whatever the origin of the losses may be (lossy dielectric or metal with conduction losses) region ②️ is now characterized by a <u>complex</u> refractive index ν_2. The Fresnel formula (5) and (6) still hold, but r and t are <u>complex</u> numbers. So is the value $\hat{\delta}$ given by (7'). In the case of a good metallic reflector like silver at wavelength 0.5 µm, for which $\nu_2 = 0.05 + i\,2.87$, $\hat{\delta}$ is found to be equal to $1.0668 + i\,0.00257$. But it is most surprising to discover that the value $\hat{\delta}$ of δ given by (7') is no longer a <u>zero</u>, but is a <u>pole</u> of the reflection coefficient $r(\delta)$.

In order to understand that, let us first remark that, as soon as δ is allowed to be complex the definitions of $\beta^{(1)}$ and $\beta^{(2)}$ given by (2) and (4) are ambiguous and the function $r(\delta)$ given by (5) is non uniform. In order to <u>get a uniform function, we</u> have to introduce cuts in the complex plane to uniformize $\sqrt{k_0{}^2 - \alpha^2}$ and $\sqrt{k_0{}^2\nu_2{}^2 - \alpha^2}$. Both roots can be written in the common form :

$$\beta^{(j)} = k_0\sqrt{\nu_j{}^2 - \delta^2} \qquad \text{with} \quad j = 1, 2 \qquad (8)$$

Thus in the complex δ plane, the cuts start at the branch points $\pm\nu_j$ and go to infinity. Since $\nu_1 = 1$, the cuts starting at $\pm\nu_1$ start from points on the real axis. When δ moves between $-\nu_1$ and $+\nu_1$ on the real axis, $\beta^{(1)}$ is real and positive. Outside the interval, $\beta^{(1)}$ is chosen purely imaginary, with imaginary part positive. For the sake of continuity, we place the cuts as indicated on Fig.2. One can verify that the argument of $\beta^{(1)}$ is increased by $\pi/2$ when one goes from A to A', following the small half circle in the lower half plane.

The second set of cuts, starting from $\pm\nu_2$ is placed in a similar way. However, for a metallic interface $+\nu_2$ and $-\nu_2$ lie far from the real axis and the cuts are never crossed in the following studies. Thus they will not play a significant role hereafter.

What is the image of the first set of cuts in the complex $\beta^{(1)}$ plane ? When $\delta = \nu_1 + i\eta$ with $\eta > 0$, $\left[\beta^{(1)}\right]^2 = k_0{}^2(\eta^2 - 2i\nu_1\eta)$ and its locus is a half parabola with axis Ox and tangent to the imaginary axis. Thus the locus of $\beta^{(1)}$ is the curve represented on fig.3, which can be easily replaced by the second bisector without any consequence. In conclusion, a unique value of $\beta^{(1)}$, as well as $\beta^{(2)}$, corresponding to a given complex value of δ is given by eq.(8) and the condition :

$$\mathrm{Im}\{\beta^{(j)}\} + \mathrm{Re}\{\beta^{(j)}\} > 0 \qquad (9)$$

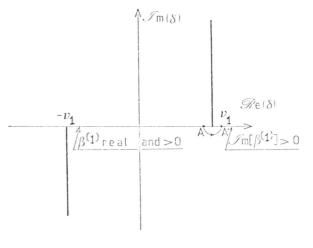

Figure 2.   Location of the cuts in the complex δ plane

The corresponding part of the $\beta^{(1)}$ complex plane is the one which has a physical interest since when δ is real, $\beta^{(1)}$ has to be real and positive.

Now what happens to the multivalued function $r(\delta)$? One can verify that, whatever the chosen determinations of the two square roots may be, two and only two values of r correspond to a given value of δ and that these 2 values are inverse each others. Thus $r(\delta)$ may be considered as a single valued function on a two sheet Riemann surface. If we come back to the more familiar language of multivalued functions, the functions $r(\delta)$ presents 2 determinations. The first determination, r_I defined by eqs.(5) and (9) is the one of physical interest. The second determination, r_{II} is deduced from r_I by replacing $\beta^{(1)}$ by $-\beta^{(1)}$ and is equal to $1/r_I$. Thus a complex value $\hat{\delta}$ of δ which is a zero of r_I is also a pole of r_{II}. Since replacing $\beta^{(1)}$ by $-\beta^{(1)}$ in (1) does not change the physical problem (only the incident and reflected waves exchange their nature) a given

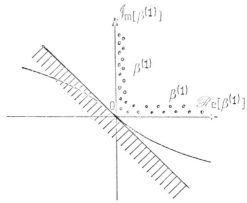

Figure 3.   Image of the cuts in the complex $\beta^{(1)}$ plane

phenomenon such as Brewster absorption may be associated to a zero or a pole, depending on the determination that we choose. Hereafter δ^z and δ^p designate the eventual zeros and poles of r_I , which implies that Brewster absorption is associated to δ^z.

We are now able to understand the reason why we find a pole instead of a zero in the case of a metallic interface. Fig.4 shows the trajectory of $\hat{\delta}$ when, starting from the refractive index of glass, ν_2 is varied to arrive at the refractive index of silver. When the trajectory of $\hat{\delta}$ crosses the cut, $\hat{\delta}$ which was a zero of r_I becomes a zero of r_{II} , i.e. a pole of r_I. Thus a metallic interface presents a pole on the sheet of physical interest. It implies the existence of a reflected and a transmitted leaky waves without an incident one. Such a solution of Maxwell equations, generally called a resonance, is attributed to collective oscillations of electrons near the surface (plasmon resonance). But since $\hat{\delta}$ has a real part greater than unity, such a resonance can never be excited by an incident plane wave, for which $\delta = \sin\theta \leqslant 1$. Thus the resonance is never seen when one studies the reflection of a plane wave on a metallic interface, and the function $r(\delta)$ is a smooth function without sharp minima or maxima. The question which arises is whether the resonance could be observed by replacing the plane interface by a grating.

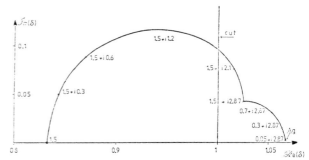

Figure 4.   Locus of $\hat{\delta}$ in the complex plane

Reflection of a plane wave on a grating

Fig.5 gives a schematic representation of a metallic grating with groove spacing d and groove depth h. The incidence is θ, the wavelength in vacuum is λ_0 and TM polarization is assumed.

In any homogeneous region, the field can be described by a superposition of plane waves :

region ① :

$$H(x,y) = \exp\left[i\left(\alpha_0 x - \beta_0^{(1)} y\right)\right] +$$
$$+ \sum_{n=-\infty}^{+\infty} B_n \exp\left[i\left(\alpha_n x + \beta_n^{(1)} y\right)\right] \quad (10)$$

with $\alpha_0 = k_0 \sin\theta$, $\alpha_n = \alpha_0 + n\dfrac{2\pi}{d}$ (11)

$$\alpha_n^2 + \left[\beta_n^{(1)}\right]^2 = k_0^2 \quad (12)$$

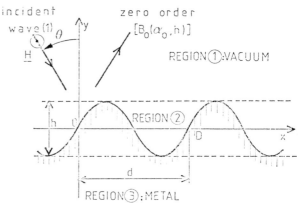

Figure 5.   Schematic representation of a grating

region (2) :

$$H(x,y) = \sum_n A_n \exp\left[i\left(\alpha_n x - \beta_n^{(2)} y\right)\right] \qquad (13)$$

$$\alpha_n^2 + \left[\beta_n^{(2)}\right]^2 = k_0^2 \nu_2^2 \qquad (14)$$

Here too, $\beta_n^{(1)}$ and $\beta_n^{(2)}$ are chosen in order to verify condition (9).

Let us choose the λ/d ratio in such a way that only one diffracted order (the zero order) propagates. The grating then acts like a mirror and the only quantity of interest for the Optician is the coefficient B_0 , which depends on α_0 and on the groove shape. If we assume the profile to be sinusoidal, and if we introduce again the normalized parameter $\delta_n = \alpha_n/k_0$, we find that B_0 is only dependent on δ_0 and h. Moreover, when $h \to 0$,

$$\lim_{h \to 0} B_0(\delta_0,h) = r(\delta_0) \qquad (15)$$

The study of the poles and zeros of the function $B_0(\delta_0,h)$ leads to the prediction of strong energy absorption phenomena.

Poles of $B_0(\delta_0,h)$

The research of poles for $B_0(\delta_0,h)$ in the complex plane implies that we now consider an electromagnetic field given by eq.(10) and (13) in which the real constant α_n have been replaced by complex constants $\hat{\alpha}_n$, and in which the amplitude of the incident term is equal to zero. Thus we search solutions of the form :

region (1) :

$$H(x,y) = \sum_{n=-\infty}^{+\infty} \hat{B}_n \exp\left[i\left(k_0\hat{\delta}_n x + \beta_n^{(1)} y\right)\right] \qquad (16)$$

with $\hat{\delta}_n = \hat{\delta}_0 + n\dfrac{\lambda_0}{d}$ and $-1 < \mathrm{Re}\{\hat{\delta}_0\} < +1$ $\qquad (17)$

region (2) :

$$H(x,y) = \sum_{n=-\infty}^{+\infty} \hat{A}_n \exp\left[i\left(k_0\hat{\delta}_n x - \beta_n^{(2)} y\right)\right] \qquad (18)$$

Contrary to what happened for a plane interface, the periodicity of the profile introduces an infinite set of possible values $\hat{\delta}_n$ for the propagation constant. When $h \to 0$, the periodicity vanishes, and among this infinite set, a particular value called δ_{n_0} must tend towards $\hat{\delta}$ of the plane. If we remember that λ/d has been chosen in such a way to have only one diffracted order when α is real, that $\mathrm{Re}\{\hat{\delta}\} > 1$ and that $\mathrm{Re}\{\hat{\delta}_0\} < 1$, we find that $n_0 = +1$. Fig.6 illustrates the location of the $\hat{\delta}_n$ for the grating and the location of $\hat{\delta}$ for the plane.

If we find values $\hat{\delta}_n$ for which (16) and (17) are solutions of the boundary value problem without any incident wave, it implies that for a given incident wave with complex constant propagation $\hat{\delta}_0$, $B_n(\hat{\delta}_0,h)$ are infinite. Thus such value $\hat{\delta}_0$ is found to be the pole $\delta P(h)$ for the function $B_n(\delta_0,h)$ and in particular, for $B_0(\delta_0,h)$. Its determination is then conducted in the following way. We first have to extend the existing grating theories [1,2] to the case for which the propagation constant is complex, and to introduce the cuts previously described.

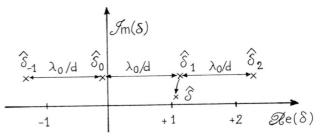

Figure 6. Location of $\hat{\delta}_n$ of the grating and $\hat{\delta}$ of the plane

As soon as $B_0(\delta_0,h)$ can be computed for complex values of δ_0 , we determine the pole $\hat{\delta}_0$ by looking for the zero of $1/B_0(\delta_0,h)$ with the use of an iterative method [4] which requires the knowledge of an approximate value of the zero as starting point. For low modulated gratings, we start the calculations from $\hat{\delta}$ of the plane, and we use the fact that $\lim_{h \to 0} \hat{\delta}_0 = \hat{\delta} - \lambda_0/d$. For high modulated gratings, one must start from the value $\hat{\delta}_0$ numerically determined for a more shallow grating.

Zero of $B_0(\delta_0, h)$

From the preceding paragraph, if we assume that δ^P is a simple pole, it follows that the function $B_0(\delta_0, h)$ may be written as :

$$B_0(\delta_0, h) = \frac{u(\delta_0, h)}{\delta_0 - \delta^P(h)} , \qquad \text{where} \quad u\left|\delta^P(h), h\right| \neq 0 \qquad (19)$$

Let us suppose that region ③ is a perfectly reflecting metal. Then $|B_0| = 1 \quad \forall \delta_0$ on the real axis. This implies the existence of a complex zero $\delta^Z(h)$, given by :

$$\delta^Z(h) = \overline{\delta^P(h)} \qquad (20)$$

In the case of a real metal, the unitarity of B_0 does not hold and (20) is no longer valid. But numerical calculations show the existence of a zero $\delta^Z(h)$ which presents a real part very close to $\mathrm{Re}\{\delta^P(h)\}$ and an imaginary part different from $\mathrm{Im}\{\delta^P(h)\}$. This zero also tends towards $\hat{\delta}$ of the plane if $h \to 0$ and its determination is conducted in the same way as for the pole $\delta^P(h)$. Thus B_0 can be written in the form :

$$B_0(\delta_0, h) = w(\delta_0, h) \left[\delta_0 - \delta^Z(h)\right] / \left[\delta_0 - \delta^P(h)\right] ,$$

where $w(\delta_0, h)$ is a complex regular function near δ^Z and δ^P and does not present any zero in their vicinity.

From (15) it follows that $\lim\limits_{h \to 0} w(\delta_0, h) = r(\delta_0)$.

Thus B_0 can be approximated near δ^P to a good degree of accuracy by :

$$B_0(\delta_0, h) \simeq r(\delta_0) \left[\delta_0 - \delta^Z(h)\right] / \left[\delta_0 - \delta^P(h)\right]$$

The zero order absolute efficiency $\mathcal{E}_0$ immediately follows :

$$\mathcal{E}_0(\delta_0, h) = |B_0|^2 \simeq R(\delta_0) \, |\delta_0 - \delta^Z(h)|^2 / |\delta_0 - \delta^P(h)|^2 , \qquad (21)$$

where $R(\delta_0)$ is the reflection factor for the energy.

Energy absorption phenomenon

It is worth recalling that from (17), $|\mathrm{Re}\{\delta^P\}| < 1$. The same condition is verified by δ^Z. Thus, when an incident plane wave falls on the grating with a real propagation constant $\delta_0 = \sin\theta$, there exists an incidence θ_R for which $\sin\theta_R = \delta_0 = \mathrm{Re}\{\delta^P\} \simeq \mathrm{Re}\{\delta^Z\}$. Thus the efficiency predicted by (21) presents a minimum given by :

$$\mathcal{E}_{0m} \simeq R(\sin\theta_R) \left[\mathrm{Im}\{\delta^Z\}/\mathrm{Im}\{\delta^P\}\right]^2 \qquad (22)$$

and the width at half depth of the absorption peak is found to be close to $2\,\mathrm{Im}\{\delta^P\}$. These predictions are confirmed by direct electromagnetic calculations on a given grating, as can be seen on fig.7. But the question which arises is whether it is possible to find a grating which gives a minimum efficiency equal to zero.

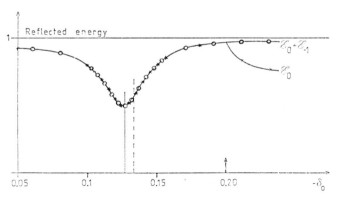

Figure 7.   Zero order efficiency of a 2400 gr/mm, 0.05 μm groove depth silver grating as function of δ_0.
$*$: rigorous calculations
o : values given by eq.(21)

Total absorption of a plane wave by a grating

From (22), $\mathscr{C}_{0m} = 0$ if $\mathrm{Im}\{\delta^z\} = 0$, i.e. if δ^z is on the real axis. Following fig.8, which gives, for a 2400 gr/mm sinusoidal silver grating and wavelength 0.5 μm, the loci of δ^p and δ^z when the groove depth is varied, δ^z crosses the real axis for $h = h_c = 0.021$ μm. It implies that the corresponding grating absorbs in totality an incident plane wave falling under the incidence θ given by $\sin\theta = \delta^z(h_c)$. This phenomenon may look very strange if we think that the silver plane has a reflectance close to 0.98 under any incidence, at $\lambda = 0.5$ μm. But it has been thoroughly confirmed by direct calculations and by experiments,[3] as can be seen on fig.9. Under conditions predicted by our theory, no significant reflected energy was seen, and the incident energy was totally dissipated in heating the grating. It is worth noticing that for groove depth higher than the one which produces the total absorption, the absorption peak grows weak, a fact which can be predicted by eq.(22) and the loci of fig.8. The conclusion is that high modulated gratings do not necessarily present a stronger absorption than low modulated ones.

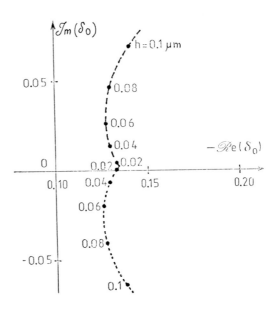

Figure 8. Locus of δ^p(‒‒‒‒‒‒) and δ^z(........) when h is varied

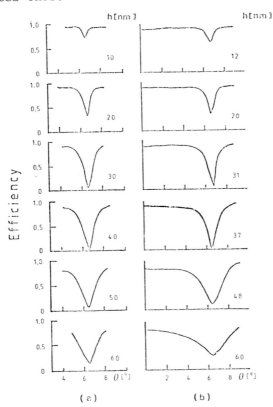

Figure 9. Theoretical a) and experimental b) zero order efficiency curves for 1800 groove/mm sinusoidal aluminum gratings of various groove depth h, used with TM polarization ($\lambda = 0.647$ μm).

Absorption of a plane wave by a dielectric coated grating used in T.E. polarization

Leaky modes of a dielectric slab bounded with metal on one of its sides

Although no surface plasmon exists for TE polarization, a surface (leaky) wave can be excited in the structure of fig.10 if the thickness e is high enough. As previously, region ① is vacuum ($\nu_1 = 1$). Region ② is a lossless dielectric (ν_2 is a real number) and region ③ is a metal (ν_3 is a complex number). The electric vector is parallel to axis Oz.

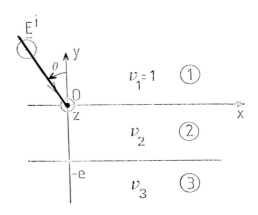

Figure 10. Dielectric slab

Here too, the boundary conditions at the two interfaces allows one to determine the reflection coefficient r(δ). Again, r(δ) presents poles $\hat{\delta}$ in the complex δ plane which are the propagation constants of the guided modes in the structure. Since ν_3 is a complex number, the modes are necessarily leaky and the values of $\hat{\delta}$ are never real. Thus their determination has to be conducted on a computer, using the iterative method described in ref.4. It is convenient to begin with the research of the real pole corresponding to a lossless structure which can support lossless guided waves. This lossless structure is derived from the one in fig.10 by taking a real value for ν_3 less than ν_2. Since only an approximate value of the pole is required, a plot of the graph of $1/r(\delta)$ on the interval $[-1,+1]$ is sufficient to show the pole location. The study may be restricted to the interval $[\max(\nu_1,\nu_3), \nu_2]$ since $\beta^{(2)}$ has to be real in order to get a guided wave, but $\beta^{(1)}$ and $\beta^{(3)}$ must be imaginary to avoid propagating waves outside the layer.

When the pole has been found on the real axis, one puts a small imaginary part for ν_3 and look for a new pole in the vicinity of the previous one. The process has to be continued until ν_3 reaches the value of the refractive index of the metal filling region ③, for which the complex pole is again designated by $\hat{\delta}$. As soon as a particular value of $\hat{\delta}$ has been found, the iterative method [4] allows us to determine the pole for different values of the parameters. Fig.11 shows as example the locus of $\hat{\delta}$ associated to TE waves guided on an aluminum plane coated by SiO when the coating thickness e is varied. The wavelength is 0.436 μm and the values of e are expressed in μm.

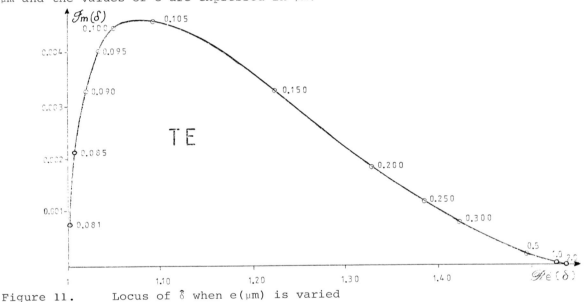

Figure 11.   Locus of $\hat{\delta}$ when e(μm) is varied

The locus of $\hat{\delta}$ meets the parallel to the imaginary axis Re{$\hat{\delta}$} = 1 for a particular cut-off thickness $e_c \simeq 0.081$ μm. Under that, no guiding phenomenon exists.

Reflection of a TE plane wave on a dielectric coated reflection grating

The study is conducted in the same way as for a bare grating used in TM polarization. Starting from the propagation constant $\hat{\delta}$ of a plane structure, we look for the complex poles $\hat{\delta}_n$ of $B_0(\delta_0,h)$ reminding that, if only one diffracted order propagates, $\lim_{h\to 0} \hat{\delta}_0 = \hat{\delta}- \lambda/d$.

The only difference is that the codes which compute $B_0(\delta_0,h)$ must include the dielectric overcoating. As previously explained, the introduction of the grating modulation implies the existence of a pole and a zero, which have different imaginary parts, and real parts very close to each other. Fig.12 shows their evolution when the dielectric

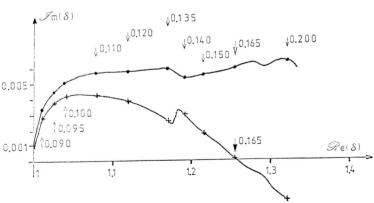

Figure 12.   Loci of pole (dots) and zero (crosses) when e is varied

thickness is varied. The results are related to a 0.37 μm groove spacing, 0.02 μm groove depth aluminum grating coated with SiO. The wavelength is 0.436 μm. One can see that for a thickness e closed to 0.165, the locus of the zero $\delta^Z(e)$ crosses the real axis. It means that a plane wave striking the structure under incidence θ given by :

$$\sin \theta = \text{Re}\{\delta^Z(0.165)\} - \lambda_0/d$$

is absorbed in totality. The prediction has been again thoroughly confirmed by experiments.[5] For any value of the dielectric thichness, the numerical values of $\delta^Z(e)$ and $\delta P(e)$ taken from fig.12 enable us to predict the location and strength of the absorption peak by the use of eq.(21). Here too, the predictions are in perfect agreement with the results given by direct computations.

In the case where the grating supports several diffracted orders, the prediction of the absorption peaks is more difficult.[6,7] Several values of $\tilde{\delta}_n$ have a real part on the $|-1,+1|$ interval, leading to several incidences θ_n which can excite the leaky modes. The knowledge of δP here too enables us to predict the location of the grating anomalies and to understand why a smooth TE efficiency curve can be turned into a very irregular one by the introduction of a dielectric overcoating, a fact already pointed out in ref.1 and 8. The problem of coupling a laser beam into an optical waveguide by means of a holographic thin film coupler can be studied in the same way. Here, one wants to maximize the energy carried by the guided wave, rather than the reflection efficiency of the grating. The study of the complex pole δP of the periodic structure again allows predicting the coupling angles θ_c via eq. $\sin \theta_c = \text{Re}\{\delta^P\} - n\frac{\lambda}{d}$, (n integer). By representing the incident beam by means of its Fourier transform, one is then able to determine the optimum beam width and the maximum coupling coefficient of the device.[4,9,10]

Conclusion

The theory that we have presented is able to predict grating anomalies in a quantitative way. When the grating only supports one diffracted order, the depth of the absorption peak can be predicted by a very simple formula. Unexpected phenomena of total absorption of light for TM and TE polarization have been predicted and confirmed by experiments.

References

1. Nevière, M., Vincent, P., Petit, R., Nouv. Rev. Optique, Vol. 5, p. 65, 1974.
2. Maystre, D., J.O.S.A., vol. 68, p. 490, 1978.
3. Hutley, M.C., Maystre, D., Opt. Commun., vol. 19, p. 431, 1976.
4. Nevière, M., Vincent, P., Petit, R., Cadilhac, M., Opt. Commun., vol. 9, p. 48, 1973.
5. Loewen, E.G., Nevière, M., Appl. Optics, vol. 16, p. 3009, 1977.
6. Maystre, D., Nevière, M., Vincent, P., Optica Acta, vol. 25, p. 905, 1978.
7. Petit, R., Electromagnetic Theory of Gratings ; Methods and Application, Springer-Verlag, Berlin, 1980.
8. Hutley, M.C., Verrill, J.P., Mc Phedran, R.C., Nevière, M., Vincent, P., Nouv. Rev. Optique, vol.6, p. 87, 1975.
9. Nevière, M., Petit, R., Cadilhac, M., Opt. Commun., vol.8, p.113, 1973.
10. Nevière, M., Vincent, P., Petit, R., Cadilhac, M., Opt. Commun., vol. 9, p. 240, 1973.

New improvement of the differential formalism for high-modulated gratings

Patrick Vincent

Laboratoire d'Optique Electromagnétique, E.R.A. du C.N.R.S. n° 597
Faculté des Sciences et Techniques, Centre de St-Jérôme, 13397 Marseille Cedex 4, France

Abstract

The differential method used to compute the field diffracted by a grating is based on a shooting method which requires the integration of a set ordinary differential equation from the bottom up to the top of the grooves. Numerical instabilities appearing for dielectric gratings with grooves deeper than the grating period are reduced if a formulation using two simultaneous counter-running numerical step-by-step integrations is used. Such a formulation is devised here for the TE, TM and general (conical diffraction) cases. Numerical examples are given to demonstrate the improvement of the method.

Introduction

Up to the last years, diffraction gratings were used only in spectrometric applications but now the situation is moving very fast. The field of application of gratings is increasing rapidly, including Integrated Optics,[1-3] Colour Imaging,[4,5] and Solar Energy.[6-8] These new applications require others devices than the classical echelette or sinusoidal Aluminum-covered gratings used by spectroscopists. For example, multilayer dielectric gratings and high modulated gratings have shown an increasing interest. The computation of their efficiencies by codes written for classical gratings is often expansive and sometimes impossible : It is the case when you try to compute the field diffracted by a very high modulated grating using the differential method.[9-11]

In this paper, a new formulation of this method is presented, which extends its range of applicability to high modulated gratings and preserves its main advantages. Numerical examples are given to demonstrate this improvement.

The differential formalism of diffraction

In the same Proceedings, a review paper by Nevière[11] gives an overview of this formalism and its applications. The classical implementation, used for echelette or sinusoidal grating is described. In our case, most of the high modulated gratings are lamellar gratings, and one must be careful to match the field with its Rayleigh expansions outside of the modulated zone. Thus we give, first, a short description of the classical shooting-method which is slightly different from the method described in Nevière's paper. The modifications of this method allowing to deal with gratings about twice higher than the preceding one are then given.

From a mathematical point of view, the diffraction problem of the grating reduces to the resolution of a set of partial derivative equations with suitable boundary condition in the planes $y = 0$ and $y = a$. (Fig.1). The projection of these equations onto an exponential basis gives a set of ordinary differential coupled equations and then, allows to use efficient integration algorithms such as Runge-Kutta or Numerov ones. It is not the case for direct methods using a 2-dimensional mesh to compute the field. On the other hand mapping techniques[7,20] where the grating is mapped onto a plane of coordinate are complicated and does not work for lamellar gratings. Thus the differential method used in our laboratory looks like an efficient compromise, often more precise and efficient than the direct integration of partial derivative equations, simpler and more versatile than the mapping technique. Nevière[9] has show that the Numerov algorithm makes this

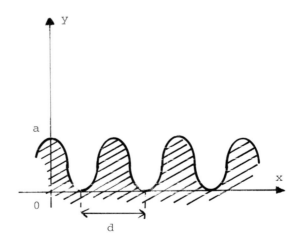

Figure 1.   Schematic representation of the grating

method useful for dielectric and conducting gratings used in P light (TE case : electric field parallel to the grating's grooves). Then it has been extended to S light (TM case : magnetic field parallel to the grooves), off plane mountings,[12] and crossed gratings.[8] It has been also used to study overcoated gratings,[13,14] corrugated waveguides[15] and gratings couplers.[16-18]

Let us recall first the main features of the differential method in the simple TE case ($\underline{E}$ parallel to the grooves).

The TE case

Assuming a time dependence in $\exp(-i\omega t)$, the electric field is represented by the complex amplitude $E(x,y)$ of its only non-zero component, that associated with Oz axis (Fig.1). The propagation equation is :

$$\Delta E + k^2 E = 0 \; , \qquad \text{with} \qquad k^2(x,y) = \omega^2 \, \varepsilon(x,y) \, \mu_0 \; . \tag{1}$$

Let us define $K = 2\pi/d$ (d is the grating period) and $\alpha_n = \alpha_0 + nK$. Using exponential basis functions, we write :

$$E(x,y) = \sum_n E_n(y) \, \exp(i \, \alpha_n \, x) \; , \tag{2}$$

$$k^2(x,y) = \sum_n (k^2)_n \, \exp(i \, nK \, x) \; . \tag{3}$$

The propagation equation (1) projected on the exponential basis leads to a set of ordinary coupled differential equations :

$$\frac{d^2 E_n}{dy^2} + \sum_m (k^2)_{n-m} \, E_m - \alpha_n^2 \, E_n = 0 \; . \tag{4}$$

Outside the modulated zone $0 < y < a$, the dielectric permittivity ε is a constant and the field can be represented by plane waves expansions ; thus :

$$\text{for} \quad y \leqslant 0 : \qquad E_n(y) = T_n \, \exp(-i \, \beta_n^t \, y) \qquad \text{with} \quad \beta_n^t = \sqrt{k_2^2 - \alpha_n^2} \; , \tag{5}$$

$$\text{for} \quad y \geqslant a : \qquad E_n(y) = A_n \, \exp(-i \, \beta_n^i \, y) + B_n \, \exp(i \, \beta_n^i \, y) \qquad \text{with} \quad \beta_n^i = \sqrt{k_1^2 - \alpha_n^2} \; . \tag{6}$$

The continuity of E and $\partial E/\partial y$ in the $y = 0$ and the $y = a$ planes implies that of E_n and dE_n/dy. Thus, we have to solve the differential set (4) with boundary conditions for $y = 0$ and $y = a$.

The classical shooting method. Numerical algorithms such as Runge-Kutta or Numerov ones allow to compute the values of $E_n(a)$ and $dE_n(a)/dy$ provided that $E_n(0)$ and $dE_n(0)/dy$ are given. Unfortunately, in our problem, the initial values in the $y = 0$ plane are unknown. Thus we must integrate numericaly eq.(4) between the two abscissa several times in order to get the right values for $E_n(0)$ and $dE_n(0)/dy$. It is possible to use an iterative method to compute the effective value of the field,[19] but it may be not convergent for high modulated gratings, when truncation and round-off errors are not negligible. Thus, we take advantage of the linearity of the relation between the field in the $y = a$ and the field $y = 0$ planes to get the solution.
Eliminating the unknown T_n coefficients between (5) and its derivative, we get :

$$\frac{dE_n}{dy}(0) = -i \, \beta_n^t \, E_n(0) \; . \tag{7}$$

Eliminating the coefficient B_n of the reflected field between eq.(6) and its derivative, we get :

$$E_n(a) - \frac{1}{i \, \beta_n^i} \frac{dE_n}{dy}(a) = 2 \, A_n \, \exp(-i \, \beta_n^i \, a) \; , \tag{7'}$$

where the A_n coefficients describing the incident field are known.

Moreover, we assume that the field is accurately represented by N components on the exponential basis, thus that all the expansions can be truncated to a finite number of terms. Consequently, on this basis, the linear relation between $E_n(0)$ and A_n can be described by a square matrix M_A with dimension N. This matrix can be computed by performing N numerical integrations of eq.(4) from $y = 0$ to $y = a$ taking N arbitrary, linearly independent values values for the field in the $y = 0$ plane. As we know that a linear application acting on a finite-dimensional space is completely determined when the image of this basis is given, the matrix M_A is fully determined.

Then, the diffraction problem is easily solved. Let ψ_i and ψ_0 be respectively the matrix representation of the incident field and the field in the $y = 0$ plane. We get :

$$\psi_i = M_A \psi_0 . \qquad (8)$$

Thus, solving the linear system (8) we get the value of the field in the $y = 0$ plane. The transmitted field, that is the T_n coefficients are immediately computed using eq.(5). A matrix M_B giving the reflection B_n coefficient can be computed simultaneously with M_A , using the same way. The only difference is that, instead of (7'), we use the relation obtained by eliminating the A_n coefficients between eq.(6) and its derivative. Thus the reflected field ψ_r is given by the matrix product :

$$\psi_r = M_B \psi_0 . \qquad (9)$$

For high modulated gratings, the important point is that the numerical integration must be performed from the bottom of the modulated zone ($y = 0$ plane) up to the top of this zone ($y = a$ plane). This can be a serious disadvantage relatively to round-off and truncation-errors introduced in the first steps of the numerical integration. To give an example, let us take a differential second-order equation with constant coefficients ; thus the solution is the sum of two exponentials with opposite exponents. Consequently, a small starting error at the beginning of the numerical integration will grow up exponentialy as the procedure is continued. When the effective solution of the problem is the decreasing exponential, it is easy to see that their exist an integration length where the off-set growing function becomes more important than the decreasing effective solution. This phenomenon is the main factor limiting the differential method for high modulated dielectric gratings.

The idea of the improvement of the shooting method is very simple : the growth of starting errors will be minimized if we perform two independent counter-running integrations starting, one from $y = 0$, the other from $y = a$, and match at the $y = a/2$ abscissa. Two effects can be expected : the exponential growth occuring over a half-length is smaller than in the previous method ; the symmetry of the procedure may induce an effect of errors-balance between the two counter-running numerical integrations.

The improved shooting method. The idea is the following : the continuity of functions E and $\partial E/\partial y$ is true, not only in the $y = 0$ and $y = a$ planes, but in any plane between these two abscissa. Thus, the matching of the values of these functions computed one from the top, the other from the bottom of the grating, can be performed at an arbitrary abscissa y_1 , usually equal to a/2.

Let us take N arbitrary linearly independent values for E(x,0) and compute $\partial E(x,0)/\partial y$ by formula (7). We are now ready to perform N numerical integrations from $y = 0$ to $y = y_1$ and compute the corresponding values of $E_n(y_1)$ and $dE_n(y_1)/dy$. Similarly to what happened in the preceding paragraph, we get two matrices M and M' relating the transmitted field to the field and its derivative at the abscissa y_1. Thus, in the exponential basis, we may write :

$$\psi_1 = M \psi_0 , \qquad (10)$$

$$\psi_1' = M' \psi_0 , \qquad (11)$$

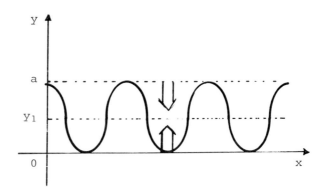

Figure 2

where ψ_0 is for the field at y = 0, ψ_1 and ψ_1' are for the field and its derivative in the y = y_1 plane.

Let us consider now the upper part of the grating (y > y_1). The field in the y = y_1 plane is the sum of the fields induced by all the different waves propagating in medium 1 (Fig.2). Each of these waves creates in the y = y_1 plane a field that we can compute by a numerical integration of eq.(4) from y = a to y = y_1. Thus, taking N arbitrary linearly independent values for the outgoing or evanescent waves, we are able to compute two matrices M_+ and M_+' relating these waves to the field in the matching plane y = y_1. Another integration performed with the incident field as initial value gives the contribution ψ_1^i and $\psi_1'^i$ of the incident wave to the field and its derivative in the matching plane (Fig.3).

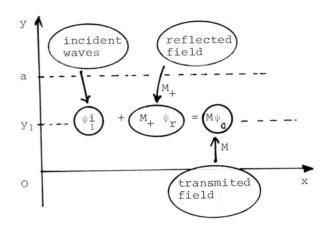

Figure 3

Thus, we write :

$$\psi_1 = M_+ \psi_r + \psi_1^i , \qquad (12)$$

$$\psi_1' = M_+' \psi_r + \psi_1'^i . \qquad (13)$$

The continuity of E and $\partial E/\partial y$ implies, using eqs.(10) to (13) :

$$\psi_0 = \psi_t$$

$$\psi_1 = M \psi_0 = M_+ \psi_r + \psi_1^i , \qquad (14)$$

$$\psi_1' = M' \psi_0 = M_+' \psi_r + \psi_1'^i . \qquad (15)$$

From eqs.(14) and (15) we deduce a linear system of 2N equations giving the reflected and transmitted fields :

$$\begin{pmatrix} M & -M_+ \\ M' & -M_+' \end{pmatrix} \begin{pmatrix} \psi_t \\ \psi_r \end{pmatrix} = \begin{pmatrix} \psi_1^i \\ \psi_1'^i \end{pmatrix} . \qquad (16)$$

The matrices M, M', M_+, M_+' are computed by numerical integration of the propagation equations as well as the two column matrices ψ_0^i and $\psi_0'^i$; the unknown vectors ψ_t and ψ_r describe respectively the transmitted and reflected fields. Thus the diffracted field is found by solving a set of 2N linear equations. This can be done by standards algorithms such as the Gauss-Jordan one. Compared to the preceding one, this method requires the resolution of a linear system twice greater, but this point is not important. The reduction of the offset at the end of the integrations due two the splitting of the integration length in two parts is far more important than the errors in solving the linear equations for high modulated dielectric gratings.

The improvement of the shooting method that we have described for the simple E // case can be transposed easily to the over cases of polarization ; we shall now give a brief description of its implementation.

The TM case

Let us take now the magnetic field parallel to the grooves of the grating (Fig.1). Thus, our diffraction problem is always a scalar one, but now the field is represented by the complex amplitude H(x,y) of the magnetic field. This function is continuous with respect to y, but, contrary to the E // case, its derivative is not a continuous function. Thus, we must use another propagation equation, where all the differential operators act on continuous quantities :

$$\text{div} \left[\frac{1}{k^2} \text{grad } \vec{H} \right] + H = 0 . \qquad (17)$$

Using the same exponential basis, we define the functions $H_n(y)$ in the same way that $E_n(y)$ and get similar plane waves expansions than (5) and (6). The shooting method used is identical to that of the TE case, the only difference is that instead dE_n/dy, the continuous functions are $k^{-2}dH_n/dy$. Column matrices ψ_i, ψ_r, ψ_t are made with the Rayleigh coefficients for the magnetic field, but eq.(9) to (16) are the same for the two cases, provided that matrices M' and M'_+ are defined with continuous functions $k^{-2}dH_n/dy$ instead of dE_n/dy. Thus the improvement acts in the same way with comparable results.

The general case

Let us now consider an incident monochromatic plane wave with wavevector $\underline{k_i}$ (α, β, γ) not perpendicular to the grooves of the grating which are parallel to z-axis (Fig.4). Thus, the electromagnetic field has an $\exp(i\gamma z)$ dependance ; γ and z look like conjugated quantities as it is the case for ω and t. Thus, we shall always use the complex representation of the field, the $\exp(i\gamma z - i\omega t)$ dependance being omitted. It can be shown[21] that, in such problems, all the components of the field can be expressed in terms of the z-components for example. Outside the modulated zone $0 < y < a$, $E_z(x,y)$ and $H_z(x,y)$ can be represented by expansions similar to (5) and (6). Eliminating the others components in Maxwell equations, we get two scalar coupled propagation equations for E_z, H_z and their partial derivatives. Projected on the $\exp(i\alpha_n x)$ basis, these equations gives a set of differential coupled equations corresponding to eq.(4) of the TE case. Runge-Kutta and Adams-Moulton algorithms are used to integrate numericaly these equations.[10] The shooting methods previously described for the TE case work here but the difference is that, instead of one component $E_n(y)$, the electric field must be represented by two functions : $E_n(y)$ and $H_n(y)$. Thus, column matrices ψ_0, ψ_r, ψ_t are now made by juxtaposing the coefficients for the electric and for the magnetic field. Consequently, the dimension of the matrices is 2N. This is the only difference in the description of the shooting-method, and it can be improved by the same technique than in the TE case.

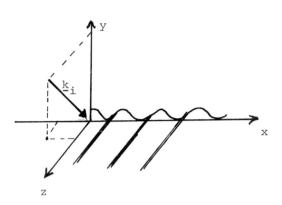

Figure 4

Applications

The differential method is useful for gratings having a period in the same range as the wavelength and made in material with moderate conductivity. For highly conducting metals, the integral formulation of D. Maystre[22] is often more reliable, but also more expansive. When the period is far greater than the wavelength and the incident wave propagating near the perpendicular to the grating plane, approximate scalar methods can be used, but in others cases, the differential method is very efficient for TE polarization, even for metals such as silver or aluminum in the visible region. For the TM polarization or the general case, one must be careful and use it only with moderate conductivity materials. The accuracy of the results can be checked by the energy balance criterion, the reciprocity relation[23] and the stability of the results when the number of terms in the expansions or the number of integration steps are varied. When it is possible, the comparison with others method and experiments is useful.

The improved shooting method has been cheched first against the classical one for gratings with shallow grooves. The results are in excellent agreement, and usually the energy balance is more precise with the new method (better than 10^{-6}).

Comparisons have been made with the integral method.[22,24] The following table gives the values of the absolute efficiencies of an echelette grating with blaze angle 54 1 degrees, period 1.6667 µm, made with aluminum ($\nu = 0.34 + i\ 4.01$), used at the wavelength 0.36 µm, with incidence angle $\theta = 10.2153°$. The differential method was performed with 19 terms in Fourier series and 2×100 integration steps. The integral method used 37 matching points on the grating profile.

Absolute efficiency Order	Differential Method	Integral Method
-5	0.0326	0.0316
-4	0.0042	0.0041
-3	0.0359	0.0316
-2	0.5107	0.5360
-1	0.2076	0.2099
0	0.0132	0.0152
1	0.0134	0.0157
2	0.0118	0.0129
3	0.0314	0.0335

Table 1. Comparisons of the efficiencies computed by the improved differential method and by the integral method of D. Maystre.[24]

Using the classical shooting-method for a symmetric dielectric lamellar grating with optical index ν = 1.5 and period d = 2 μm, good results are obtained when the groove height is equal to the period but the computed values are not accurate when it is increased up to 2.4 μm. The result is that, for normal incidence, the -1 and +1 orders exhibit different efficiencies. It is not the case when the improved shooting-method is used. This method gives reasonable results up to a groove height about 4. μm. The figures 5a and 5b give the TE and TM transmitted efficiencies under normal incidence computed with 13 Fourier components and 2 × 100 integration steps for the grating describe in Fig.5.

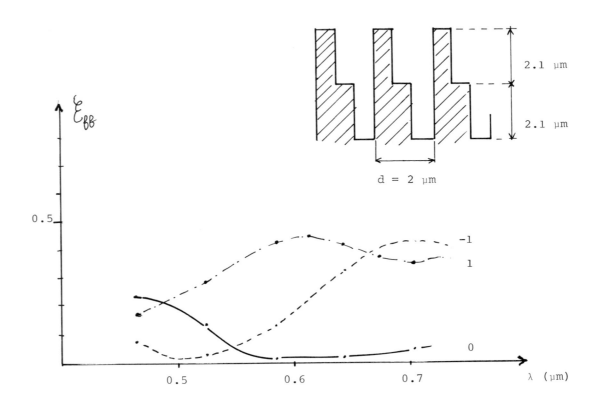

Figure 5-a.   Efficiency u.s. wavelength in TM polarization

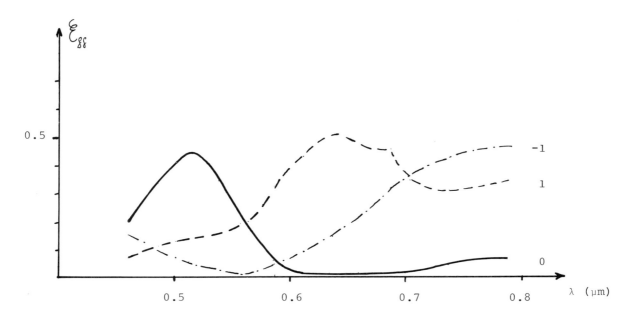

Figure 5-b.   Efficiencies v.s. wavelength in TE polarization for the
              same grating as in Fig.5-a

Figure 6 gives the efficiencies of a sinusoidal gratings with the groove depths greater
than the period in TE polarization, used in normal incidence.

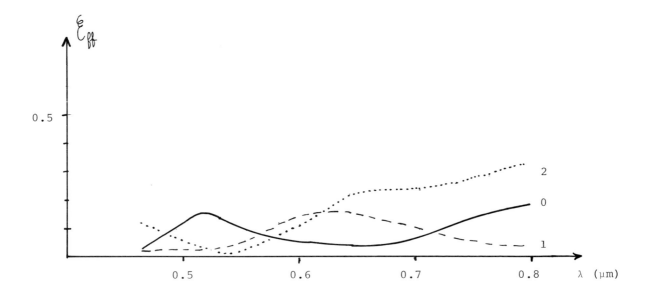

Figure 6.   Transmitted efficiencies in various order for a sinusoidal dielectric
            grating (a = 2.1 μm,   d = 2 μm,   ν = 1.5)

Conclusion

The use of counter-running integration steps has shown a substantial enhancement of the field of application of the differential method to dielectric gratings with deep grooves. However, the reader should keep in mind that special profiles such as lamellar ones can be treated by special methods such as modal developments, which are usually more efficient than general ones. The main interest of the differential method is that it is very easy to change the profile of the grating studied, which is defined only by a set of Fourier coefficients. The improvement of the shooting-method gives reasonable computed efficiencies for grating as high as two times of the grating period. This can be a useful tool to study transmission gratings proposed for coloured image recording [4],[5] It works for TE, or TM polarization, and will certainly gives the same effects in the general case of conical diffraction.

References

1. T. Tamir, Integrated Optics, Springer-Verlag, 1975.
2. M., K. Barnosky, Introduction to Integrated Optics, Plenum Press, 1974.
3. H. Kogelnik, C.V. Shank, "Coupled-Wave Theory of Distributed Feedback Lasers", J. Applied Phys., vol. 43, pp. 2327-2335, 1972.
4. K. Knop, Applied Optics, vol. 17, pp. 3598-3603, 1978.
5. K. Knop, J. Opt. Soc. Am., vol. 68, pp.1206-1210, 1978.
6. C.M. Horwitz, Opt. Commun., vol. 11, p. 210, 1974.
7. G.H. Derrick, R.C. Mc Phedran, D. Maystre, M. Nevière, "Crossed gratings : a theory and its applications", Appl. Phys., vol. 18, p. 39-52, 1979.
8. P. Vincent, "A finite difference method for dielectric and conducting crossed gratings" Opt. Commun., vol. 26, p. 293-296, 1978.
9. M. Nevière, P. Vincent, R. Petit, Nouv. Rev. Opt., vol. 5, pp.65-67, 1974.
10. P. Vincent , "Differential Methods", in "The Electromagnetic Theory of Gratings : Methods and Applications", R. Petit, ed., Springer-Verlag, Heidelberg, New-York, 1980.
11. M. Nevière, Proceedings of the SPIE's 24th Annual Technical Symposium, July 28-August 1, 1980, San Diego, Ca., U.S.A.
12. P. Vincent, M. Nevière, D. Maystre, Nuclear Instruments and Methods, vol. 152, pp. 123-126, 1978.
13. D. Maystre, M. Nevière, R. Petit, "Experimental Verification and Applications", in "The Electromagnetic Theory of Gratings : Methods and Applications", R. Petit, ed., Springer-Verlag, Heidelberg, New-York, 1980.
14. M.C. Hutley, J.P. Verrill, R.C. Mc Phedran, M. Nevière, P. Vincent, Nouv. Rev. Optique, vol. 6, pp. 87-95, 1975.
15. P. Vincent, M. Nevière, Appl. Phys., vol. 20, pp. 345-351, 1979.
16. M. Nevière, R. Petit, M. Cadilhac, Opt. Commun., vol. 8, pp. 113-117, 1979.
17. M. Nevière, P. Vincent, R. Petit, M. Cadilhac , Opt. Commun., vol. 9, pp. 48-53, 1973.
18. M. Nevière, P. Vincent, R. Petit, M. Cadilhac, Opt. Commun., vol. 9, pp. 240-245, 1973.
19. J.P. Hugonin, Private Communication.
20. J. Chandezon, D. Maystre, G. Raoult, J. Optics, submitted for publication.
21. A.C. Hewson, An introduction to the theory of electromagnetic waves, Longman Group Ltd., London, 1970.
22. D. Maystre, "A new general integral theory for dielectric coated gratings", J. Opt. Soc. Am., vol. 68, pp. 490-495, 1978.
23. P. Vincent, M. Nevière, "The reciprocity theorem for corrugated surfaces used in conical diffraction mountings", Optica Acta, vol. 26, pp. 889-898, 1979.
24. D. Maystre, "Integral Methods", in "The Electromagnetic Theory of Gratings : Methods and Applications", R. Petit, ed., Springer-Verlag, Berlin, Heidelberg, New-York, 1980.

Fresnel images, coherence theory and the Lau effect

Ronald Sudol

The Institute of Optics, College of Engineering and Applied Science
University of Rochester, Rochester, New York 14627

Abstract

The Lau effect is observed when a double-grating pair is illuminated with spatially incoherent light. It would be of interest to generalize this experiment to include illuminating fields which have some degree of spatial coherence. To gain some insight into the changes the intensity distribution undergoes as the degree of spatial coherence is varied, we consider the special case of completely coherent illumination. This particular problem is easily discussed in terms of Fresnel images where it is found that the Fresnel image planes are defined by distances which are identical to the allowed grating separations in the incoherent Lau experiment. This suggests an interesting relationship between the nature of the fields in the incoherent Lau experiment and those in the coherent formation of Fresnel images.

Introduction

Recently,[1-3] there has been renewed interest in an interference phenomenon known as the Lau effect.[4] The Lau experiment, shown in Figure 1, consists of transilluminating an identical pair of separated line gratings and making an observation in the far-field of the second grating. The field incident on the first grating is assumed to be spatially incoherent. Figure 2 shows an example of these Lau fringes. By examining the coherence properties of the field incident on the grating G_2, it can be shown that the existence of high contrast fringes corresponds to a matching between the periods of the grating G_2 and the coherence function.

An analysis of this matching principle shows that Lau fringes are observed whenever the grating separation, z_o, is a rational multiple of the grating period squared divided by the wavelength, λ, i.e. $z_o = \alpha p^2 / \beta \lambda$ where α and β are integers. This family of separations contains the set originally discussed by Lau (i.e. $z_o = np^2/2\lambda$, where n is an integer) but is not limited to that set and, hence, may be considered as an extension or generalization of the Lau effect.

To further generalize the Lau experiment, it is natural to consider the effect of illuminating the grating-pair with fields having an arbitrary degree of spatial coherence. This consideration is motivated by the possibility of applying this particular optical system to coherence measurements as well as to the determination of rough surface parameters. To develop some indication of the changes the Lau fringes would undergo as the coherence properties of the illuminating field are changed, we consider the limit of completely spatially coherent illumination.

It is found that the periodic nature of the system lends itself to an analysis based on the theory of Fresnel images.[5] In this limit, for a normally incident plane wave on the first grating the intensity distribution in the far-field of the second grating reduces to the far-field diffraction pattern of this grating. What is interesting though, is the fact that the condition for the location of distinct Fresnel images for coherent illumination is identical to the condition on grating separation for the observation of high contrast Lau fringes in incoherent illumination. A comparison of the fields in these two cases shows a certain reciprocity between them. This leads us to consider the interesting question of whether or not it is possible to define a coherent experiment which is "equivalent" to the original incoherent Lau experiment.

In this paper the Lau effect is reviewed using optical coherence theory, indicating how the condition on grating separation arises naturally in this analysis. This also serves the purpose of describing the nature of the fields in the original experiment which will be used for later comparisons. We next turn our attention to a discussion of the Lau experiment in the coherent limit based on the theory of Fresnel images. This ultimately leads us to a comparison between the nature of the fields in these cases thereby enabling us to point out certain interesting reciprocities and equivalences.

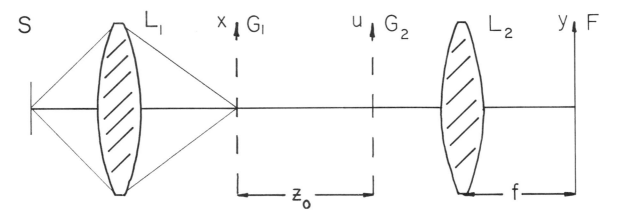

Figure 1. The Lau experiment. The extended source is imaged by the lens L_2 onto the line grating G_1 which illuminates a second identical grating G_2. High contrast fringes appear in the rear focal plane F of the lens L_2.

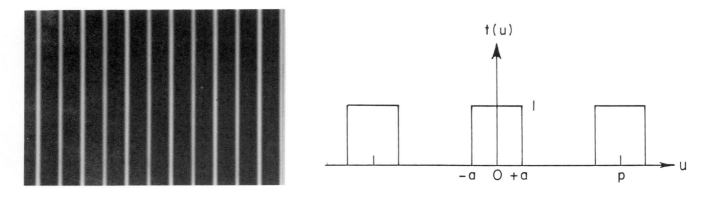

Figure 2. Lau fringes for grating separation z_o equal to $p^2/2\lambda$.

Figure 3. Grating amplitude transmittance.

Coherence theory and the Lau effect

In the theory of partial coherence, propagation of the second order correlation of the optical field is of central interest. As applied specifically to the Lau experiment, the correlations of the field incident on the second grating G_2 provide the key to explaining the presence of the high contrast Lau fringes. This may be seen by the discussion that follows. Before proceeding, however, it should be noted that only an outline of this coherence theory interpretation is presented since a detailed analysis has been presented elsewhere.[2,3,6]

The field incident on the grating G_1 in Figure 1 is essentially spatially incoherent and hence the field incident on the second grating G_2 can be considered as being produced by an array of spatially incoherent sources. The complex degree of spatial coherence[7] of this field is found by an application of the van Cittert-Zernike theorem.[8] Since this theorem represents a Fourier transform relation between the source intensity distribution (in this case the grating G_1) and the coherence function, the complex degree of spatial coherence produced in the field by propagation from the first grating will be "periodic" with a period equal to $\lambda z_o/p$. The second grating, then, will be illuminated by a beam of uniform intensity and whose field correlations are "periodic." Since the coherence function period depends on the grating separation, z_o, its period can be made to match that of the second grating by adjusting this separation. Assuming the gratings have a binary amplitude transmittance (see Figure 3) with sufficiently narrow slit-widths, high contrast fringes will be observed whenever the coherence function period is a rational multiple of the grating period.

On a more formal basis, the complex degree of spatial coherence, $\mu_{G_2}^{(-)}(u_1, u_2, \omega)$, of the field incident on the grating G_2, in a one-dimensional analysis, can be shown to be given by the expression

$$\mu_{G_2}^{(-)}(u_1, u_2, \omega) = \sum_{n=-\infty}^{\infty} C'_n \; e^{i2\pi n u_2/p} \; e^{i\pi\lambda z_0 (n/p)^2} \delta(\sigma - z_0\lambda n/p) \; , \tag{1}$$

where p is the grating period, C'_n is proportional to the Fourier coefficients, C_n, of the grating amplitude transmittance and $\sigma = u_1 - u_2$, i.e. the coordinate difference of two points in the plane of the grating G_2. Clearly, this function has a spacing between peaks equal to $z_0\lambda/p$. If we restrict the grating separation, z_0, to the following set of values

$$z_0 = \frac{\alpha}{\beta} \frac{p^2}{\lambda} \; ; \quad \alpha, \beta \text{ are integers} \; , \tag{2}$$

then Eq. (1) reduces to

$$\mu_{G_2}^{(-)}(u_1, u_2, \omega) = \sum_{n=-\infty}^{\infty} C'_n \; e^{i2\pi n u_2/p} \; e^{i\pi\frac{\alpha}{\beta} n^2} \delta(\sigma - \alpha n p/\beta). \tag{3}$$

By plotting Eq. (3) as a function of z (i.e. α and β), it is possible to demonstrate the significance of this set of grating separations. An example is shown in Figure 4 for the cases where (b) $\alpha = 1$ $\beta = 3$ and (c) $\alpha = 2$, $\beta = 3$. Figure 4a locates the centers of the transmitting elements of the grating G_2. The phase factor in u_2 has been neglected in the diagrams since it is not essential for the present discussion. Referring to the solid arrows in Figure 4, it is to be observed that corresponding points in slits which are separated by a distance equal to p are coherent with respect to each other (although not completely coherent since $|C'_n| < 1$. In addition, these points are 180 degrees out of phase with respect to each other whenever the product $\alpha\beta$ is odd (see Figure 4b). Figure 5 illustrates a case where $\alpha\beta$ is even and corresponding points in the slits are in phase. Thus, the lens L_2 effectively "sees" α independent gratings each having a period αp and each illuminated by a partially coherent field such that corresponding points in every slit have some degree of coherence with respect to each other. This behavior in the field correlations at the second grating is a direct result of the field having propagated a distance equal to a rational multiple, α/β, of the quantity p^2/λ. The Lau fringes corresponding to the cases of $\alpha = 1$, $\beta = 3$ and $\alpha = 2$, $\beta = 3$ are shown in Figure 6a and b respectively. They were obtained using a low pressure mercury arc filtered for the .5461 μm line.

The resulting intensity distribution has the following Fourier series representation

$$I_F(y, \omega) = A \sum_{n=-\infty}^{\infty} C'_n \; (-1)^{n\alpha\beta} \; e^{-i2\pi n y/\frac{p}{\beta}(f/z_0)} \; ; \text{ for } z_0 = \alpha p^2/\beta\lambda \; , \alpha, \beta \text{ integers}, 2a < p/2\beta, \tag{4}$$

where A is a proportionality constant taking into account geometrical factors and the original source intensity and where the effect of an odd $\alpha\beta$ - product is a half period shift in the intensity pattern relative to the pattern produced when $\alpha\beta$ is even. This distribution, known as the Lau fringe pattern, can be recognized as being a series of triangular shaped pulses of period, p', and width w

$$p' = \frac{p}{\beta} \cdot \frac{f}{z_0} = \frac{\lambda f}{\alpha p} \; ; \quad w = (2a) \cdot \frac{2f\lambda}{p^2} \cdot \frac{\beta}{\alpha} \; . \tag{5}$$

The condition that the slit width, $2a$, be less than the grating period p divided by 2β insures non-overlapping of these triangular pulses.

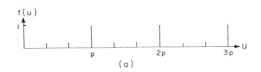

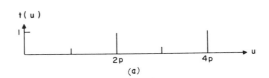

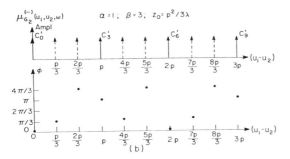

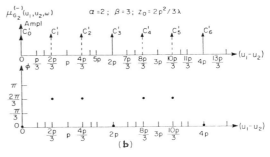

Figure 4. Illustration of the matching principle for $z_0 = p^2/3\lambda$. (a) relative position of the centers of the elements of grating G_2 (b) amplitude and phase of complex degree of coherence.

Figure 5. Illustration of the matching principle for $z_0 = 2p^2/3\lambda$. (a) relative position of the centers of the elements of grating G_2 (b) amplitude and phase of complex degree of coherence.

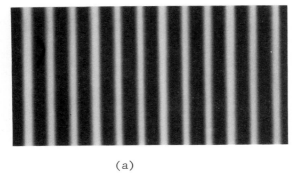

(a)   (b)

Figure 6.   Lau fringes for   (a) $z_0 = p^2/3\lambda$ and   (b) $z_0 = 2p^2/3\lambda$.

Completely coherent "Lau" experiment

We now turn our attention to the other extreme, that is when the Lau experiment operates under completely spatially coherent light. In particular, consider the system illustrated in Figure 7. A monochromatic point source located on axis in the front focal plane of the lens L_1 produces a normally incident plane wave on the grating G_1. The field incident on the second grating G_2 is thus produced by a plane periodic object illuminated by a spatially coherent field. Fields of this nature can be described by the theory of Fresnel images. Though previously developed[5], we have chosen to repeat some of the main features of this theory which allow information about certain properties and conditions on the Fresnel images to be derived analytically, thereby making the theory immediately applicable to the coherent Lau experiment.

Basically, the theory of Fresnel images of plane periodic objects is concerned with the images formed in the near-field of a plane periodic object illuminated by spatially coherent, quasimonochromatic light. Specifically, the theory deals with fields, $U(u,z)$, which are represented by integrals of the form

$$U(u,z) = \int_{-\infty}^{\infty} t(x) e^{ik(x-u)^2/2z} dx \quad , \tag{6}$$

when the function $t(x)$ is periodic. Equation (6) can be considered, for example, as expressing the field produced in the u-plane by a unit amplitude plane wave normally incident on an object (whose amplitude transmittance is $t(x)$) located a distance z away in the x-plane. Obviously, Eq.(6) is the Fresnel diffraction integral and, thus, its magnitude is referred to as the Fresnel diffraction pattern or Fresnel transform of the object transmittance $t(x)$.

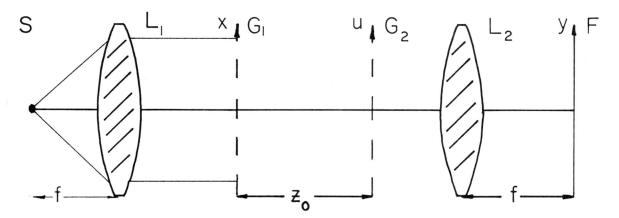

Figure 7. Completely coherent "Lau" experiment.

If the object transmittance has a periodic structure, the resulting Fresnel diffraction pattern will be a sharply defined pattern which resembles the original object t(x) in some sense. These particular reconstructions are referred to as Fresnel images. In fact, if the propagation distance, z, is given by

$$z = 2\frac{p^2}{\lambda}(n + R) \quad , \tag{7}$$

with $0<R<1$ and $R = M/m$ where M,m, and n are integers, then well-defined Fresnel images are observed. For this reason, these distances given by Eq.(7) are called Fresnel image distances and are the ones which are of interest here. Noting the similarity to the Lau grating separations, we equate Eqs.(2) and (7) arriving at the relation

$$\frac{\alpha}{\beta} = 2(n + R) \quad , \tag{8}$$

which establishes the correspondence between the Fresnel image distances and the grating separations in the Lau experiment.

Consider, then, the wavefield expressed by Eq.(6) when the object has an amplitude transmittance t(x) which is periodic with period p and has a single element transmittance $t_p(x)$. The overall transmittance t(x) can be written as

$$t(x) = t_p(x)*A(x) \quad , \tag{9}$$

where

$$A(x) = \sum_{n=-\infty}^{\infty} \delta(x-np) \quad , \tag{10}$$

and $\delta(\ )$ is the one-dimensional Dirac delta-function, while * denotes convolution. In view of Eq.(9), Eq.(6) for the field, U(u,z), may be written as

$$U(u,z) = t_p(u)*T(u) \quad , \tag{11}$$

where

$$T(u) = A(u)*e^{iku^2/2z} \quad . \tag{12}$$

Equation (11) is the heart of the theory. It expresses the diffracted field or alternatively the resulting Fresnel image as a convolution between the single element transmittance $t_p(u)$ and the array function T(u). This formula for the diffracted field is particularly useful since it indicates how the periodic elements in the original object are replicated during the process of propagation. It would be even more useful if the array function T(u) could be transformed into a weighted delta-function series since this would make obvious the changes that the periodic elements undergo. Winthrop and Worthington[7] have, in fact, shown two methods for performing this transformation. Here, we outline one of these procedures.

We begin the process by expressing the array, A(u), by its Fourier series representation

$$A(u) = \frac{1}{p} \sum_{h=-\infty}^{\infty} e^{i2\pi hu/p} \quad , \tag{13}$$

in which case Eq.(12) yields the array function ,T(u),

$$T(u) = \frac{1}{p} \sum_{h=-\infty}^{\infty} e^{-i\pi h^2 \lambda z/p^2} e^{i2\pi hu/p} \quad , \tag{14}$$

where we have neglected unessential constants. Now making use of Eq.(7) the integer h is transformed to the integers k and v via

$$h = mk + v \quad , \quad \text{where } 0 < v < m-1 \quad . \tag{15}$$

Substituting Eq.(15) into Eq.(14), T(u) becomes

$$T(u) = \frac{1}{p} \sum_{k=-\infty}^{\infty} e^{i2\pi um/p} \sum_{v=0}^{m-1} e^{-i2\pi Mv^2/m} e^{i2\pi kv/m} \quad . \tag{16}$$

The summation in k is recognized as the Fourier series representation of a delta-function series. Thus T(u) becomes

$$T(u) = \sum_{k=-\infty}^{\infty} C(k,M,m)\delta(u-kp/m) \quad , \tag{17}$$

where

$$C(k,M,m) = \frac{1}{m} \sum_{v=0}^{m-1} e^{-i2\pi Mv^2/m} e^{i2\pi kv/m} \quad . \tag{18}$$

Consequently, the Fresnel image field, as expressed by Eq.(12) can be rewritten as

$$U(u,z) = \sum_{k=-\infty}^{\infty} C(k,M,m)t_p(u-kp/m) \quad . \tag{19}$$

Therefore, the Fresnel image of a periodic object whose single element transmittance is $t_p(x)$ is again an array of the same single elements but with a period p/m and weighted by the coefficient C(k,M,m). The period of the image depends on the actual propagation distance,z, (i.e. via m). It follows from Eq.(19) that, in order to observe distinct Fresnel images, for a given propagation distance, the width of the single element, $t_p(x)$, must be less than p/m. The coefficients, C(k,M,m), are ,in general, complex. This leads to images which have single elements out-of-phase with respect to each other. However, we see from Eq.(18) that if R=1, then C(k,1,1) = 1, in which case the Fresnel image is an exact replica of the original object. These special Fresnel images are referred to as Fourier images[9,10] and the corresponding distances, $z = 2np^2/\lambda$, are called self-imaging distances.

For the Lau experiment, $t_p(x)$ represents the transmittance of a single grating slit. In Figure 8 the field incident on the grating G_2 as described by the Equation (19) is shown for $z_0 = p^2/2\lambda$ ($\alpha=1,\beta=2$ or n=0,M=1,m=4), Fig.(8b) and $z_0 = p^2/3\lambda$ ($\alpha=1,\beta=3$ or n=0,M=1,m=6), Fig.(8c). From these typical fields we make the following general observations: 1.) as long as the slit widths are sufficiently narrow non-overlapping images of the grating slits will be incident on the grating G_2, 2.) when the product $\alpha\beta$ is even the Fresnel image of the grating G_1 will be aligned with the grating G_2 such that every slit is illuminated coherently and in-phase, and 3.) when the $\alpha\beta$ product is odd the Fresnel image is shifted by half a period and consequently none of the slits in the grating G_2 will be illuminated.

The result is obvious. For the coherent "Lau" experiment where the grating G_1 is illuminated by a normally incident plane wave, the intensity distribution in the rear focal plane of the lens L_2 will be either a.) a dark field or b.) the far-field diffraction pattern of the second grating G_2 corresponding to $\alpha\beta$ odd and $\alpha\beta$ even, respectively.

Using these results, the field in the rear focal plane of the lens L_2 may be written as

$$U(y) = \int_{-\infty}^{\infty} \left[t(u) \int_{-\infty}^{\infty} t(x) e^{ik(x-u)^2/2z_0} dx \right] du \quad , \tag{20}$$

where the quantity in brackets has the the following property

$$t(u) \int_{\infty}^{\infty} t(x) e^{ik(x-u)^2/2z_0} dx = \begin{cases} t(u) & \text{for } \alpha\beta \text{ even} \\ 0 & \text{for } \alpha\beta \text{ odd} \end{cases} \tag{20a}$$

and unessential constants have been neglected.

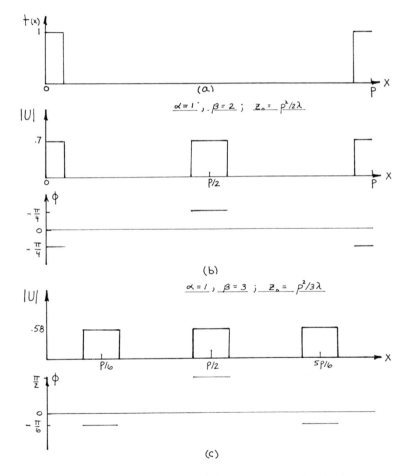

Figure 8.   Illustration of Fresnel images. (a) one period of the grating amplitude transmittance, (b) and (c) show magnitude and phase of Fresnel images for $z_0 = p^2/2\lambda$ and $z_0 = p^2/3\lambda$, respectively.

Reciprocal and equivalent nature of the fields

We have just seen, then , that in the completely coherent "Lau" experiment where the first grating is illuminated by a normally incident plane wave the observed intensity distribution is either a dark field or the far-field diffraction pattern of the grating G_2. The particularly interesting result associated with the coherent case, though, is obtained if we refer to Equation (8). Basically, this equation indicates that the distances which define the Fresnel image planes in the case of coherent illumination are precisely the same distances which define the grating separations for which high contrast fringes are observed in the original incoherent Lau experiment. This curious finding is a direct result of the field in the coherent case and the cross-spectral density function in the Lau experiment both being periodic and satisfying Helmholtz equations. The field satisfies a single Helmholtz equation while the cross-spectral density function satisfies coupled Helmholtz equations.

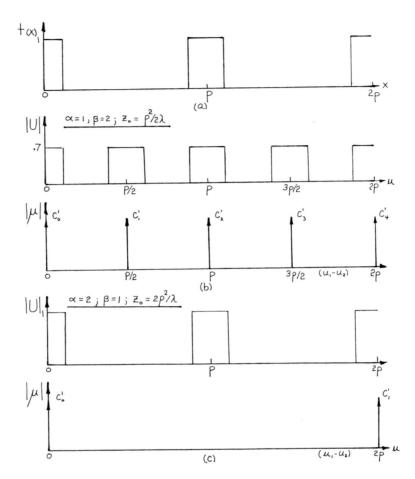

Figure 9. Comparison of coherent and incoherent Lau experiments. (a) two periods of the grating G_2; (b) and (c) show magnitudes of the Fresnel images and the complex degree of spatial coherence of the fields incident on grating G_2 for $z_0 = p^2/2\lambda$ and $z_0 = 2p^2/\lambda$, respectively.

In addition, a comparison of the fields incident on the grating G_2 for these two extreme cases reveals an interesting relationship between them. It follows from Equations (3) and (19) that in the original Lau experiment the field incident on grating G_2 has a periodic degree of spatial coherence and is of uniform intensity while in the coherent case the field has a periodic intensity distribution and is of uniform coherence. Thus there exists a reciprocity between the coherence function and the intensity distribution for these two cases. It is the periodic nature of these fields incident on the periodic grating G_2 which is responsible for the periodicity observed in the intensity distribution at the rear focal plane of the lens L_2. For a given grating separation, z_0, however, the two cases will, in general, produce patterns of different periods, though there do exist separations for which the periods will be the same. The situation is depicted in Figure 9 where (a) shows two periods of the grating G_2 and (b) and (c) show the magnitude of the Fresnel image and the complex degree of spatial coherence of the fields incident on the grating G_2 for two different grating separations. In the completely coherent case, as shown earlier, the diffraction pattern of the grating G_2 is observed (when $\alpha\beta$ is even). Hence, the period of the pattern will always be inversely proportional to the grating period, p. From the figure we see that in the case of incoherent illumination (refer to the complex degree of spatial coherence μ) that for certain separations ($\alpha=1, \beta=2$, for example) corresponding points in every slit are coherent with respect to each other. This will result in a fringe pattern having a period inversely proportional to p. However, for other separations (such as $\alpha=2, \beta=1$) this is not the case.

This comparison, though, does raise an interesting question and that is, is it possible to define a completely coherent experiment whose intensity distribution in the rear focal plane of the lens L_2 behaves in the same manner as the fringe pattern in the incoherent case   when the position of the first grating is changed. The answer is yes. The "equivalent" coherent experiment is shown in Figure 10. A point source located on axis a

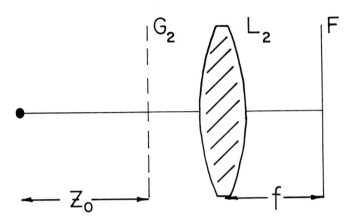

Figure 10. "Equivalent" coherent experiment.

distance z_0 in front of the grating G_2 provides coherent illumination. The intensity distribution due to this point source is a Fresnel image of the grating G_2 where the field is described by Eq.(19) with u replaced by yz_0/f. We note that this configuration is actually a special case of the original Lau experiment. In fact, Lohmann[1] used this system to determine the intensity distribution in the original Lau experiment by first finding the intensity due to a single point source on the grating G_1 and performing an incoherent sum over this grating. We are, however, taking a somewhat different point of view in that we are attempting to find an "equivalent" coherent experiment which in a sense simulates the original Lau experiment. Figure 11 shows the intensity distribution for the original Lau experiment and the equivalent coherent experiment shown in Fig.10 for two values of z_0. It is important to note in these figures that although the exact structures are not the same, their variation with z_0 is the same. That is, as the separation, z_0, changes the fringe width and period change in the same manner for both the incoherent Lau experiment and the "equivalent" coherent experiment. It is in this sense that the two experiments are equivalent.

To determine the basis of this equivalence, the field incident on the grating G_2 in the case of the original Lau experiment is of uniform intensity over a small angle and it's complex degree of spatial coherence is periodic with a quadratic phase. The grating G_2 can be thought of as imposing a periodic intensity distribution on this field. Hence, the grating G_2 can equivalently be thought of as a source whose far-field intensity distribution interests us. This source has a periodic intensity distribution and a degree of spatial coherence which is periodic with a quadratic phase. Thus the "equivalent" coherent experiment would in some sense have to produce a similar type source. For the equivalent coherent system shown in Figure 11 the field incident on the grating G_2 is completely coherent, is essentially uniform in intensity over a small angle, and has a quadratic phase. Here, too, the grating G_2 imposes a periodic intensity distribution. The grating G_2 now viewed as a source has a periodic intensity distribution with a quadratic phase and uniform coherence. though not entirely identical there is a certain equivalence between these two "sources" as well as some reciprocity between the nature of the fields. It is the similar nature ot these two sources which results in the identical behavior of their corresponding far-field intensity patterns, while the fact that they are not completely identical results in the different detailed structure observed.

Conclusion

In this article we have reviewed the Lau effect from an optical coherence theory point of view showing how the condition on grating separation arises naturally in this analysis. To develop a feeling for the changes the fringes would undergo as a function of spatial coherence we considered the case of completely coherent illumination in which case the Lau fringes reduced to the far-field diffraction pattern of the second grating. In analyzing this particular case we found that the periodic nature of the optical system gave rise to a periodic nature in the field which ultimately led us to certain reciprocities and the consideration of equivalent experiments.

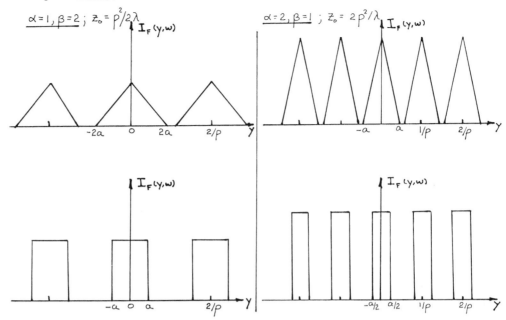

Figure 11. Intensity distributions in the Lau experiment and the "equivalent" coherent experiment. Top figures show Lau fringes while lower figures show "equivalent" coherent results for (a) $z_0 = p^2/2\lambda$ and (b) $z_0 = 2p^2/\lambda$.

Acknowledgements

The author wishes to thank Dr. Brian Thompson for his encouragement and support. The research was supported by U.S. Army Contract DAAK70-77-C0190.

References

1. Jahns, J. and Lohmann, A. W.,"The Lau effect(A diffraction experiment with incoherent illumination)," Opt. Commun., Vol.28,pp. 263-267. 1979.

2. Gori, F."Lau effect and coherence theory", Opt. Commun., Vol. 31, pp. 4-8, 1979.

3. Sudol, R. and Thompson, B. J.,"An explanation of the Lau effect based on coherence theory," Opt. Commun., Vol. 31, pp. 105-110. 1979 .

4. Lau, E.,"Interference phenomena on double grating," Annalen der Physik, Vol. 6, pp. 417-423. 1948.

5. Winthrop, J. J. and Worthington, C.R.,"Theory of Fresnel images.I.plane periodic objects in monchromatic light," J.Opt.Soc.Am., Vol. 55, pp. 373-381. 1965.

6. Sudol, R. and Thompson, B. J.,"The Lau effect: theory and experiment," submitted to Applied Optics.

7. Mandel, L. and Wolf, E.,"Spectral coherence and the concept of cross-spectral purity," J.Opt.Soc.Am., Vol. 66, pp. 529-535. 1976.

8. Born, M. and Wolf, E.,Principles of Optics, fifth edition, Pergamon, Oxford and New York, 1975, chapter 10.

9. Talbot, H. F.,"Facts relating to optical science," Phil. Mag., Vol. 9, pp.404. 1836.

10. Cowley, J. M. and Moodie, A. F.,"Fourier images:I- the point source," Proc.Phys.Soc., Vol. B70, pp. 486-496. 1957.

One hundred percent efficiency of gratings in non-Littrow configurations

Magnus Breidne*, Daniel Maystre**

*Department of Physics II and Institute of Optical Research
Royal Institute of Technology, S-100 44 Stockholm, Sweden

**Laboratoire d'Optique Electromagnétique, E.R.A. au C.N.R.S. n° 597
Faculté des Sciences et Techniques, Centre de St-Jérôme, 13397 Marseille Cedex 4, France

Abstract

The possibility of perfect blazing for infinitely conducting gratings used in non-Littrow mountings has been shown recently. This study is devoted to a thorough investigation of this interesting property for sinusoidal and echelette gratings. We show that most commercial gratings can exhibit this property.

Introduction

It is generally believed that infinitely conducting gratings can have a 100 % efficiency only in the Littrow configuration.[1] However, a recent paper[2] has shown the theoretical possibility of obtaining, for TM polarized light, all the incident energy in a non-zero diffracted order even in non-Littrow mountings. The aim of this study is to make a thorough investigation of this property for commercially available gratings (holographic sinusoidal and ruled echelette gratings).

Phenomenological theory of gratings

By using the well known analytical properties of gratings, such as the energy conservation theorems[3] and the reciprocity theorem,[4] the phenomenological theory shows that all the interesting quantities, efficiency and phase of the diffracted orders, can be deduced from four fundamental parameters, if only two orders propagate. Two of these parameters define the "electromagnetic center of the groove" and have no influence on the efficiency. It can be shown that the efficiency in the −1 order depends on the two remaining fundamental parameters according to the formula :

$$E = \sin^2 \rho \cdot \cos^2 \phi \cdot \tag{1}$$

The parameters ρ and ϕ are both functions of the grating period d, the depth h, the wavelength λ, the polarization and the type of profile. However, they change much more regularly, than e.g. the efficiency when the depth or period of the grating is changed. To calculate the values of ρ and ϕ numerical computer programs have to be used.

For the Littrow mounting, $\phi \equiv 0$. In this case $\lambda = 2d \sin \theta$, where θ is the angle of incidence, and formula (1) is simplified into :

$$E = \sin^2 \rho \cdot \tag{1'}$$

To obtain an efficiency of 100 % it is necessary that :

$$\rho = \frac{\pi}{2} + n\pi, \quad \text{with} \quad n \in \mathbb{Z} \cdot$$

From formula (1) it can be seen that for a non-Littrow mounting two conditions must be fulfilled to have $E = 1$:

$$\rho_n = \frac{\pi}{2} + n\pi,$$

$$\phi_m = m\pi, \quad \text{with} \quad n, m \in \mathbb{Z} \cdot$$

However, in this case we have one additional parameter to vary, because there is no longer any relation between λ and θ.

So, for a fixed groove profile, keeping the angular deviation D constant and varying the wavelength, we obtain a curve in the ρ-ϕ-plane. If this curve passes through any of the points (ρ_n, ϕ_m) we have a perfect blaze. This is illustrated in Fig.1.

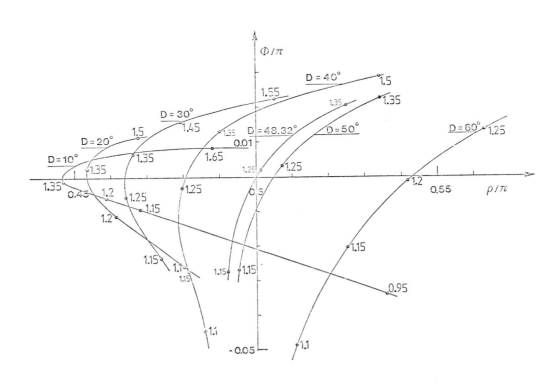

Figure 1. Parametric curves giving ρ/π and ϕ/π as functions of λ/d for various deviations D (TM polarized light). The curve corresponding to the Littrow mount coincides with the ρ/π axis. For D = 48.32°, the curve crosses the point (ρ_0, ϕ_0) and the efficiency is equal to 1 for $\lambda/d \simeq 1.25$. The grating is sinusoidal with h/d = 0.35.

Perfect blazing for sinusoidal gratings, TM-case

In Fig.2 the result of a systematic investigation of non-Littrow perfect blaze is presented. On the abscissa we have the depth-to-groove spacing ratio of the sinusoidal grating. Two ordinates have been drawn. The one to the left gives the deviation and that to the right gives the wavelength-to-groove spacing ratio at which a perfect blaze occurs in the TM-case. The dots correspond to the deviation, while the squares correspond to the ratio λ/d.

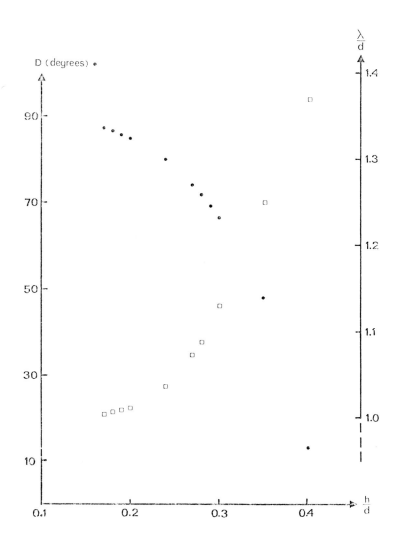

Figure 2. Values of D (filled circles) and λ/d (squares) for which a perfect blaze is obtained for a sinusoidal grating used for TM polarized light

As h/d → 0, the deviation curve tends to 90°, but we have not found any perfect blaze for h/d < 0.17. It is to be noted that, for Littrow mountings, the perfect blaze also disappears, for h/d < 0.25.

For two values of the deviation, the corresponding efficiency curves for the related value of h/d is plotted, Fig.3a and 3b.

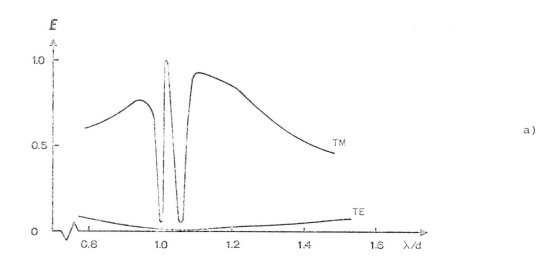

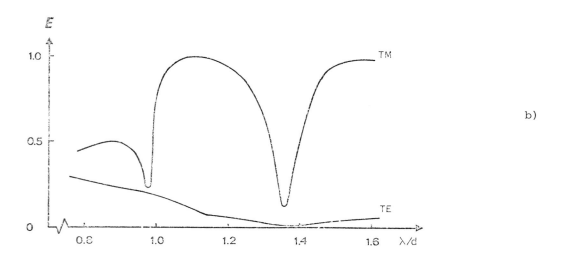

Figure 3. Efficiency curves in constant deviation mounting for two infinitely conducting sinusoidal gratings.

a) D = 87°,   h/d = 0.17 ,

b) D = 69°,   h/d = 0.29 .

Perfect blazing for echelette gratings, TM-case

The same investigation as above has been done for echelette gratings, the blaze angle α being varied (apex angle 90°). In contrast to the sinusoidal gratings, we find that the interval where non-Littrow perfect blaze can be obtained is divided into two, Fig.4. Notice that there is no perfect blaze effect between α = 25° and α = 32°.

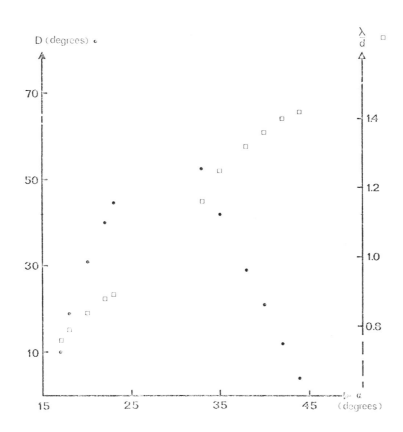

Figure 4. Same as Figure 2, but for echelette grating

No empiric rule for when we obtain a non-Littrow perfect blaze has been found. Notice, that the intuitive approach of geometrical optics is not valid since, at the perfect blaze position, the incident and diffracted directions are not symmetrical with respect to a facet.

In Fig.5a and 5b efficiency curves for two points, one from each of the intervals, are plotted.

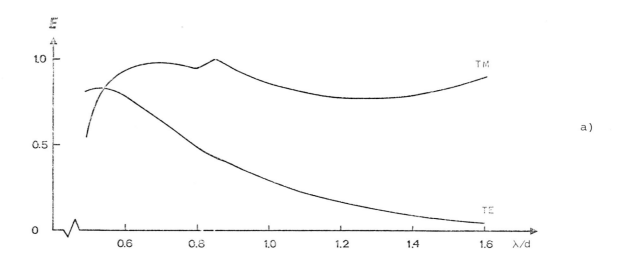

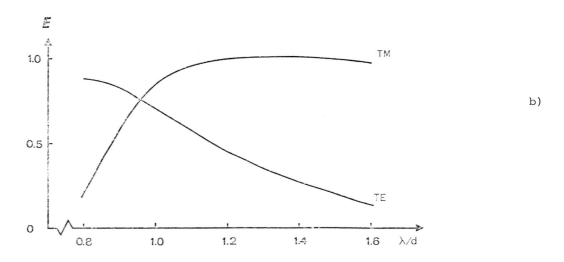

Figure 5. Efficiency curves in constant deviation mounting for two infinitely conducting echelette gratings :

a) _D = 31° , α = 20° ,

b) D = 21° , α = 40° .

Conclusions

We have discussed an interesting phenomenon, that of 100 % efficiency in non-Littrow
mountings for TM polarized light. Interesting from the theoretical point of view, but also
of large interest for applications where the deviation is imposed. If the deviation can be
chosen arbitrarily the Littrow configuration still seems to be the best mounting.

In the TE-case we have not been able to find, neither theoretically nor by calculations,
this phenomenon.

References

1. Hessel A., Schmoys J., Tseng D.Y., J. Opt. Soc. Am. 65, 380 (1975)
2. Maystre D., Cadilhac M., Chandezon J., submitted for publication in Optica Acta
3. Petit R., Revue d'Optique 45, 249 (1966)
4. Maystre D., Mc Phedran R.C., Opt. Commun. 12, 164 (1974)

Echelle efficiency and blaze characteristics

M. Bottema

Ball Aerospace Systems Division, Ball Corp., P.O. Box 1062, Boulder, Colorado 80306

Abstract

On the basis of a simple model for diffraction by the groove, we show that the efficiency of an echelle is high in the Littrow arrangement, but becomes markedly less if the angle of deviation between the incident and the diffracted beams is increased. We also show that the height and width of the spectral image change if the directions of incidence and diffraction are interchanged, but that the efficiency remains the same. In general, the direction of maximum efficiency does not coincide with the blaze direction, defined as the maximum of the envelope of the diffracted-light distribution function. The difference depends on the mode of operation of the spectrograph (scanned or stationary spectrum) and typically changes from a fraction of a degree at high orders in the UV to several degrees at low orders in the IR. Experimental evidence agrees qualitatively with these calculations, but a more refined diffraction model may be needed for a quantitative interpretation.

Introduction

Echelles are diffraction gratings with deep triangular grooves, as shown in Figure 1. The light, diffracted by the grooves, is concentrated by high-order interference in sharply defined directions that obey the relation

$$\sin\alpha + \sin\beta = m\lambda/d, \tag{1}$$

where α is the angle of incidence, relative to the grating normal, β the angle of diffraction, m the spectral order (i.e. order of interference), λ the wavelength and d the groove spacing (grating constant). The angle between the reflecting groove facet and the grating surface is called the echelle blaze angle, θ. The diffracted-light distribution is represented by the well-known equation

$$I(\alpha,\beta) = I_e \sin^2\{\tfrac{1}{2}Nkd(\sin\alpha + \sin\beta)\}/\sin^2\{\tfrac{1}{2}kd(\sin\alpha + \sin\beta)\}, \tag{2}$$

where I_e is the diffraction function of a single groove and $k = 2\pi/\lambda$. The diffracted light is concentrated in narrow spectral images of height $N^2 I_e$ and width (full width at half maximum)

$$\Delta\beta = \lambda/Nd\cos\beta_m), \tag{3}$$

where β_m is the angle of diffraction for the center of the peak (Figure 2). The integrated flux in the spectral image can readily be derived by integration of Eq. (2) in a narrow interval around β_m and is found to be

$$I_{int}(\alpha,\beta_m) = I_e \{N\lambda/(d\cos\beta_m)\} \tag{4}$$

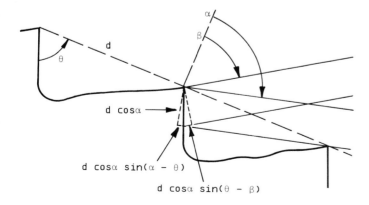

Figure 1. Echelle profile with light beams used in mathematical model $(\alpha > \beta)$

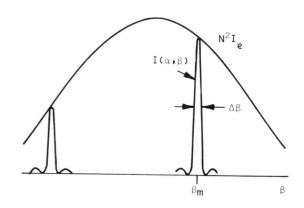

Figure 2. Spectral images and envelope function $N^2 I_e$

Of interest to the throughput of a spectrograph is the ratio of the integrated flux in the spectral image to the incident flux of the same wavelength. This ratio is called the "absolute efficiency", E_a. It is equal to the product of the reflectivity of the groove facet and the so-called "relative efficiency", E_r, i.e. the ratio of the flux in the spectral image to that in the image of a mirror of the same reflectivity as the facet.

The purpose of this paper is to present a theoretical derivation of the relative efficiency in various spectrograph configurations and to compare these predictions with some experimental data. Although the subject of grating efficiency in general is well covered in the literature,[1,2] little attention seems to have been given to the special case of the echelle.

Relative efficiency

As is evident from Eq. (4), the prediction of the efficiency hinges entirely on the calculation of I_e. To treat diffraction by the groove rigorously, we should model the groove as a slit of width d in the plane of the grating, followed by a mirror image of the same slit in the groove facet. We will return to this later. For the present, we will use a much simpler model, by considering the unvignetted part of the reflecting facet as the diffracting aperture. This model is commonly used in the analysis of reflective gratings with small blaze angles.

We assume that the directions of incidence and diffraction lie in the plane of dispersion and consider separately the situations in which $\alpha > \beta$ or $\alpha < \beta$.

In case $\alpha > \beta$, the width of the illuminated part of the facet is approximately $d \cos \alpha$. The angle of incidence is $\alpha - \theta$ and the angle of diffraction $\beta - \theta$. We apply the well-known scalar theory for a long narrow slit. If the length of the groove is h, the flux in the diffracted beam, for unit flux density in the incident beam, is equal to

$$I_e(\alpha, \beta) = R(hd^2 \cos^2\alpha / \lambda) \operatorname{sinc}^2\left[\tfrac{1}{2}kd\cos\alpha\{\sin(\alpha-\theta) + \sin(\beta-\theta)\}\right], \qquad (\alpha > \beta) \qquad (5)$$

where R is the facet reflectivity.

This equation represents the flux in the beam, integrated in the direction perpendicular to the dispersion, as a function of α and β. In Eq. (5), the obliquity factors, associated with non-normal incidence, are approximated by $\cos(\alpha-\theta) \simeq 1$ and $\cos(\beta-\theta) \simeq 1$.

The absolute efficiency is now found immediately by substituting Eq. (5) in Eq. (4) and dividing by the incident flux. Since we assumed unit flux density in the incident beam, the incident flux is equal to $Nhd\cos\alpha$. Hence,

$$E_a(\alpha, \beta_m) = R(\cos\alpha / \cos\beta_m) \operatorname{sinc}^2\left[\tfrac{1}{2}kd\cos\alpha\{\sin(\alpha-\theta) + \sin(\beta_m-\theta)\}\right] \qquad (\alpha > \beta_m) \qquad (6)$$

This equation applies to discrete spectral images for monochromatic radiation.

With the help of Eq. (1) we can eliminate λ and d and write E_a as a continuous function of α, β and spectral order m. After division by R, the relative efficiency is then

$$E_r(\alpha, \beta, m) = (\cos\alpha / \cos\beta) \operatorname{sinc}^2\left[m\pi\cos\alpha\sin\{\tfrac{1}{2}(\alpha+\beta)-\theta\}/\sin\{\tfrac{1}{2}(\alpha+\beta)\}\right] \qquad (\alpha > \beta) \qquad (7)$$

In case $\alpha < \beta$, the efficiency can be derived quite similarly, except that now only part of the incident beam falls on the reflecting facet. Furthermore, obstruction of the diffracted beam limits the effective width of the facet to $d \cos \beta$. The diffraction function now becomes

$$I_e(\alpha, \beta) = R(hd^2 \cos^2\beta / \lambda) \operatorname{sinc}^2\left[\tfrac{1}{2}kd\cos\beta\{\sin(\alpha-\theta) + \sin(\beta-\theta)\}\right] \qquad (\alpha < \beta) \qquad (8)$$

and the relative efficiency, as a function of α, β and m, becomes

$$E_r(\alpha, \beta, m) = (\cos\beta / \cos\alpha) \operatorname{sinc}^2\left[m\pi\cos\beta\sin\{\tfrac{1}{2}(\alpha+\beta)-\theta\}/\sin\{\tfrac{1}{2}(\alpha+\beta)\}\right] \qquad (\alpha < \beta) \qquad (9)$$

Efficiency and spectrograph configuration

The echelle can be mounted in the spectrograph either in an "in-plane" or an "off-plane" arrangement. In the former, the incident and diffracted beams lie in the plane of dispersion, in the latter on opposite sides. In either arrangement the echelle may remain fixed ("stationary echelle") or be rotated ("scanning echelle"). We concentrate here on the "in-plane" geometry and discuss consecutively the stationary and the scanning situation.

Stationary echelle

The echelle can be illuminated either with $\alpha > \theta$ or $\alpha < \theta$. We define as the blaze direction the angle $\beta_0 = 2\theta - \alpha$. If $\alpha > \theta$, I_e reaches its maximum at $\beta = \beta_0$ (Eq. 5). It can readily be shown that $N^2 I_e$ at β_0 is equal to the peak value, I_{max}, of the diffraction function of a mirror of the same reflectivity as the groove facets. If we introduce the "normalized envelope function" $I_n = N^2 I_e / I_{max}$, then $I_n(\beta_0) = 1$. If $\alpha > \theta$, the relative efficiency at β_0 is given by (Eq. 7)

$$E_r(\alpha, \beta_0) = \cos\alpha / \cos\beta_0 . \qquad (\alpha > \theta) \qquad (10)$$

This result can be interpreted as representing the width of the spectral image, $\lambda / (N d \cos\beta_0)$, as a fraction of the width of the mirror image, $\lambda / N d \cos\alpha)$.

If the directions of incidence and diffraction are interchanged $(\alpha < \theta)$, I_e is reduced by a factor $(\cos\beta / \cos\alpha)^2$, as follows immediately from Eqs. (5) and (8). This phenomenon has been observed by Burton and Reay,[3] who also presented the explanation in terms of losses in the incident beam, given above. However, the relative efficiency remains exactly the same, as follows from Eqs. (7) and (9). This is in full agreement with the general principle of reciprocity in optical imaging, and has convincingly been confirmed by echelle efficiency measurements at Bausch and Lomb.[4] The explanation is that while I_e is lower when $\alpha < \theta$, the spectral image is wider. Examples of I_e and E_r as a function of β are shown in Figure 3. In each case ($\alpha > \theta$ and $\alpha < \theta$), the range is divided in sections for $\alpha > \beta$ and $\alpha < \beta$. In an in-plane configuration, part of one section may be blocked, but with off-plane optics, the full range should be accessible.

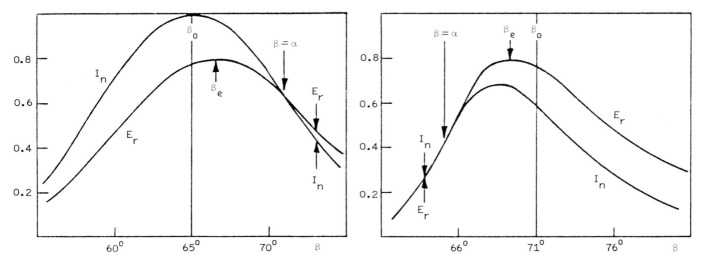

Figure 3. Normalized envelope function, I_n, and relative efficiency, E_r as functions of angle of diffraction, β, in stationary spectrum. Echelle blaze angle $\theta = 68°$. Left diagram: $\alpha = 71°$ ($\alpha > \theta$), blaze direction $\beta_0 = 65°$. Right diagram: $\alpha = 65°$ ($\alpha < \theta$), blaze direction $\beta_0 = 71°$. In both diagrams, Eqs. (5) and (7) apply to the left of $\beta = \alpha$, Eqs. (8) and (9) to the right.

As is evident from Figure 3, the maximum relative efficiency does not occur at the blaze direction β_0, but at a larger angle β_e if $\alpha > \theta$ and a smaller angle if $\alpha < \theta$. The deviation $\beta_e - \beta_0$ decreases with increasing spectral order. For large values of m, $\beta_e - \beta_0$ is approximately given by

$$\beta_e - \beta_0 \simeq 6 \tan\beta_0 (\sin\theta / \cos\alpha)^2 / (m^2 \pi^2) \qquad (\alpha > \theta, m \text{ large}) \qquad (11)$$

$$\beta_e - \beta_0 \simeq -6 \tan\beta_0 (\sin\theta / \cos\beta_0)^2 / (m^2 \pi^2) \qquad (\alpha < \theta, m \text{ large}) \qquad (12)$$

However, we must point out that the simple model for groove diffraction, used here, may not warrant precise predictions of such small effects as the difference between β_e and β_0. This will be discussed in more detail below.

Scanning echelle

We introduce the echelle scan angle

$$\gamma = \tfrac{1}{2}(\alpha + \beta) \qquad (13)$$

and the deviation angle

$$\delta = \tfrac{1}{2}|\alpha - \beta| \tag{14}$$

Both Eq. (7) and Eq. (9) then reduce to

$$E_r(\gamma,\delta) = \{\cos(\gamma+\delta)/\cos(\gamma-\delta)\}\,\mathrm{sinc}^2\left[m\pi\cos(\gamma+\delta)\sin(\gamma-\theta)/\sin\gamma\right] \tag{15}$$

Clearly, in a scanning spectrograph, the orientation of the echelle ($\alpha > \beta$ or $\alpha < \beta$) is immaterial as far as the efficiency of the echelle is concerned. Other differences remain, however. If the spectral resolution is limited by the pixel size of the detector or the width of the exit slit, the resolution can be higher if $\alpha < \beta$, since the dispersion,

$$d\beta/d\lambda = m/(d\cos\beta), \tag{16}$$

is larger. On the other hand, a wider entrance slit is acceptable if $\alpha > \beta$, as the image of the slit is demagnified at the detector by a factor $\cos\alpha/\cos\beta$. This allows a higher spectrograph throughput without loss of spectral purity.

An example of the relative efficiency as a function of γ is shown in Figure 4. The scan angle, corresponding to the blaze direction, is $\gamma = \theta$. In the model, used here, the angle of maximum efficiency, γ_e, is found to be somewhat smaller. In first-order approximation, the difference is

$$\gamma_e - \theta \simeq -\,(3/2)\left[\{\sin^2\theta\sin 2\delta\}/\{\cos^3(\theta+\delta)\cos(\theta-\delta)\}\right]/(m^2\pi^2). \qquad \text{(m large)} \tag{17}$$

As in the stationary spectrum, the predicted difference decreases quadratically with spectral order. A comparison with experimental data is made below.

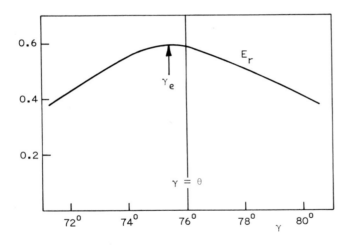

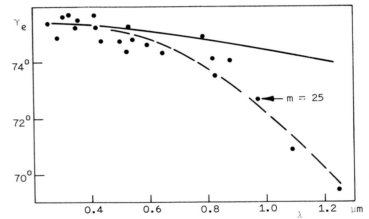

Figure 4. Relative efficiency, E_r, as function of scan angle, γ, predicted by Eq. (15). The maximum efficiency lies at a scan angle γ_e, smaller than the blaze scan angle, $\gamma = \theta$. Echelle data: $\theta = 76°$, $\delta = 3.75°$, d = 12.565 μm. Order m = 25.

Figure 5. Scan angle, γ_e, for maximum relative efficiency as function of wavelength. Broken line: best fit to experimental data. Solid line: γ_e, predicted by Eq. (17). Echelle data same as in Figure 4.

Both when $\alpha > \beta$ and $\alpha < \beta$, the blaze efficiency decreases rapidly if the deviation angle is increased. This effect is especially noticeable at large echelle blaze angles. Some examples are given in Table 1. Only in the off-plane configuration is it possible to select $\alpha \simeq \theta \simeq \beta_o$ and achieve $E_r \simeq 1$. This maybe a distinct advantage in some applications.

Table 1. Effect of blaze angle, θ, and deviation angle, 2δ, on blaze efficiency

Blaze angle		Relative blaze efficiency			
tan θ	θ	$2\delta = 0$	$2\delta = 5°$	$2\delta = 10°$	$2\delta = 15°$
2	63.4°	1	0.84	0.70	0.58
4	76.0°	1	0.70	0.48	0.31
6	80.5°	1	0.58	0.31	0.12

Measured maximum relative efficiencies of four different echelles are presented in Table 2. For three of these, the data were provided by Bausch and Lomb.[5] The measurements were done in an in-plane scanning arrangement with a deviation angle $2\delta = 7.5°$ and $\alpha < \beta$. With the exception of a few data points, the maximum relative efficiencies for the r/2 echelles ($\theta = 63.4°$) lie consistently between 80% and 90% of maximum possible values, calculated from Eq. (15). The data for the r/4 echelle ($\theta = 76.0°$) are even higher, but may have been affected by decreased reflectivity of the comparison mirror.[5] The data refer to odd-generation replicas and indicate excellent groove shape.

The fourth example in Table 2 refers to an echelle, ruled by Bausch and Lomb for the International Ultraviolet Explorer. The UV efficiency was measured at the Johns Hopkins University.[6] These measurements were also done in an in-plane scanning arrangement. The deviation angle in this case was $2\delta = 17.4°$, with $\alpha > \beta$. The data indicate a relative blaze efficiency between 260 nm and 300 nm of 77% of the theoretically possible value. This is still very high, considering the increased effects of small groove imperfections and surface roughness at short wavelengths.

As pointed out in the preceeding section, the scan angle for maximum efficiency, γ_e, is smaller than the blaze scan angle, which is equal to θ. This effect has, indeed, been observed at Bausch and Lomb. However, the experimental difference $\theta - \gamma_e$ is much larger than that predicted by Eq. (17). An example is shown in Figure 5. Undoubtedly, the lack of agreement is entirely due to the limitations of the theoretical model, especially at low spectral orders. One might expect that a theory, which predicts γ_e correctly, would also predict slightly higher values of the efficiency at γ_e. The performance of the echelles would then be somewhat less than presented in Table 2.

Table 2 Comparison of measured and theoretical maximum relative efficiencies

Blaze angle	Ruling frequency (mm⁻¹)	Deviation angle	Wavelength (nm)	Order	Maximum relative efficiency		
					Measured	Theoretical	Percentage
63.4°	79	7.5°	300	75	0.64	0.77	83%
			600	38	0.71	0.77	92%
			900	25	0.67	0.77	89%
63.4°	316	7.5°	300	19	0.70	0.77	91%
			600	9	0.65	0.77	85%
			900	6	0.52	0.77	68%
76.0°	79	7.5°	600	41	0.56	0.59	95%
			800	31	0.55	0.59	94%
			1000	25	0.54	0.59	91%
48.1°	63	17.4°	260-300	78-90	0.54	0.71	77%

Alternate groove model

An effort was made to treat the diffraction by the groove more rigorously by means of the model, shown in Figure 6. The groove is equivalent to two slits of width d, which include an angle 2θ. The incident light is diffracted in two consecutive steps. The diffraction function of each of these can readily be derived, but at the time this paper was submitted, we had not succeeded in convoluting the two functions.

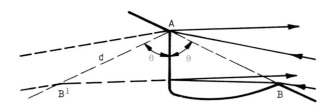

Figure 6. Groove model consisting of two tilted slits of width d (grating constant). The included angle is twice the echelle blaze angle, θ.

One might argue that for angles of incidence, sufficiently larger than the blaze angle, the interception of the wavefront by the second slit could be ignored. The relative efficiency in the in-plane scanning configuration is then found to be

$$E_r = \left[\cos^2(\theta+\delta)\cos^2(\gamma-\theta)/\{\cos(\gamma+\delta)\cos(\gamma-\delta)\} \right] \text{sinc}^2 \left[m\pi \cos(\theta+\delta)\sin(\gamma-\theta)/(\sin\gamma\cos\delta) \right] \qquad (18)$$

The efficiency at the blaze setting ($\gamma = \theta$) is the same as predicted by Eq. (15), but the predicted maximum efficiency occurs at a scan angle, larger than θ. Remarkably, the magnitude of $\gamma_e - \theta$, calculated from Eq. (18), agrees well with the Bausch and Lomb data, but the sign is wrong. This shows the single-slit model to be inadequate. However, the demonstrated effect of the choice of the model on the magnitude and sign of $\gamma_e - \theta$ suggests that a satisfactory explanation of the experimental data might be feasible, once a correct theoretical model has been constructed.

Acknowledgements

This work was undertaken as part of the design of the High Resolution Spectrograph for the Space Telescope, being built by the Ball Aerospace Systems Division for the Goddard Space Flight Center.[7] Experimental data for this paper were provided by R.K. Dakin at Bausch and Lomb and by W.G. Fastie at the Johns Hopkins, with cooperation from G.H. Mount, University of Colorado.

References

1. R.P. Madden and J.D. Strong, Appendix P to J.D. Strong, "Concepts of Classical Optics", Freeman and Comp, San Francisco, 1958.

2. G.W. Stroke, "Diffraction Gratings", in Encyclopedia of Physics, S.F. Flügge, ed., Springer Verlag, Berlin, 1967, Vol. 29.

3. W.M. Burton and N.K. Reay, Applied Optics 9, 1227 (1970)

4. E.G. Loewen, Bausch and Lomb, Rochester NY, personal communication

5. R.K. Dakin, Bausch and Lomb, Rochester NY, personal communication

6. G.H. Mount and W.G. Fastie, Applied Optics 17, 3108 (1978)

7. J.C. Brandt e.a., Proc. SPIE, 172 254 (1979)

PERIODIC STRUCTURES, GRATINGS, MOIRÉ PATTERNS AND DIFFRACTION PHENOMENA

Volume 240

SESSION 5

FABRICATION TECHNIQUES, APPLICATIONS I

Session Chairman
Carey L. O'Bryan III
Air Force Weapons Laboratory

Power losses in lamellar gratings

A. Gavrielides, P. Peterson
AFWL/ARLO, Kirtland AFB, New Mexico 87117

Abstract

Power losses in lamellar gratings per groove length are obatined by integrating the square of the tangential component of the magnetic field, obtained from the infinite conductivity solutions, along the grating profile. The groove fields for the perfectly conducting grating are generated by matching a superposition of diffracted plane waves above the grating to an exact solution of the groove boundary value problem for each polarization. Diffraction effects in the groove energy density and in the power losses are clearly evident.

Introduction

We begin by assuming a superposition of diffracted plane waves and match these, via Maxwell's boundary conditions, to an exact solution of the groove vector boundary value problem. This then yields the field solutions in and above the groove for the two polarization states as shown in Figure 1. It is well known that once the solutions for the perfect conductor are known one may obtain a first-order approximation to the power loss/unit area (dp/da) for a conductor with conductivity σ by using [2]

$$\frac{dp}{da} = \frac{R_s}{2} \left| H_{||} \right|^2 \tag{1}$$

where $R_s = (\omega\mu_g/2\sigma)^{1/2}$ is the surface resistance, ω is the angular frequency, μ_g is the grating permeability and $H_{||}$ is the tangential component of the magnetic field for the perfect conducting case evaluated at the surface. In this approximation one assumes the conduction current is much larger than the displacement current.

Perfectly conducting case

In order to obtain solutions to Maxwell's equations for a bare, perfectly conducting lamellar grating, as shown in Figure 1, we assume plane wave solutions above the grating ($z > 0$) and solve the Helmholtz equation, subject to the appropriate boundary conditions in the groove for the plane of incidence perpendicular to the groove profile.

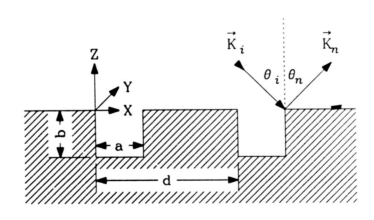

Figure 1. Groove profile and plane of incidence

These solutions are then matched across the (x, y, z = 0) surface by using Maxwell's boundary conditions. This leads to the two polarization cases: Transverse Electric (TE), the electric field parallel to the groove, and Transverse Magnetic (TM), the magnetic field parallel to the grooves. The expressions which satisfy these conditions are Incident plus diffracted TE or TM

$$E_o \, e^{i\beta_i x} \, e^{-i\gamma_i z} + \sum_n A_n \, e^{i\beta_n x} \, e^{i\gamma_n z} \tag{2}$$

GROOVE TE

$$E_g = \sum_{m=1}^{\infty} B_m \frac{\sin m\pi x}{a} \left[e^{-i\lambda_m z} - e^{2i\lambda_m b} \, e^{+i\lambda_m z} \right] \tag{3}$$

GROOVE TM

$$H_g = \sum_{m=0}^{\infty} B_m \frac{\cos m\pi x}{a} \left[e^{-i\lambda_m z} + e^{2i\lambda_m b} \, e^{+i\lambda_m z} \right] \tag{4}$$

where

$$\beta_n = K\sin\theta_n = \beta_i + 2n\pi/d$$

$$\gamma_n = (K^2 - \beta_n^2)^{1/2} = K\cos\theta_n \tag{5}$$

$$\lambda_m = (K^2 - (\frac{m\pi}{a})^2)^{1/2}$$

with γ_n, λ_m being either real or imaginary. E_0 and A_n are electric fields for TE polarization and magnetic fields for TM polarizations. We also adopt the notation that d = groove period, b = depth, a = width, θ_i the angle of incidence and θ_n the diffraction angle with K(= $2\pi/\lambda$) the free space propagation constant. Note that $\gamma_n(\lambda_m) \to i\gamma_n'(i\lambda_m)$ for $K^2 < \beta_n^2$ ($K^2 < (\frac{m\pi}{a})^2$).

The coefficients A_n, B_m for both the TE and TM cases are obtained by joining these solutions, Eq. (2) and Eq. (3) or Eq. (2) and Eq. (4), across the (x, y, z = 0) surface. This yields the following matrix equation for the diffracted amplitudes

$$A_n - \sum_{s=-\infty}^{+\infty} T_{sn} A_s = R_n \tag{6}$$

and the groove amplitudes

$$B_m = \sum_n A_n Q_{nm} + (-1)^p X_m \tag{7}$$

Since the definitions of the quantities contained in Eqs. (6) and (7) are very lengthy we have omitted them and refer the reader to reference 1.

Equation (6) represents an infinite number of complex linear equations with A_n as unknowns and the coefficients T_{sn} being infinite sums over the groove index m. Our method of solution is to truncate arbitrarily the number of equations, as well as the sum over m, then invert the matrix. The size of the array and the number of terms in T_{sn} are determined by the convergence of A_n to within .001. The groove solutions are then obtained by inserting A_n into Eq. (7). Also $\sum_{\gamma_n \, real} |A_n|^2 \, \gamma_n/\gamma_0 \simeq 1$ to within 10^{-5}.

These equations are general and apply to all lamellar gratings under any operating conditions i.e. angle of incidence θ_i and wavelength λ. As an example we choose a grating with a period 6.7μm, width 0.67μm and depth 2.0μm. Fig (2) shows the n = 0 diffracted power for both polarizations, as a function of wavelength at an angle of incidence of $\theta_i = 39^0$. Figure 3 shows the diffracted power, again for both polarization in the same order as a function of angle of incidence θ_i for $\lambda = 9.28$μm. In both figures the incident field is normalized to $|E_0|^2 = 1$.

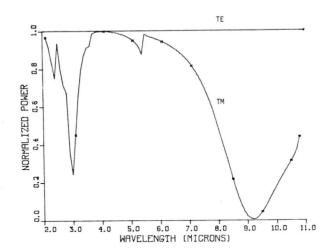

Figure 2. P_0 vs wavelength for TM and TE.

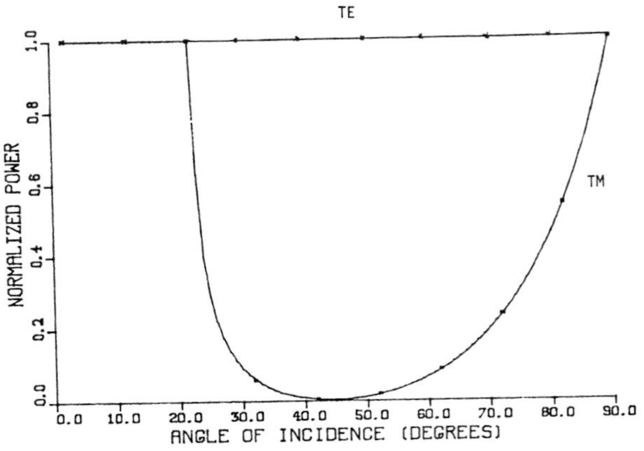

Figure 3. P_0 vs angle of incidence for TM and TE

Once the diffracted amplitudes A_n are known the groove solutions are obtained by using Eq. (7). From this the energy density in the grooves can be obtained by using

$$\text{energy density} = \frac{1}{2} \left(\varepsilon E_g^2 + \mu H_g^2 \right) \tag{8}$$

Figures (4) and (5) show contour plots of the groove energy density, for both polarizations, at the operating point $\Theta_i = 39^\circ$, $\lambda = 9.28\mu m$, with the normalization $\left| E_0 \right|^2 = 1$.

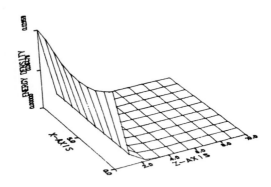

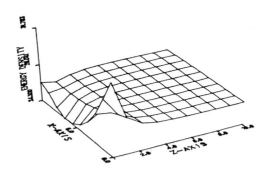

Figure 4. Groove energy density for TE polarization with width and depth scaled to $10\mu m$

Figure 5. Groove energy density for TM polarization with width and depth scaled to $10\mu m$

The dominate feature in Fig. (4), where $P_0 = 1$, is the rapid exponential attenuation reminiscent of the skin-depth attenuation in good conductors. However, Fig. (5) displays a diffraction peak at the (0, 0, 0) top corner which is a manifestation of the non-specular nature of the TM diffraction. A comparison between these two figures shows the dominance of the TEM (m = 0) mode present in TM polarization.

Finite conductivity power losses

To find the power loss/unit groove length (dP/dy) we do one integration of Eq. (1) with the grating area (da) equal to (dydx) on the groove botton ($0 \leq x \leq a$, z = -b) and top $(a < x \leq d, z = 0)$, and equal to $(dy \, dz)$ on the walls $(-b \leq z \leq 0, x = 0, a)$. The resulting equations are then normalized to the incident power/unit groove length obtained by integrating the normal component of the Poynting vector in the interval $0 \leq x \leq d$ for $z = 0$.

In carrying out these integrations one must be cognizant of the complexity of γ_n, λ_m and special cases where the integrated functions become infinite. For TE polarization we must obtain the magnetic field from $\underline{H} = (i/\omega\mu) \, \nabla \times \underline{E}$ and for TM polarization the magnetic field is given by Eq. (4).

An example of these integrations is presented here for the power loss per unit groove length on the bottom. Returning to Eq. (4) and integrating over x for $z = 0$ gives

$$\frac{dP}{dy} = \frac{a \, R_s}{(\omega\mu)^2 p} \sum_{m=p}^{\infty} \left| B_m \right|^2 e^{i(\lambda_m - \lambda_m^*)b} (\lambda_m \lambda_m^*)^p \tag{9}$$

All the power loss equations are then normalized to ,

$$\frac{dP}{dy} = \frac{\gamma_0}{\omega\mu} \left| E_0 \right|^2 d \qquad \text{TE} \tag{10}$$

$$\frac{dp}{dy} = \frac{\gamma_0}{\omega\varepsilon} \left| E_0 \right|^2 d \qquad \text{TM} \tag{11}$$

for $\left| E_0 \right|^2 = 1$. The remaining equations for losses along the top and walls are contained in reference 1.

The use of the power loss equations can be illustrated by taking the conductivity to be that of bulk gold in the IR region where

$$\sigma(\lambda) = \frac{1.829 \times 10^6 \, (\lambda + 3)^2}{\lambda} \qquad 1\mu m \leq \lambda \leq 10\mu m \tag{12}$$

in MKS units. We note that for wavelengths less than $0.5\mu m$ in gold films the index of refraction and the extinction coefficient are much less than the values implied by Eq. (12). In a more general treatment the attenuation of the conductor fields have to be included. This could be accomplished with modified exact solutions matched to the conductor solutions in an iterative manner. However, as an example, we use Eq. (12) since it gives losses which are consistent with experimental results. Figures 6-9 show power losses over the same range of θ_i and λ as before.

We begin our discussion of the losses by considering the TE polarization losses as a function of λ, shown in Fig. (6). Figure 2 shows that the diffraction is specular, hence, the loss is dominated by the top with the wavelength dependence given by λ^{-1} in the conductivity. As a function of angle of incidence θ_i the diffraction is again specular, so the power loss shown in Fig (7) is still dominated by the top with the angular dependence of $\cos\theta_i$. Both features are characteristic of power losses from infinite planar surfaces.

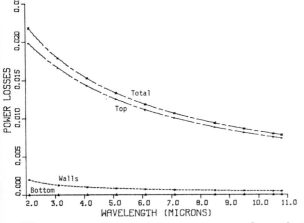

Figure 6. Power losses vs wavelength for TE polarization with $\left| E_0 \right|^2 = 1$.

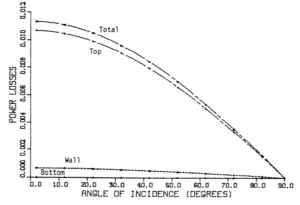

Figure 7. Power losses vs angle of incidence for TE polarization with $\left| E_0 \right|^2 = 1$.

The remaining figures depict losses for TM polarization. We consider first the losses as a function of wavelength as shown in Fig. (8). This graph has two peaks at 3 and 9μm corresponding to an increase in the n = -1 diffraction component as is shown in Fig. (2). Some of the features in the TE case are also present here; namely, when the n = 0 diffraction order is dominate (3.5μm - 6μm) the losses decrease and are dominated by the top which displays the $(\lambda\sigma(\lambda))^{-1/2}$ dependence as before. Expectedly, at the power loss spikes (3μm, 9μm) the major contributor is due to the walls with spiking also occurring in the bottom losses. The relative magnitudes of these contributions are difficult to intuite since they are strongly dependent on $B_m|(\lambda)|^2$ and this can be understood only after the solution for the A_n's and B_m's have been obtained.

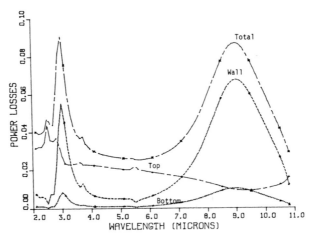

Figure 8. Power losses vs wavelength for TM polarization with $|E_0|^2 = 1$.

The final figure, Fig. (9), shows the TM losses as a function of θ_i for $\lambda = 9.28$μm. Returning to Fig. (3) we see that P_0 begins a rapid decrease at $\theta_i = 22°$. Accordingly, in Fig. (9), we see a local minimum at the same angle. Since the diffracted power at this angle is relatively small (30%) we see a larger difference between the bottom and wall losses than was present in Fig (7) when P_0 was about 96%. The only new feature is the cross-over near $\theta_i = 35°$ between the top and bottom losses. This is in the region where the bottom corner (a, 0, -b) becomes visible to the incident beam, and hence an increase in the field strengths on the bottom.

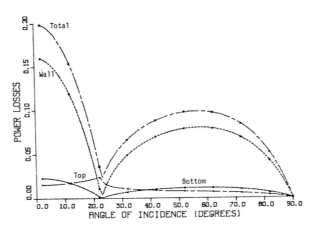

Figure 9. Power losses vs angle of incidence for TM polarization with $|E_0|^2 = 1$

Conclusion

The extension from infinite conductivity to power losses in a finite conducting grating, for $\sigma/\omega\varepsilon \gg 1$, has been straight forward albeit the rather extensive and tedious algebra.[1] An advantage of our approach has been the ability to predict separately the power losses on the top, walls, and bottom of the grating. This information, in conjunction with the groove energy density contour plots, gives a fairly complete theoretical picture of what is happening in the grooves.

For arbitrary $\sigma/\omega\varepsilon$ equation 1 is generalized to $\dfrac{dP}{da} = \dfrac{\mu_g \omega}{16\pi} \text{Re}\left[\bar{\delta}(1-i)\right]\left|H_{||}\right|^2$

where $\bar{\delta} = \alpha + i\beta$

$$\alpha = \pm\left[\frac{A}{2}\left(1 + (1 + B^2)^{1/2}\right)\right]^{1/2}$$

$$\beta = \frac{AB}{2\alpha}$$

with $A = \dfrac{2c^2}{4\pi\sigma\omega\mu_g} \dfrac{1}{\left(1 + \left(\frac{\omega\varepsilon}{4\pi\sigma}\right)^2\right)}$

$$B = \frac{\omega\varepsilon}{4\pi\sigma} \quad .$$

The analysis used to obtain these equations is a generalization of the work done in reference 2 to include the displacement current.

References

1. A. Gavrielides and P. Peterson, Appl. Opt. 18, 1468 (1979).

2. J. D. Jackson, Classical Electrodynamics (John Wiley and Sons, 1975).

Spectral shaped aperture component

Changhwi Chi
Hughes Aircraft Company, High Energy Laser Laboratory
Culver City, California 90230

Norton B. James III, Peter L. Misuinas
Department of the Air Force, Air Force Weapons Laboratory Office
Kirtland Air Force Base, Albuquerque, New Mexico 87117

Abstract

A dichroic (or multichroic) beam splitter operating in the high energy laser HEL environment is called a shared aperture component and is becoming an increasingly desirable component in the design of HEL systems. At present, there are four basic types: Buried Short Period (BSP) grating, Buried Long Period (BLP) grating, Dichroic Beam Splitter (DBS) and Component Interlaced (CI) grating. The BLP and CI gratings are new types of grating that have recently been proposed and are currently under development. The four basic shared aperture components and their characteristics, their design and fabrication issues, and present technology status are described in this paper.

Introduction

As high energy laser (HEL) systems begin to require high precision alignment and tracking and concentration of maximum energy density on the target objects, it becomes more desirable to use the shared aperture approach in which the HEL beam, the infrared (IR) beam returning from the target object, and the alignment and tracker beams all share the same optical path. A dichroic (or multichroic) beam splitter operating in a HEL environment, called a "shared aperture component", is used to inject the test beams into or split them from the HEL beam.

A spectral shared aperture concept offers several advantages. First, since all beams share the common optical path, the tracking, alignment, and beam control performance are immune from the jitter, misalignment, and aberrations occurring within the common optical path. Second, there is no need for a second set of alignment or pointing optics, reducing system size, weight, and complexity. Third, the IR radiation returning from the object can provide aberration information along the actual beam path, allowing detection of beam induced aberrations as well as those caused by the atmosphere.

The HEL beam is a single line or sometimes a multiline unpolarized beam with more than 1 µm spectral range. The IR radiation from the target object is incoherent and very low power and ranges over the 3 to 12 µm spectral band. The spectral band of 8 to 12 µm is called the long wave infrared (LWIR). The alignment and tracker beams originate from the lasers on board of moderate power level.

The performance requirements imposed on a shared aperture component are formidable. The HEL beam should be reflected with high efficiency and sampled with low efficiency, and the IR and alignment beams should be sampled or combined with high efficiency. The component should handle very high, medium and very low power beams simultaneously, while maintaining precise alignment among the beams and a good optical figure.

A government contract was awarded to Hughes Aircraft Company to develop shared aperture components. The material presented in this paper includes some of the results of the effort accomplished under the Cooled Spectral Shared Aperture Concept, or COSSAC, program.

Buried Short Period (BSP) Grating

The construction of a buried short period (BSP) grating is shown in Figure 1. The term "short period grating" refers to the ordinary diffraction grating, but this nomenclature is used to distinguish it from the "long period grating" — a new type of component discussed later in this paper. The grating is diamond ruled in gold and covered by a burying material, such as ZnSe, which transmits long wave infrared (LWIR) beams. Since this burying material follows the grating profile, its top surface is polished flat. A multilayer dichroic filter is then deposited on top of the burying layer. This filter is designed to reflect the laser beam but transmit the low power beam.

BSP grating samples have been fabricated and tested in the past. A BSP grating is, therefore, a practical and workable approach; it is used in a double rhomb configuration (Figure 2) to eliminate the spectral dispersion and shear, and obtain the desired LWIR wavefront form for interferometric sensors.

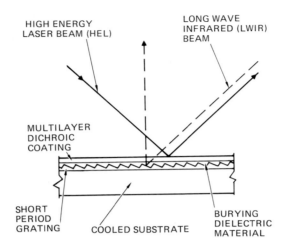

Figure 1. Buried short period (BSP) grating.

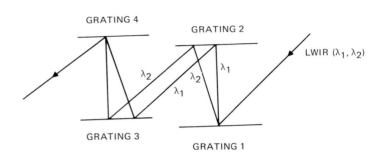

Figure 2. Double grating rhomb configuration which corrects spectral dispersion.

The diffraction grating is designed to diffract the LWIR beam into first order with high efficiency. To aid in this design, a computer code developed by D. Maystre[*] has been used. This code predicts diffraction efficiencies in the presence of dielectric layers covering the grating. The grating parameters for a typical design are listed below.

1. Incidence Angle   -   30 degrees
2. Grating Blaze Angle -   3.5 degrees
3. Grating Apex Angle -   90 degrees
4. Grating Period   -   34 μm

The grating period here was chosen to be large to keep grating dispersion low, which simplifies design of dispersion correcting optics.

Typical results for the first order diffraction efficiency are shown in Figures 3 and 4. The diffraction efficiency at a given wavelength varies considerably with burying layer thickness. The efficiency at 10 μm (for a grating which is blazed for 10 μm) is shown in Figure 3a as the burying layer thickness varies up to 3 μm (measured from the top of the grating groove). Both TE and TM polarization are shown. The variation of efficiency with burying layer thickness for various wavelengths is shown in Figure 3b. The results for TE and TM polarization have been averaged for each wavelength. The diffraction efficiency versus wavelength for much larger burying layer thicknesses (up to 50 μm) is shown in Figure 4.

For many applications, the quantity of interest is the average diffraction efficiency over the entire 8 to 12 μm band rather than the efficiency at any particular wavelength. When the efficiency is thus averaged for each polarization, the averaged quantity varies only a few percent over the range of thicknesses examined above. Thus in constructing a buried grating component, the exact thickness of the burying layer is not critical in achieving high efficiency. The actual value of the average efficiency is typically 75 to 90 percent, depending on grating design and on the transmission of the dichroic filter in the 8 to 12 μm wavelength band. Also, since the average efficiency was insensitive to small variations in groove apex angle and blaze angle, manufacturing tolerance was relaxed somewhat. This result is important for ruled gratings, since such gratings often depart significantly from the intended groove profile.

A dichroic reflector filter is an essential part of several of the concepts for shared aperture components. Such filters must be designed carefully to ensure their optimum performance. This care applies not only to the theoretical design of the filter but also to the fabrication, since survivability of dielectric coatings under intense laser illumination is still a problem with present day components. Breakdown mechanisms in such films are not well understood and are a topic of ongoing research.

---

[*]D. Maystre, "A New General Integral Theory for Dielectric Coated Gratings," J. Opt. Soc. of Am., Vol. 68, p. 490, 1978.

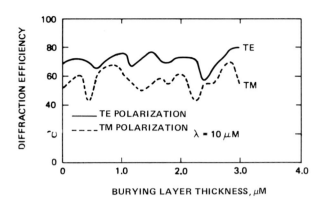

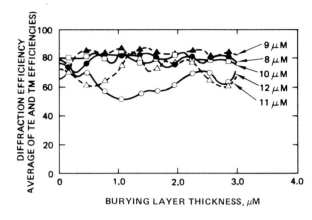

a. Burying layer thickness up to 3 μm

b. Burying layer thickness for various wavelengths

Figure 3. Diffraction efficiency versus burying layer thickness.

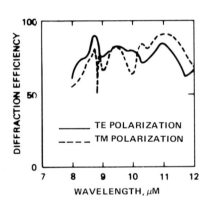

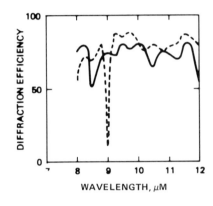

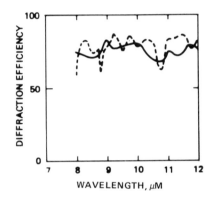

a. Burying layer thickness = 5 μm

b. Burying layer thickness = 25 μm

c. Burying layer thickness = 50 μm

Figure 4. Diffraction efficiency versus wavelength for various burying layer thicknesses.

As an example of the optical performance of a filter designed for a shared aperture component, Figure 5 shows the transmission of three similar filters that were prepared by several laboratories. These filters are composed of alternating layers of ThF_4 and ZnSe and were designed to be highly reflective in the DF laser wavelength band (3 to 4 μm) and to be transparent in the 8 to 12 μm wavelength band. The transmission curve for a simple filter composed of 10 pairs of alternating quarter wave layers of high index and low index material is shown in Figure 5a. The filter in Figure 5b, (designed by Optical Coating Laboratories, Inc. (OCLI)) shows considerable improvement over Figure 5a in the 8 to 12 μm wavelength transmission. This improvement was obtained by altering the thickness of some of the filter layers. The filter in Figure 5c, (designed by Perkin-Elmer) shows further improvement in the 8 to 12 μm transmission, giving greater than 98 percent transmission over most of this band. This filter is also a modified form of the simple quarter-wave stack filter.

High electric field strengths in the dichroic filter layers may adversely affect coating survivability. The field distribution which will occur in a particular filter design can be computed to determine whether excessive field strengths will exist. For reflector filter designs, the field strength decreases rapidly in successive layers of the filter, so that only the topmost layers are exposed to the high incident field strengths. The electric field distribution in the OCLI filter design, normalized to unity incident field strength is shown in Figure 6. In this example, the peak field strength occurring in any layer is approximately 1.2 times the incident field.

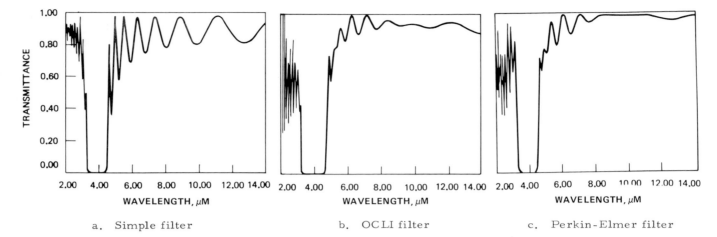

a.   Simple filter   b.   OCLI filter   c.   Perkin-Elmer filter

Figure 5.   Transmission of multilayer dichroic coating (incident angle = 0 degree).

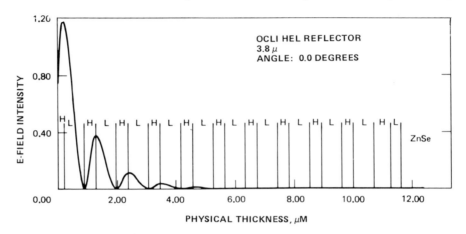

Figure 6.   Dichroic filter E-field distribution.

The packaging of BSP grating rhombs is a formidable task in terms of both the structural rigidity required (to hold four gratings aligned) and the available space to install the gratings (while remaining outside the HEL and LWIR beam path).   The skew grating configuration, shown in Figure 7, provides some relief in this respect.   In this configuration the grating is rotated so that the groove is not perpendicular (in Figure 1, it is parallel) to the incident beam, and the diffracted beams are in the direction out of the plane of incidence.   Consequently, the space outside the plane of incidence, not available in the ordinary grating configuration, becomes available for packaging.   The grating analysis shows that the diffraction efficiency remains approximately the same when the skew configuration is implemented.

Although the BSP grating is a workable approach, it has shortcomings also.   The BSP grating is sensitive to polarization and wavelength, has large spectral dispersion and a limited spectral range, requires four gratings, and the total efficiency is significantly reduced by the time the LWIR beam diffracts through the four gratings.

Present activities concerning BSP grating technology include design optimization of a short period grating and dichroic coating, alternative methods to double grating rhombs, and the investigation of improved coating materials and coating techniques.

Buried Long Period (BLP) Grating

The buried large period (BLP) grating, a new shared aperture component, is considered to be an important technique because it eliminates most of the shortcomings of BSP gratings.

In the BLP grating shown in Figure 8, the period is so large (on the order of a few millimeters) that the grating facet becomes a mirror for the wavelength of interest (10 μm in the present case), and consequently the diffraction effect is minimal.   The large period grating is placed on a cooled molybdenum substrate, buried with a dielectric material, and overcoated with a multilayer dichroic coating that reflects the high energy laser (HEL) beam and transmits the LWIR beam.

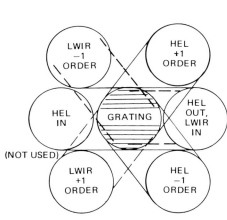

TOP VIEW SHOWING BEAM SEPARATION

Figure 7. Skew oriented grating configuration (incident HEL and LWIR beams are parallel to the grating grooves).

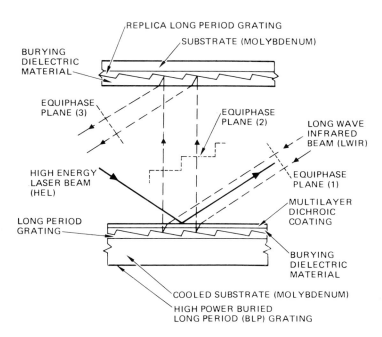

Figure 8. Buried long period (BLP) grating pair.

Since the grating facet is a mirror, the BLP grating exhibits the following desirable characteristics:

1. High efficiency for LWIR (mirror has good reflectivity)
2. Insensitive to the polarization
3. Insensitive to the wavelength
4. Works for a wide spectral range
5. Minimal spectral dispersion
6. Easy to align, as is the mirror alignment
7. Requires no double rhomb but a BLP grating pair (one is a low power replica).

A small spectral dispersion of the BLP grating originates from the wedge shaped burying material that acts like a prism but is compensated by the BLP grating replica.

When a LWIR plane wave is incident on one BLP grating, the reflected wavefront has the phase discontinuity, as shown in Figure 8, because the rays reflected by different facets have traveled different path lengths. The beams having a discontinuous wavefront are detrimental to the interferometric sensors, especially when the discontinuity is larger than the coherence length of the LWIR. To recover the original wavefront and recover the phase information, a second BLP grating (a replica of the first) is used. The replica BLP grating is a low power unit and does not need a cooled substrate.

The BLP grating pair thus constructed recovers the original wavefront by introducing the extra beam path length to the ray that traveled less in the first BLP grating, thus compensating for the optical path difference (OPD) introduced by the first BLP grating. Introduction of the replica BLP provides other advantages. The fabrication tolerance of the first BLP grating parameters, such as the blazed angle, is significantly reduced because the error in the first BLP grating is also present in the replica BLP but in the opposite sense, thereby compensating the effect of fabrication errors. The spectral dispersion introduced by the first BLP grating is also compensated by the replica and converted to a lateral spectral shear of minimal amount.

An error in the BLP grating parameters and an imperfect replica will result in a wavefront OPD coming out of the BLP pair. To determine the fabrication tolerance, ray tracing through the BLP grating pair was performed. The parameter notation and tolerance analysis results are shown in Figure 9 and Table 1, respectively. The intra-facet parameters (between facets within one BLP grating) have a large tolerance; thus, the fabrication tolerance for the first BLP grating is quite loose. The inter-element parameters (between the mating facets of two BLP gratings) have relatively tight tolerances, which means that a reasonably good replica is needed. The BLP grating pair fabrication is within the present precision machining capability.

The BLP grating operates in the regime where the diffraction effect is negligible. To determine the acceptable range of the BLP grating period, a diffraction calculation was performed. It is interesting to find the diffraction effect as the LWIR beam travels from the high power BLP grating to the replica BLP

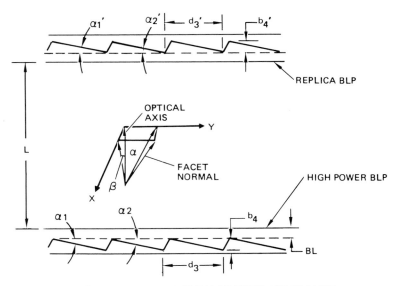

ANGLE β IS IN THE PLANE NORMAL TO THE PLANE OF PAPER

Figure 9.   BLP grating parameter notation.

Table 1.   Fabrication parameter tolerance errors for BLP grating pair.

a. Intra-facet Parameters (Between facets within one BLP grating)		
Parameter Ratio	Sensitivity Derivatives	Parameter Error that Produces 1.0 μm OPD
OPD/Δ	0.38606 CM/Rad	53.5 Arc Seconds
OPD/Δ	0.42880 CM/Rad	48.2 Arc Seconds
OPD/Δb	10^{-11} μM/μM	-
OPD/ΔL	10^{-15} μM/μM	-
OPD/ΔBL	10^{-15} μM/μM	-
OPD/Δd	10^{-13} μM/μM	-

Negligible (for OPD/Δb through OPD/Δd)

b. Inter-element Parameters (Between the mating facets of two BLP gratings)		
Parameter Ratio	Sensitivity Derivatives	Parameter Error that Products 1.0 μM OPD
OPD/Δ/Length*	5.127 μM/μM-Rad	4.0 Arc Seconds
OPD/Δ/Length**	4.781 μM/μM-Rad	0.22 Arc Seconds
OPD/Δb	4.781 μM/μM	0.21 μM
OPD/ΔL	10^{-15} μM/μM	-
OPD/ΔBL	10^{-15} μM/μM	-
OPD/Δd	0.38746 μM/μM	2.6 μM

Negligible (for OPD/ΔL and OPD/ΔBL)

*Assume facet width of 1.0 CM
**Assume facet length of 20 CM

grating.   The diffraction effect is determined by the Fresnel number N, $N = d^2/4\lambda L$ where d is the width of equivalent slit representing one facet as shown in Figure 10, L is the grating separation, λ is the wavelength. An equivalent slit is used in Figure 10 to represent the aperture size of the facets.

The LWIR energy diffracted outside the geometrical shadow and thus lost is shown in Figure 11.   It indicates that the diffraction loss is low when the Fresnel number is larger than 4.

When the LWIR wavefront coming from the BLP grating is focused, the resulting airy disks for different Fresnel numbers are shown in Figure 12.   The airy disk degradation is minimal even for the extreme case of N = 1; therefore, the image resolution and efficiency are not degraded by the diffraction effect.

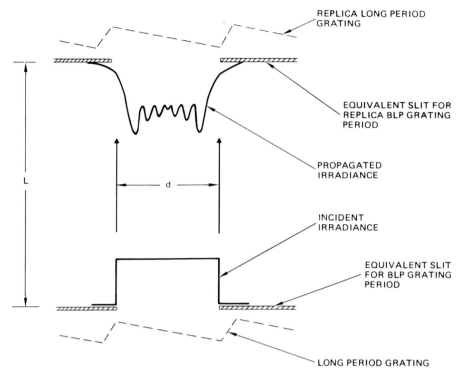

Figure 10. Equivalent geometry for diffraction analysis of single period of BLP grating.

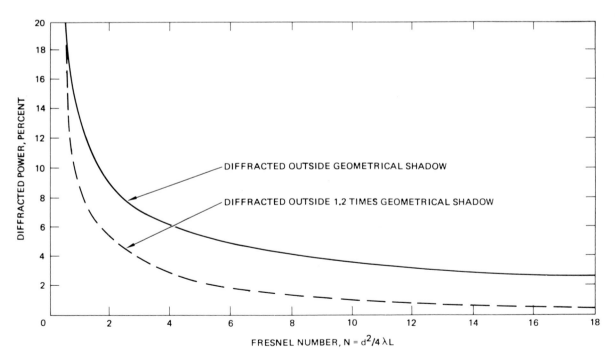

Figure 11. Power diffracted outside geometrical shadow of slit.

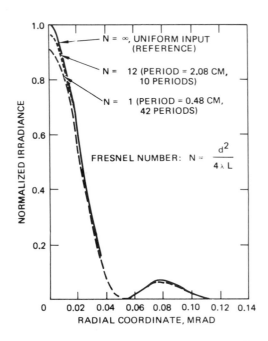

Figure 12. Point spread function of BLP grating pair.

When performing the diffraction calculation, the amplitude distribution of the wavefront emerging from the BLP grating pair contains a significant function due to edge effect. It should be remembered that the phase distribution, important parameter, is hardly degraded; therefore, a good airy disk is obtainable even with a Fresnel number of 1.

As the BLP grating period is increased, the average distance from the multilayer coating to the cooled substrate increases. Eventually the cooling effect disappears, and a failure will result. The failure mechanism of a BLP grating is not well understood at this time. To gain an insight into the failure mechanism, an investigation is in progress concerning the electromagnetic field behavior and the thermal/mechanical stress and deformation within the dielectric media.

Thermal/mechanical analysis within one period was performed on the model shown in Figure 13. It represents one period in which a uniform thermal irradiation is incident on the dichroic coating. The bottom of the surface is held at a temperature T_O, and the same boundary conditions are imposed at both the vertical boundaries since the BLP grating is periodic.

The thermal and structural analysis also predicted the stresses in the CVD ZnSe burying layer. The maximum values of the stresses (compression, shear and tension) for the principal axes are plotted as a function of period, d, in Figure 14. The breaking stress for CVD ZnSe is on the order of 5500 psi. For a safety factor of 5, the stresses must be kept at or below 1000 psi. This criterion is satisfied for a period of 1 cm where the compression is just over 700 psi.

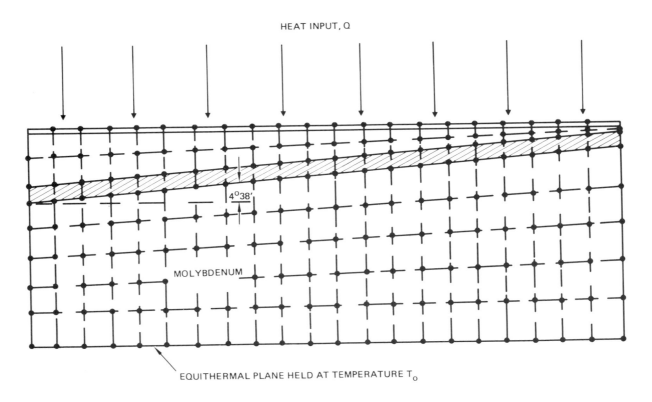

Figure 13. Model used for thermal/mechanical analysis within one BLP grating period.

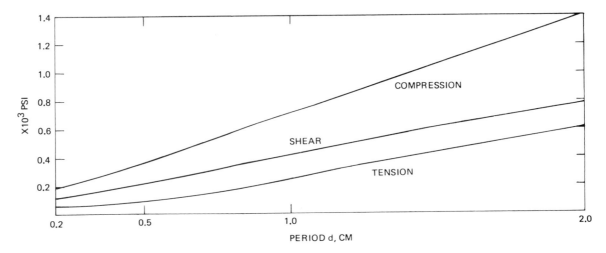

Figure 14. Maximum stresses, about principal axes, resulting from temperature gradients.

The locations of the maximum temperature gradient, distortion and stress are indicated in Figure 15. The thermal conductivity of CVD ZnSe is one-eighth that of molybdenum. The maximum temperature should occur in the vicinity of the thickest CVD ZnSe, i.e., vertically above the valley of the molybdenum facet (Point A). The minimum dichroic surface occurs in the vicinity of the region above the minimum vertical CVD ZnSe thickness, i.e., the peak of the molybdenum facet (point B). Points A and B are close together, and the maximum temperature gradient is located between these two points. Hence the maximum stresses are also located in this region (C). By lowering the facet slope in the peak-to-valley transition region, as indicated by the dotted lines, the thermal gradient and material stresses should be decreased. The thermal expansion of ZnSe is 1.7 times that of molybdenum, and the thicker the ZnSe the larger the vertical distortions. Hence the maximum out-of-plane distortions occur near point A or equivalent point A'. These out-of-plane distortions tend to force the ZnSe wedge (shaded region) in a horizontal direction, which results in maximum in-plane distortions at point D, or equivalent near point B, or equivalently to the left of C.

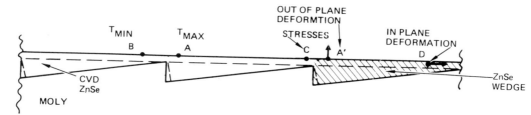

Figure 15. Locations of surface temperature gradient, distortion and stress extrema.

The present effort on the BLP grating is centered on the investigation of various fabrication techniques, which includes the diamong turning, sputtering, electro forming, replication process, room temperature coating and low absorption coating.

Dichroic Beam Splitter (DBS)

The DBS, Figure 16, is a more traditional approach to the shared aperture component. In both the rotating and stationary approach, the multilayer dichroic coating is designed to reflect the HEL beam and transmit the LWIR.

In the rotating DBS, the substrate is larger than the HEL beam (typically six times); therefore, the HEL beam energy absorbed at the dichroic coating is spread over an annular region larger than the HEL beam footprint and the effective density of the absorbed energy is reduced (by a factor of six). The DBS is cooled by gas jets impinging on the DBS within the stationary cooling shroud. For most applications, the DBS substrate rotational speed is not critical and can be as slow as seven hundred rpm.

The stationary DBS is a much simpler configuration in which the substrate is the same size as the HEL beam footprint. The coolant gas is blown by the jet nozzles located outside the path of HEL and LWIR beams. In the stationary DBS (unlike the rotating DBS), the absorbed HEL beam is not spread over a wider area and the cooling capacity is much less.

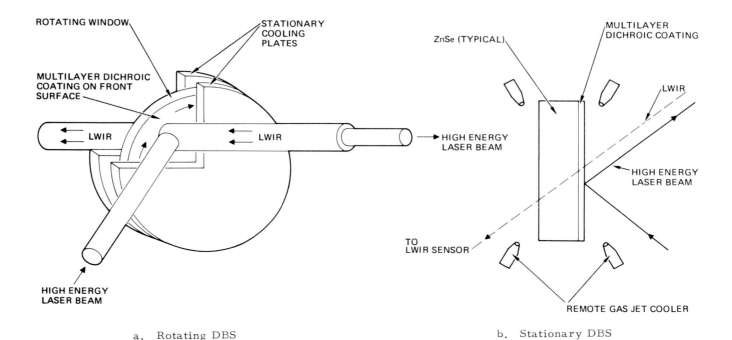

ROTATING WINDOW

STATIONARY COOLING PLATES

MULTILAYER DICHROIC COATING ON FRONT SURFACE

LWIR

LWIR

HIGH ENERGY LASER BEAM

HIGH ENERGY LASER BEAM

ZnSe (TYPICAL)

MULTILAYER DICHROIC COATING

LWIR

HIGH ENERGY LASER BEAM

TO LWIR SENSOR

REMOTE GAS JET COOLER

a. Rotating DBS

b. Stationary DBS

Figure 16. Dichroic beamsplitter (DBS).

The rotating DBS is the simplest aperture sharing concept if the cooling system is ignored. It is basically a large dichroic filter that provides good isolation between the HEL and LWIR, has high efficiency, is polarization insensitive, and exhibits no spectral dispersion. This concept does not require unconventional optical finishing, spectral dispersion correction elements, or replicated correction elements as do the other concepts. However, proper design of the cooling system is a significant task. Issues include the amount required and subsequent disposal of the coolant gas, fogging of substrate, and overcooling, which is as undesirable as the overheating. Also, the rotating mechanism adds complexity in that a finite start-up time will be required, and dynamic steering errors due to window wobble and substrate wedge may require correction.

The stationary DBS offers the advantage of the configuration simplicity. Since it has no moving parts, all the auxiliary subsystems attached to the rotating mechanism are eliminated. Its much smaller size impacts optic size, fabrication techniques, mounting structures and overall weight. Its usage is limited, however, by its low cooling capacity.

The material window, used to separate two environments and allow transmission of the HEL beam, employs a similar configuration. Significant differences exist between the material window and DBS. Since the material window is a refractive element (transparent plate) with an antireflection coating on the substrate for the HEL beam, the substrate deformation results in a minimal degradation of the HEL beam quality. The DBS, on the other hand, is a reflective element (mirror) for the HEL beam and a refractive element for the LWIR beam; therefore, the HEL beam is highly sensitive to the DBS substrate deformation. DBS requires more stringent optical figure control, a more careful cooling design, and an optical compensation system to correct for the substrate deformation.

A simplified one-dimensional analysis and heat balance equations for DBS plate are useful for determining the performance bounds of various DBS configurations. The parameters of interest include the mechanical stress that determines the failure threshold, the DBS substrate deformation that degrades HEL beam quality, and OPD for LWIR beam.

Mechanical stress is caused by gravity, centrifugal force (for rotating DBS), coolant gas impingement pressure, and the thermal gradient induced stress. The dominant source of the mechanical stress is the thermal gradient in a typical rotating DBS configuration, which consists of a ZnSe substrate, having 60 cm diameter, 3 cm thickness, rotation speed of 500 rpm, and the gas impingement cooling from the jet nozzle of a cooling shroud providing the heat transfer coefficient (H) of 0.1 watt/cm^2/°C.

The CVD processed ZnSe substrate has the flexural strength of 7500 psi when four-point loaded. Allowing a design safety factor equal to 5, the allowable stress becomes 1500 psi. The thermal stress factor of ZnSe is 80 psi/°C. Therefore, the allowable temperature rise is 20°C. Since the absorption of HEL beam occurs on one face of the ZnSe plate, thereby causing expansion of that face, the plate exhibits a bowing deformation and assumes a curvature. The OPD caused by the substrate deformation for the LWIR beam results from the variation of the index of refraction with the temperature and the increased path length caused by the material expansion. The nominal values of the material constants of substrate materials are given in Table 2.

Table 2. Substrate Material Constant

Material	Thermal exp Coefficient, α	Poisson Coefficient, ν	dn/dT	Index of refraction, n
ZnSe	$8.0 \times 10^{-6}/^{\circ}C$	0.3	$6 \times 10^{-5}/^{\circ}C$	2.4
CaF$_2$	$2.0 \times 10^{-5}/^{\circ}C$	0.3	$-5 \times 10^{-5}/^{\circ}C$	1.3
K Cl	$3.4 \times 10^{-5}/^{\circ}C$	0.3	$-3 \times 10^{-5}/^{\circ}C$	1.45

The effectiveness of the cooling technique is expressed by the thermal transfer coefficient (H) in watts of heat removed per area for each degree centigrade difference between the heated surface and the cooling gas temperature (watt/cm^2/$^{\circ}$C). In a rotating DBS, the cooling shroud containing the gas cooling jet nozzles provides H = 0.1 watt/cm^2/$^{\circ}$C. The cooling nozzle parameters are

Orifice diameter   = 0.08 cm
Orifice density   = 3/cm^2
Separation between nozzle substrate plate = 0.13 cm

In the stationary DBS, the separation between the substrate and the nozzle is greater. Assuming this separation to be 15 cm, the thermal transfer coefficient (H) is 0.006 watt/cm^2/$^{\circ}$C. The thermal transfer coefficient (H) of the free air convection cooling is 0.0006 watt/cm^2/$^{\circ}$C.

The required amount of coolant gas is also a parameter of interest. With N$_2$ as the coolant gas and assuming a complete thermal transfer, the required amount of gas is 4 lb/sec for removing 1 kW heat when the temperature difference between the heated surface and coolant gas is 1°C. In some HEL applications, the amount of coolant can be considerable.

It is useful to find the conditions necessary to reach the thermal equilibrium. In the rotating DBS, the following condition must hold: (Power Density Absorbed) (HEL Duty Cycle) = (Thermal Transfer Coefficient) (Temperature Difference Between Coolant Gas and Substrate).

Case 1. In a rotating DBS, the size of HEL is one sixth of the total substrate area, and the cooling occurs over three-fourths of the substrate area. For a cryogenic temperature cooling gas (80°K), a room temperature DBS plate (300°K), and a heat transfer coefficient of 0.1 watt/cm^2/$^{\circ}$C (the upper limit for current gas impingement cooling technology), the thermal equilibrium equation becomes Q(1/6) = (0.1) (3/4) (300-80), or the allowable thermal input (Q) = 100 watt/cm^2 maximum. Complications arise in the implementation of this technique to carefully match cooling to the DBS substrate heating so that deleterious performance effects due to over-cooling do not occur.

Case 2. When a cooling gas temperature of 0°C and the DBS substrate temperature of 22°C are assumed in the system of Case 1, Q = 10 watts/cm^2. This configuration has the advantage of using a chilled coolant instead of the cryogenic coolant.

Case 3. In a stationary DBS, the HEL beam size, substrate cooled are approximately the same. For a chilled coolant gas (0°C) a room temperature DBS substrate (25°C), and a heat transfer coefficient of 0.006 watt/cm^2/$^{\circ}$C (for coolant jet nozzles located outside the beam path), Q = (0.006)(25) = 0.15 watt/cm^2. Although cooling capacity is reduced significantly, the simplicity of this technique provides an overwhelming advantage for those situations where small cooling is adequate.

Present state-of-the-art techniques for high-power dichroic coating fabrication and cooling make the rotating DBS a feasible approach for the shared aperture component. The stationary DBS has none of the complexity of the rotating DBS, but has a reduced cooling capacity which makes this approach unsuitable for present HEL applications. The realization of stationary DBS approach awaits further technology development in the dichroic coating design and fabrication that provides better than 0.1 percent absorption currently possible and simpler and more efficient substrate cooling techniques.

Compound Interlaced (CI) Grating

Various types of CI grating are shown in Figure 17. This type of grating is much simpler to fabricate than other types of shared aperture components. It has a superior overall performance and has no survivability problems. However, a CI grating works only for cases where the HEL wavelength is much larger (2 to 3 times) than the test beam wavelength (alignment, tracking or target object radiation).

An example of the application is when the HEL beam is 10.6 μm, the alignment beam is 0.6238 μm, and target object radiation is 3 to 4 μm. Another example is when HEL is 3 to 4 μm (chemical lasers), alignment beam is 0.6238 μm, and active tracker is 1.06 μm.

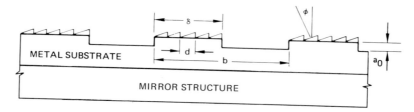

a. Partial compound interlaced (CI) grating

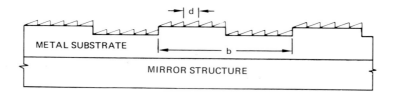

b. Full CI grating

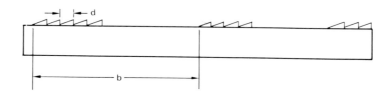

c. Degenerate CI grating

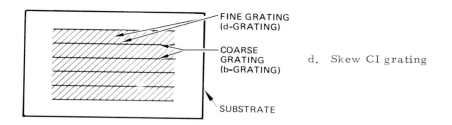

Figure 17. Compound interlaced (CI) grating types.

The CI grating has two gratings: one of period b, called a b-grating, for the diffracting long wavelength (i.e., 10.6 μm when the HEL is a CO_2 laser for instance), and one of period d, called a d-grating, for diffracting short wavelength (visible or near IR). When a 10.6 μm beam impinges on the CI grating, it does not "see" the d-grating because its period is much smaller than the wavelength. The d-grating, however, will contribute to the scatter of the 10.6 μm beam and also tilt it slightly toward the blazed direction.

When the short wavelength (visible) beam impinges on the CI grating, it will be diffracted both by the d-grating in the manner and direction desired and by the b-grating, which generates multiple ghost orders between the main d-grating orders. The amplitudes of the ghost orders can be minimized by properly designing the groove shape, period and duty cycle.

Different types of CI grating are shown in Figure 17. The d-grating diffraction efficiency for a full CI grating is larger than for a partial CI and degenerate CI grating. In a skew CI grating, the b-grating and d-grating are not parallel so that the diffractions of two gratings do not lie in the same plane (as shown in Figure 18).

A partial CI grating is obtained when the d-grating (fine) is ruled first and the b-grating (coarse) is ruled second. A full CI grating is obtained when the b-grating is ruled first and the d-grating is ruled second. A skew grating is obtained when the d-grating is ruled unparallel to the b-grating.

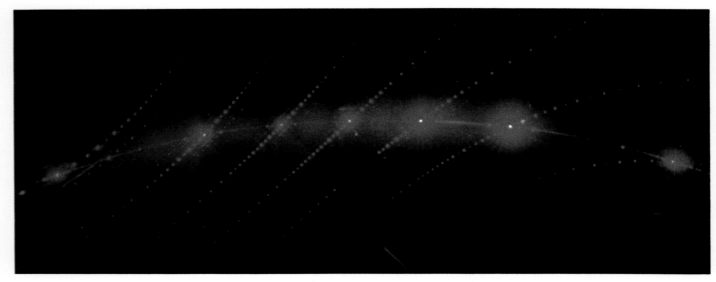

Figure 18. Diffraction pattern of a skew CI grating.

Two methods. for ruling gratings are presently in use: (1) "planning" and (2) "embossing". The planning method involves pushing the diamong tool across the substrate to "gouge" out the groove. This method actually removes material from the substrate. Embossing involves pulling the diamond across the surface to plow a groove. In this pulling method, the material is pushed aside, but none is removed. A number of small (1 by 1 inch) samples and one large prototype CI grating (4 by 5 inches) on a 9-inch cooled mirror substrate have been fabricated to verify feasibility and evaluate the performance. The embossing method was used in the prototype fabrication. A typical design for the prototype unit is shown in Figure 19, in which HEL beam is high efficiency reflected and also low efficiency diffracted by the shallow b-grating, and the visible (0. 6 μm) and mid IR (3-4 μm) are high efficiency diffracted.

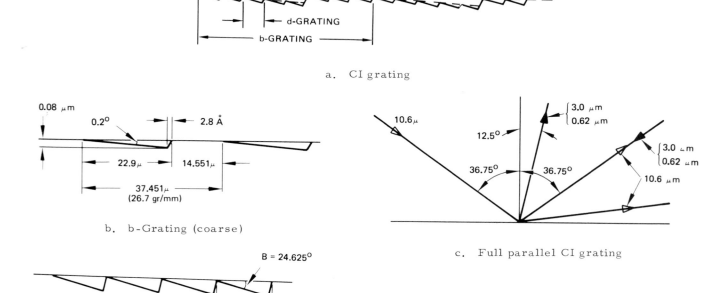

Figure 19. CI prototype grating design.

The scalar wave diffraction theory for the CI grating was formulated, assuming the substrate having infinite conductivity and shallow grooves (i.e., the multiple reflections within a groove are ignored). Figure 20 is a typical diffraction energy distribution between the major and minor orders. Figure 21 indicates that the CI grating diffraction efficiency is approximately 10 percent below the ordinary grating efficiency for the prototype CI grating design and that the experimental data with HeNe laser beam show a good agreement with theory.

Present activities include optimization of design using a more comprehensive theoretical analysis, fabrication and groove measurement techniques, coating effect, and large scale unit fabrication.

Acknowledgement

The authors wish to acknowledge the valuable assistance and contributions made by the following colleagues: Dr. T. Holcomb (BSP grating), Dr. J. Reeves (BLP grating), Mr. R. Loveridge (BLP grating), Mr. D. Sullivan (DBS), Dr. A. Lau (optical analysis), Dr. R. Holman (DBS).

The Buried Long Period (BLP) grating and the Compound Interlaced (CI) grating were originally invented by Dr. C. Chi, who also conducted the initial investigations.

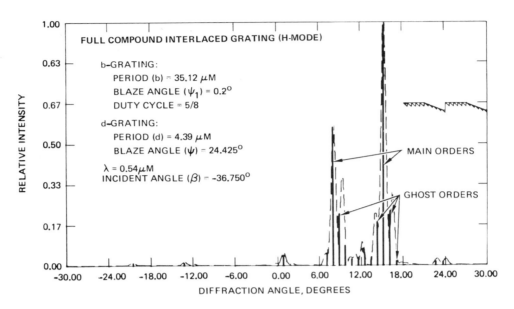

Figure 20. CI grating diffraction pattern (theoretical).

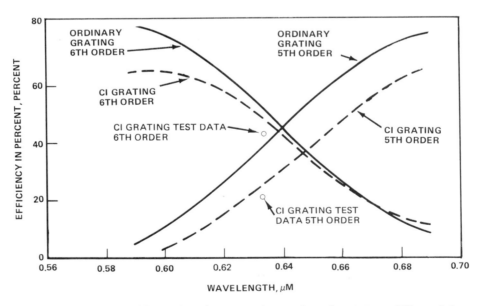

Figure 21. CI grating theory and test data (prototype TE mode).

Ruling on unusual surfaces

Edward Leibhardt
Diffraction Products, Inc., P.O. Box 645, Woodstock, Illinois 60098

The ruling of a diffraction grating on two unusual surfaces will be described, and the techniques used are dealt with in depth.

Introduction

Diffraction gratings can be ruled on many different surfaces. The usual substrates which come to mind are: cylindrical, toroidal, ellipsoidal, spherical and plane, to mention a few. For all these surfaces, a conventional ruling engine can be readily adapted with a minimum of tooling.

This paper deals with two gratings that cannot be ruled without elaborate tooling; both were ruled on cylindrical rods. The first one was ruled on a 1.5 mm-diameter sapphire rod with grooves running around its circumference. The second grating was ruled on a hardened stainless steel rod .450" in diameter with grooves along the axis of the cylinder.

The sapphire rod

For the first grating, a centerless ground sapphire rod 1.5 mm in diameter and 65 mm long, was lapped with a two-piece brass lap, which could be adjusted using push-pull screws. The lap was 3/4 the length of the rod, or approximately 48 mm long. It was grooved internally both radially and longitudinally, also a short lap, approximately 8 mm long, was made, to size the rod and insure that the diameter was uniform throughout the length of the rod.

The sapphire rod was held in the collet of a lathe and rotated at 500 R.P.M. and the two brass laps were used with 1/10 micron diamond paste until all irregularities were removed. A new lap was made from lead and now using diamond paste, the rod was polished until no pits could be seen under a 1000 power microscope.

In order to evaporate an aluminum film onto the finished rod, it was necessary to rotate the rod above the evaporating source at approximately 30 R.P.M.. This was accomplished by installing a rotating vacuum-sealed shaft centered in a stainless steel plate, which replaced one of the port windows of the vacuum chamber. On the outside, an electric motor was installed and on the inside was a collet-type chuck for holding the work.

The most difficult task was cleaning the rod and then inserting it in the collet inside the vacuum chamber. Aluminum film was deposited onto the rod; no undercoating was necessary to insure adherence of the film.

A headstock from a jeweler's lathe was obtained and the hardened steel, double-cone bearings were relapped using 1-micron diamond paste to the point at which there was no measurable play detectable, using an electronic gauging head reading to 1/100,000 inch. A 1.5 mm collet was fitted into the headstock and was ground to run dead-true by using diamond-plated grinding wheels. The headstock was fastened to a reference bar located parallel to the ways of the engine carriage via a magnetic base plate. In this way, the headstock, with its sapphire rod, could be removed and replaced during the setting up period. This made it convenient to check the grating under the microscope without difficulty. The rod was now ready for ruling 1000 gr/mm, blazed at 6238 Angstroms which were to be ruled around the circumference, similar to a screw having zero lead. The jeweler's headstock was now mounted on the ruling engine grating carriage and was driven via a rubber O-ring from the slave carriage using a phonograph motor which was quiet and vibration-free.

The actual ruling was accomplished similar to ruling a plane grating with the exception that the diamond-carriage remained stationary and its only function was to lift and lower the diamond on the rotating sapphire rod. A speed of 70 R.P.M. was selected so that several rotations of the rod were made for each groove and in this way no line of demarcation could be detected where the diamond started or was lifted from the rod. After a groove was ruled the diamond was lifted, the engine was indexed 1/1000 mm, and the process was repeated. Again, the most difficult part was removing the finished grating and devising a method for shipping to the ultimate user.

The stainless steel rod

The second grating was also ruled on a cylindrical substrate, except it was ruled with grooves parallel to the axis of the cylinder. Four such gratings were ruled with different blaze angles and spacings, the finest having the equivalent of 300 grooves/mm.

The stainless steel rods, .450" in diameter by 2 inches long, were turned on centers and left .005" oversize, then hardened and later ground. Two brass laps were fabricated similar to the ones used on the sapphire rod. The lapping was done with 1-micron diamond paste until the rods were perfectly uniform in diameter. The final finish was obtained using a pitch lap charged with 1/10 micron sapphire powder. For a lubricant, water was mixed with a small amount of mucilage, to prevent seizure of the pitch on the polished stainless steel surface. The quality of the polish was examined under a 1000 power microscope. The rods were aluminized using the same technique as on the sapphire rods.

The ruling of the rods was performed on a Gaertner Dividing Engine, with an accuracy of one second of arc. The rods were mounted in a special fixture which held them both vertically and centered on the axis of the engine. The regular scribing mechanism was removed and a diamond holder with dampened cross-springs was designed to rule in a vertical direction, ruling on the up-stroke. The indexing was accomplished on the down-stroke with the diamond clear of the cylinder. Four different spacings, with diamonds having 135° and 165° included symmetrical angular sides. To produce a symmetrical groove form on a conventional plane grating is very difficult, but on these gratings it was simply a matter of ruling 2 or 3 times around the cylindrical surface. The grooves coincided exactly every time the rod rotated during the ruling.

A special holder was constructed to enable checking the grating under the microscope and the symmetry of the grooves was checked with an interferometric microscope. One advantage these gratings have is that they are easier to pack and ship.

Measurements of diffraction grating efficiencies in the vacuum ultraviolet

W. R. Hunter

Naval Research Laboratory, Washington, D.C. 20375

Abstract

A technique to measure the efficiency of diffraction gratings in the vacuum ultraviolet spectral region at any angle of incidence has been developed and will be discussed briefly. The measurements show that concave gratings do not have a uniform efficiency over their surfaces, primarily because of the change in blaze angle across the surface. Examples will be shown of results obtained using both conventionally ruled, and holographically recorded, gratings. A short discussion of stray light will be given, followed by a description of the type of contamination found in laboratory instruments and its effect on the efficiency.

Introduction

The usefulness of a diffraction grating in the vacuum ultraviolet (VUV) is critically dependent on the grating efficiency and stray light. As the efficiency decreases, or the stray light increases, or both, the usefulness of the grating decreases. Extrapolation of grating performance from the visible to the VUV spectral region is not valid because the optical properties of coating materials can undergo large changes between these two regions. To be sure of the performance of a grating in the VUV, its efficiency and stray light must be measured at those wavelengths. It is not safe to assume that the efficiency of a grating remains constant after it is put in use. Contamination, especially in oil-pumped vacuum systems, can cause large changes in efficiency. Consequently if the speed of the system appears to decrease, the efficiency of the grating should be measured to verify the loss in speed.

Efficiency measurements

Because concave grating, unlike plane gratings, do not have a uniform efficiency across their surfaces, a technique similar to that of image dissection is necessary to obtain accurate efficiency measurements. By using a small radiation beam cross section so that only a small portion of the grating is illuminated at a time, an "efficiency map" can be built up showing the change in efficiency over the grating surface. A special apparatus, the Optical Grating Reflectance Evaluator (OGRE) has been developed that can measure grating efficiencies at almost any angle of incidence from normal to grazing. For details on its construction and use the reader should consult Ref. 1.

An example of an efficiency map is shown in Figure 1. The efficiency varies from almost zero at the left of the ruled area to about 43% at the right. Such a variation is due to the change in blaze across the ruled area [2]. Usually the ruling diamond is set for the correct blaze angle at the center of the area to be ruled, consequently, as the spherical surface moves under the diamond during ruling, the direction of the local normal of the sphere changes with respect to the groove facet normal which is fixed. This particular grating has a 4° 45' blaze angle and was ruled without changing the diamond orientation. For smaller blaze angles, or a smaller radius of curvature, the groove facet could be parallel to the spherical surface at some location, in which case the blaze would coincide with the zero order; an undesirable condition. Such a problem is avoided by ruling multipartite gratings; that is, by ruling the grating in small areas, or panels, and re-orienting the diamond to the proper blaze angle at the center of each panel. This procedure serves to increase the average efficiency across the surface but it does reduce the resolving power by the number of panels; i.e., if there are three panels, the resolving power is 1/3 of what it would be if there were only one panel the full width of the surface.

Figure 2 shows efficiency maps in the zero and strong first order, at 1216A, of a tripartite grating. The total width of the ruled area is 80 mm but it was ruled 80/3 mm at a time, in three panels, so that the groove facets would never be approximately parallel to the spherical surface at any location. The abrupt changes in efficiency at the junction between the panels are quite obvious. The center panel shows a smooth change in efficiency from one side to the other, almost monotonic, but the other panels do not show the same smooth change in efficiency. The different rates of change in efficiency in the outer panels indicate change in the effective blaze angle due to unknown causes, perhaps metal piling up on the edge of the ruling diamond to produce a non-sawtooth profile.

Concave holographic gratings with sinusoidal groove profiles generally have a quite uniform efficiency over their surfaces [3]. Variations can be kept within 5% and sometimes less. Blazed concave holographic gratings, however, show some change in blaze across their ruled area, but to a lesser extent than ruled gratings because the standing wave fronts used to record the grating are formed from spherical waves. Figure 3 [4] compares the ruled tripartite grating of Figure 2 with a concave, blazed, holographic grating. The zero and strong and weak 1st orders are shown. On either side of the zero orders are very sharp efficiency maxima which are actually specular reflections from the non-ruled portions of the gratings. Note that in the negative first order the center panel of the ruled grating has its large efficiency value to the left at 562A whereas, in Figure 2 (1216A), the large efficiency for the same order occurs to the right. This means that the blaze wavelength is between these two wavelengths. The nominal blaze wavelength given by the manufacturer is 900A. In principle, a well blazed grating has an efficiency maximum in the center of the panel at the blaze wavelength (2) and, as the measuring wavelength departs from blaze, the efficiency maximum will move off the panel leaving an almost monotonic change in efficiency across the grating, as in Figures 2 and 3.

To obtain the grating efficiency as a function of wavelength, the characteristic of importance to spectroscopists, the average value of efficiency across the surface is plotted against wavelength. Figure 4 (4) compares the efficiency of the conventional grating of Figure 2 and the concave blazed holographic grating shown in Figure 3, both measured at 10° angle of incidence. The two strong 1st and zero orders only are shown. The large maximum at about 550A for the 1st order of the holographic grating is caused, in part, by the reflectance spectrum of the gold coating. It is also partly due to the blaze being approximately at 550A. At the short wavelengths, the holographic grating has an unusually large efficiency, being about 0.7% at 300A and measurable almost to 200A. In contrast, the 1st order efficiency of the conventional grating is about 0.1% at 300A and, toward longer wavelengths, remains somewhat less than that of the holographic grating until, at about 700A, they are approaching the same value. To longer wavelengths, the conventional grating has a larger efficiency. Both zero orders are small at the short wavelengths but that of the holographic grating begins to increase just short of 500A and continues to increase toward longer wavelengths, a trend not expected in blazed gratings and, at present, not understood.

If the ratio of grating efficiency/coating reflectance is plotted, the effect of the coating reflectance is eliminated and one has the "groove efficiency" which should reach a maximum at the blaze wavelength. However, since the groove efficiency changes over the grating surface, primarily because of the blaze change, there will not necessarily be a clear indication of the location of the blaze, as would be the case for a plane grating. Thus Figure 4 does not show the effect of blaze clearly for either of the two gratings.

Figure 5 (5) shows efficiency maps at 192A of a conventional ruled grating and a holographic grating with sinusoidal groove profiles used in grazing incidence. This ruled grating is monopartite. The average value of the first order is about 18% with a maximum value of about 24%. The average value of the second order is about a factor of ten less. In contrast, the 1st and 2nd orders of the holographic grating are about the same at this wavelength. They, and the zero order, are non-uniform to an extent surprising for sinusoidal groove profiles. $T_a$lystep measurements (6), however, showed that the amplitude of the grooves decreased by a factor of two from the left to the right of the map which accounts for the non-uniform efficiency.

Figure 6 (5) shows the efficiency vs wavelength of the holographic grating of Figure 5. The two sets of data points delineating the shaded areas represent measurements made at the same angle of incidence (80°) on either side of the grating normal. The reason for an extra set of measurements was to determine whether or not the groove profiles were symmetrical, and the conclusion is that they are. Of all the orders shown, the zero order is the most efficient, a characteristic common to most grazing incidence mountings. Only the first three orders are shown but the fourth and fifth could also be measured.

Figure 7 (5) shows the efficiency vs wavelength of the conventional grating of Figure 5, and compares it with that of the holographic grating of Figure 6. For this grating the blaze can be clearly seen at about 200A because at large angles of incidence the reflectance spectrum of the coating material (gold) is fairly uniform as a function of wavelength, and because the grating is monopartite. The 2nd order, where shown, is always less than 10% of the 1st order and is increasing toward shorter wavelengths as would be expected.

Stray light

Stray light from grating surfaces may have two distinct forms. First, general stray light is scattered in all directions from the grating. The intensity of this type of stray light is usually small enough to be negligible and will not be discussed further. The second form is called focused stray light (FSL), or sometimes grass, because it appears as a background continuum in spectroscopes and is confined to the plane of dispersion. Figure 8 (7) is a photograph of the VUV emission spectrum of helium obtained with two gratings, grating 1 having excessive FSL and grating 2 with a much lower level of FSL. These spectra were obtained using a slitless spectrograph so that the grating imaged a resolution mask on the film. An aluminum filter (8) was used to remove all wavelengths longer than 800A and its absorption edge at 170A is responsible for the sharp cut-off of the FSL below 256A in grating 1. Therefore all the FSL recorded in these photographs is VUV stray light and not visible stray light from the zero order.

Stray light intensities can be measured and an example of such a measurement is shown in Figure 9 (1). This measurement was made at near-normal incidence with the detector and source approximately on the Rowland circle. Note that the intensity scale is in intensity units/angstrom to allow for the width of the aperture over the detector.

In an attempt to compare FSL in the VUV from a conventional grating, that of Figure 2, and a holographic grating with sinusoidal groove profiles, Namioka and Hunter (9) used a scanning monochromator designed for 1 meter radius-of-curvature gratings, and recorded the HeI spectrum from a dc glow discharge in helium for each grating with a Galileo channeltron as a photon counter. The results are shown in Figure 10. In the vicinity of the very intense 584A line, the conventional grating appears to have slightly more FSL than the holographic grating. At other wavelengths the FSL is about the same for either grating. These results appear to contradict the claims of manufacturers of holographic gratings, that holographic gratings have much less FSL than conventional gratings. The usual demonstration employed by the manufacturers to assert their claim is to illuminate a small portion of a conventional and a holographic grating with visible laser light to demonstrate the large amount of grass produced by the conventional grating while practically none is produced by the holographic grating. One must bear in mind, however, that the glow discharge in He used for the Namioka-Hunter experiment was not a coherent source. If a coherent VUV source is used, then the holographic grating may have a definite superiority over the conventional grating in having less FSL, but for the usual laboratory source, including synchrotron radiation, a good conventional grating should be equivalent to a holographic grating in its FSL level.

The effect of coherence of the source on the FSL of a conventional grating is shown in Figure 11 (10) which is the spectrum of a Xe laser tuned to about 1720A. The grating had 600 grooves/mm, a radius-of-curvature of 1 meter, and was coated with Pt. The zero orders are the bright lines at the center, the remainder of each spectrum is grass. It is interesting to note that the grass intensity peaks in the region of the blaze wavelength, 900A for this grating, rather than near the zero order. Also that the grass intensity is greater on the side of the strong 1st order than on the side of the weak 1st order. No satisfactory explanation for this phenomena has been advanced.

Large stray light levels cause complications in the measurement of specular reflectances or transmittances. Figure 12 (3) compares some early (1961) normal incidence reflectance measurements of an Al + LiF mirror. The measurements were made with two gratings, one had a high FSL level and was coated with Al + MgF_2, and the other was a gold grating with a low FSL level. The source was a glow discharge in hydrogen. When the grating with the high FSL level was used, the uppermost curve was obtained (diamond data points) which gave a reflectance value close to 40% at 920A. When the gold grating was used the true reflectance values were obtained (circular data points), that at 920A being close to zero. The square data points show the reflectance obtained using the grating with the high FSL level after correcting the data for the stray light. This correction can be done by measuring the direct and reflected intensities at a spectral line and then measuring another set of direct and reflected intensities at an intensity minimum between lines. The latter set of measurements are subtracted from the corresponding values in the previous set before the ratio is calculated. If the source of VUV radiation is a continuum, such as a synchrotron or a quasi-continuum of very closely spaced emission lines, this type of correction is not possible.

Contamination

Most VUV monochromators use oil pumps to achieve the vacuum required for operation. Although the system may have a liquid nitrogen trap, there is always some backstreaming of oil vapor into the monochromator that gets onto the grating and may be polymerized by the VUV photons or by charged particles from the source entering the monochromator through the entrance slit. It appears to be impossible to predict just how the grating performance will be affected, but specific cases may be worth recounting as a guide to others.

In March 1970, a 1-m radius-of-curvature grating with 600 grooves/mm was installed in a MacPherson model 225 scanning monochromator in the author's laboratory. The grating was ruled in a continuous ruling over a width of 50 mm. No overcoating was deposited on the aluminum surface supplied by the manufacturer (Bausch & Lomb, Inc.) before the grating was put into use. The nominal blaze of the grating was 2750A in the Littrow mount.

The monochromator is used for measuring the reflectances of surfaces, transmittances, and grating efficiencies. For such applications a beam of small divergence is most useful, therefore the monochromator is equipped with an adjustable aperture (11) located about 1/3 of the distance from the entrance slit to the grating. It controls the angular divergence of the exit beam by limiting the area of the grating surface that is illuminated. For the purposes described above, the illuminated area is usually approximately 1 cm high (parallel to the grooves) by 5 mm wide or less.

A windowless dc glow discharge is used for a light source, and hydrogen, neon, and helium are the gases used. A Freon-refrigerated trap is used above the oil diffusion pump and is run continuously as long as the pump is in operation. Convoil 20 is the pumping fluid. No trap is used in the roughing or backing lines. There have been no accidents that would cause a sudden backstreaming of oil vapor in large quantities. What contamination was found represented a gradual accumulation over 8 years - from the time of installation until July 1978 when the grating was removed and efficiency measurements were made.

Visual inspection of the contaminated grating revealed a peculiar marking at its center, which is the area exposed by the aperture to the entrance slit, that was difficult to see unless the illumination and viewing angles were carefully chosen. Figure 13 is a photograph of the surface in white light. Colors could not be seen, so if the marking is a layer, it is not thick enough for interference effects at visible wavelengths. Also, no scattering from the mark could be seen with the unaided eye.

Figures 14 and 15 (12) show efficiency maps, in the strong first order, at different wavelengths from 1216A to 4000A for the contaminated grating (solid lines) and a clean sister grating from the same master (dashed lines). The angle of incidence was 10°. Both gratings presumably would have the same efficiency when clean. Although the wavelengths longer than 2000A are not in the VUV spectral region, the measurements were extended to 4000A as a matter of interest.

At 1216A and 4000A the two gratings have practically the same efficiency distribution. At intermediate wavelengths, however, the contaminating layer gives rise to considerable differences in efficiency. At 1434A, for example, the contaminated grating has about 1/3 to 1/2 the efficiency of the clean grating and has a slight concavity at the center where the grating was illuminated. At 1580A, the contaminated grating has almost zero efficiency outside the illuminated area and a peak efficiency of about 7.5% within that area. In contrast, the clean grating shows a uniform change in efficiency across the surface from 45% on the left to 6% on the right, caused by the change in blaze. Note that for 1580A, the ordinate for the dashed line is from 0 to 100% (left) and that for the solid line is from 0 to 10% (right). As the wavelength increases, the efficiency of the central exposed portion of the contaminated grating appears to oscillate with respect to the efficiency of the unexposed portion. The general depression of efficiency between 1216A and 4000A shows that the contaminant is an absorbing material, and the oscillations of the central exposed area from 2000A to longer wavelengths show that interference effects must also play a part.

Gillette and Kenyon (13) have shown that polymerized pump oil, if it is carbon-based, can be removed by exposing the contaminated object in an rf-excited oxygen discharge. Such equipment is not available in the author's laboratory. Finally, the contaminated grating and the clean grating were overcoated with gold and their efficiency maps were compared. Within the limits of error of the measurements, there were no differences. Thus the efficiency of a grating contaminated by pump oil can be restored by overcoating it. Eventually the accumulated coatings, polymerized oil and restoring overcoating, become rough enough so that scatter renders the grating useless.

Conclusions

This brief exposition on grating efficiencies, stray light, and contamination gives results from just a few gratings. However, many other measurements made at the NRL provide sufficient information to justify the following conclusions.
1) Holographic gratings with sinusoidal groove profiles have a very uniform efficiency distribution over their surfaces, and a quite uniform efficiency as a function of wavelength. Their FSL level, as compared with that of a good conventional grating, is about equivalent with incoherent sources but much better for coherent sources. They have the disadvantage, however, of supporting many orders all comparable in magnitude so that order sorting with such a grating may be quite difficult.
2) Conventional gratings are usually more efficient that the holographic gratings mentioned in 1) in the region of the blaze wavelength. In addition orders other than the blaze orders are fairly strongly suppressed, making order sorting less of a problem. They have the disadvantage of a pronounced non-uniform efficiency across their surfaces, sometimes enough so that the resolving power may be affected. They may also have a restricted spectral range in that the groove efficiency decreases as the wavelength departs from the blaze wavelength. Sometimes this restricted wavelength range can be enlarged by proper

choice of the coating material. For coherent sources their FSL is larger than that of holographic gratings.

3) Blazed, concave, holographic gratings retain most of the characteristics of conventional gratings and, in addition, have a much more uniform efficiency across the surface. One would expect the FSL to be comparable with that of holographic gratings with sinusoidal groove profiles.

4) Generally contamination causes the VUV efficiency of gratings to be reduced. The loss in efficiency can be quite large even though the contamination cannot be observed visually. Sometimes overcoating the grating can restore the lost efficiency.

References

1. D. J. Michels, T. L. Mikes & W. R. Hunter, Appl. Optics 13, 1223 (1974), and W. R. Hunter & D. K. Prinz, Appl. Optics 16, 3171 (1977).

2. D. J. Michels, J. Opt. Soc. Am. 64, 662 (1974).

3. W. R. Hunter, in "Proceedings of the Fourth International Conference on Vacuum Ultraviolet Radiation Physics", (Eds. E-E. Koch, R. Haensel & C. Kunz), p. 683, Pergamon-Vieweg, Hamburg, 1974.

4. W. R. Hunter, M. C. Hutley, P. R. Stuart, D. Rudolph, & G. Schmahl, presented at the Fifth International Conference on Vacuum Ultraviolet Radiation Physics, Montpellier, France, September 1977.

5. W. R. Hunter, A. J. Caruso, & J. G. Timothy, presented at the Fifth International Conference on Vacuum Ultraviolet Radiation Physics, Montpellier, France, September, 1977.

6. The author is indebted to R. J. Speer and members of his group at the Imperial College of Science and Technology, London, for this measurement.

7. J-D. F. Bartoe, Naval Research Laboratory, Washington, D. C., private communication.

8. W. R. Hunter, in "Physics of Thin Films", (Eds. G. Hass, M. H. Francombe, & R. W. Hoffman), Vol. 7, p. 43, Academic Press, New York, 1973.

9. T. Namioka and W. R. Hunter, Optics Comm. 8, 229 (1973).

10. A. Theocarous, Optics Group, Imperial College of Science and Technology, London, private communication.

11. W. R. Hunter & R. K. Chaimson, Appl. Optics 13, 2913 (1974).

12. W. R. Hunter & D. W. Angel, Appl. Optics 18, 3506 (1979).

13. R. B. Gillette & B. Kenyon, Appl. Optics 10, 545 (1971).

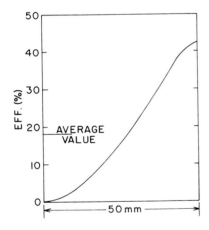

Fig. 1 Efficiency map for 1608A. 2nd order, 1 meter radius-of-curvature, 600 grooves/mm, coating of Al + MgF$_2$, blaze angle of 4° 45, angle of incidence is 0°.

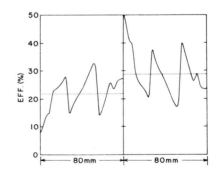

Fig. 2 Efficiency maps at 1216A in the zero (right) and strong 1st (left) orders. 1 meter radius-of-curvature, 1200 grooves/mm, Pt coating, blaze angle 2° 45', angle of incidence is 10°.

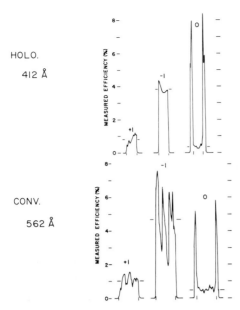

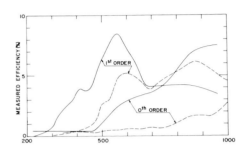

Fig. 3 Efficiency maps of a concave, blazed, holographic grating (top) at 412A, and of the conventional grating of Fig. 2 (bottom) at 562A. Both gratings are coated with gold.

Fig. 4 A comparison of the strong 1st and zero orders of the gratings of Fig. 3 as a function of wavelength.

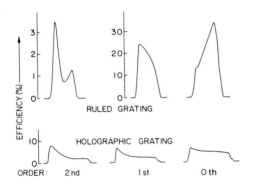

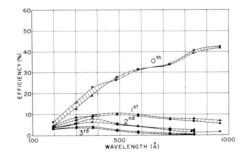

Fig. 5 Efficiency maps at 192A of a conventional (ruled) grating and a holographic grating with sinusoidal groove profiles. The angle of incidence is 80° and the coatings are gold.

Fig. 6 The efficiency in the zero, 1st, 2nd, and 3rd orders as a function of wavelength for the holographic grating of Fig. 5.

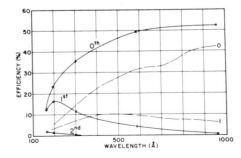

Fig. 7 The efficiency in the zero, 1st, and 2nd orders of the conventional grating of Fig. 5. The dashed lines show the zero and 1st orders of the holographic grating for comparison.

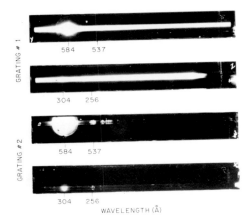

Fig. 8 Focused stray light (FSL) from two gratings. Grating 1 has a high FSL level while that of grating 2 was acceptable for VUV solar spectroscopy.

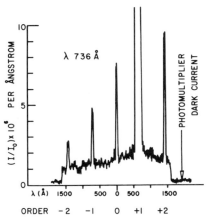

Fig. 9 Example of FSL measurements at 736A

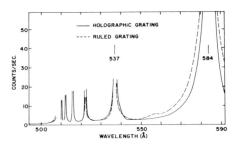

Fig. 10 Comparison of FSL from a holographic grating with sinusoidal groove profiles and from a conventional grating.

Fig. 11 FSL from a conventional grating illuminated by a coherent VUV source, a Xe laser tuned to about 1720A.

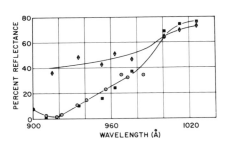

Fig. 12 Effect of FSL on reflectance measurements. The upper curve was obtained with a grating with a high FSL. The lower curve shows the true reflectance values.

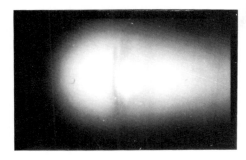

Figure 13 Photograph of a concave grating surface in white light showing contamination due to diffusion pump oil. The rulings are parallel to the short dimension.

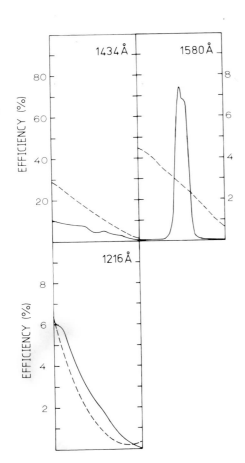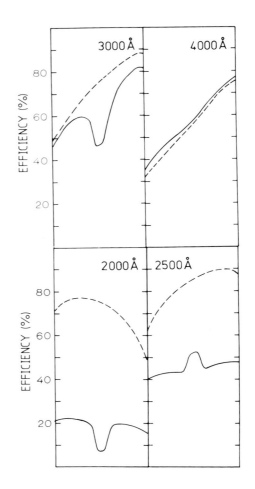

Fig. 14 & 15 Efficiency maps of a concave grating surface at different wavelengths from 1216A to 4000A showing the effect of oil contamination on efficiency. Solid lines are data for the contaminated grating, and dashed lines show the efficiency of a clean sister grating for comparison. The abscissa represents the distance across the ruled area, 50 mm. Note that for 1580A, the ordinate for the dashed line is from 0 to 100% (left), and that for the solid line is from 0 to 10% (right).

Buried long period grating for laser applications

C. H. Chi, J. M. Reeves, L. N. Au, K. D. Price
Hughes Aircraft Company, Culver City, California 90230

P. L. Misuinas
Air Force Weapons Laboratory, Kirtland Air Force Base, New Mexico 87116

Abstract

One type of aperture-sharing device, the buried long period grating (BLPG), is described in this paper. The BLPG functions as a buried segmented mirror whose primary function is to spatially redirect, by reflections, an antiparallel laser beam and its corresponding low power broad band LWIR (long wavelength infrared) return beam. The aperture sharing unit consists of a pair of BLPGs, the second BLPG being used to restore spatial coherence across the LWIR wavefront. Other system functions of the BLPG such as autoalignment and beam sampling are discussed.

The optical performance of the device is discussed in terms of energy losses due to material dispersion diffraction loss, degradation of resolution resulting from diffraction, segment fabrication tolerances, and thermally induced structural deformations due to laser beam heating. Both transient and steady-state thermal and structural analysis were performed on the device. One result from the analysis was the value of the burying dielectric thickness above the segment tips that minimized the stress within the device under laser beam irradiation.

Fabrication consisted of separately tooling segmented surfaces in the cooled substrate and in the burying dielectric (CVD ZnSe and ZnS) followed by application of a metallic coating. These two segmented surfaces were joined with an appropriate bonding agent. The critical fabrication step is to use the bonding materials that have the following properties: (1) approaches full cure during fabrication, (2) minimum of outgassing with temperature and with time, (3) stable with aging (minimize surface distortion), (4) high thermal conductivity, and (5) flexible bond line to absorb thermal expansion mismatch between the dissimilar substrate materials. The exposed dielectric surface is polished flat, vacuum baked, and dichoric coated to reflect a laser beam and transmit a LWIR beam. Some top surface distortion is introduced during vacuum baking and during dichroic coating. Present work is towards reducing these fabrication temperature deformations by accurately mixing the bonding components to give different compositions. Preliminary work indicates good device performance in a laser environment.

Introduction

A key role of aperture-sharing systems is to spatially separate two colinear beams. One of the beams is the outgoing high-intensity monochromatic laser beam and the other is the return long wavelength low-intensity wide bandwidth long wave infrared (LWIR) beam originating from the hot spot created by the focused laser beam. When the aperture-sharing element separates these two beams, it must not introduce appreciable wavefront distortions into the beams nor degrade the intensity of the beams. A candidate device to perform this function is a buried long period (BLP) grating.

A high power buried segmented mirror consists of a burying dielectric (approximately 1 mm thickness) covering metallic reflecting segments that are inclined with respect to the dielectric surface. The top exposed dielectric surface is polished flat and a multi-layer dichroic reflective coating deposited over it. The flat dichroic stack reflects the laser beam and transmits the LWIR beam which is then reflected by the buried inclined segments. The LWIR beam is once more transmitted by the dichroic stack. The laser beam and the LWIR beam are now propagating in different directions. The LWIR beam has an incoherent wavefront because of the step phases introduced by reflections from adjacent buried segments. One way to compensate these step phases across the LWIR wavefront is to reflect the LWIR beam from an aligned and matched low-power buried antireflection (AR) coated segmented mirror as shown in Figure 1. The step phases will be eliminated from the twice reflected LWIR wavefront to the extent that the reflecting segments of the two (high- and low-power) buried substrates are matched and aligned.

The advantages of using the BLPG are

1. High efficiency for LWIR (mirror has good reflectivity)
2. Insensitivity to polarization
3. Insensitivity to wavelength

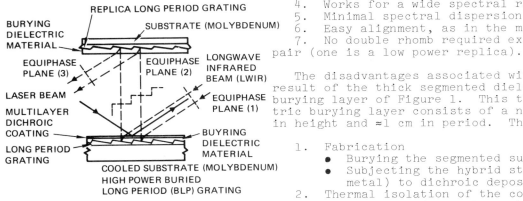

Figure 1. Buried long period (BLP) grating pair.

4. Works for a wide spectral range
5. Minimal spectral dispersion
6. Easy alignment, as in the mirror alignment
7. No double rhomb required except a BLP grating pair (one is a low power replica).

The disadvantages associated with the BLPG are a result of the thick segmented dielectric (ZnSe or ZnS) burying layer of Figure 1. This thick segmented dielectric burying layer consists of a number of wedges ≈1 mm in height and ≈1 cm in period. The problems are

1. Fabrication
   • Burying the segmented substrate
   • Subjecting the hybrid structure (dielectric/metal) to dichroic deposition temperatures
2. Thermal isolation of the cooled substrate
   • Transient due to laser beam turn on may subject the dichroic stack to damage
   • Laser beam induced operational temperatures degrade the optical performance of the transmission media and the reflecting surfaces

These disadvantages are thought to be solvable by the present technology.

System functions

An example BLP grating pair showing the laser beam and LWIR beams is given in Figure 2. As shown, the primary function of the BLPG pair is to spatially separate the antiparallel laser and LWIR beams. The redirected LWIR beam is detected by the sensor packages for wavefront correction and tracking. The high power substrate also samples the laser beam for wavefront corrections and alignment and also samples a previously injected HeNe beam (#1) for autoalignment purposes, as shown in Figure 3a. The BLP grating pair injects a HeNe (#2) beam antiparallel to the LWIR beam for further alignment purposes as shown in Figure 3b. The index dispersion of the burying material will only laterally shift the HeNe beam with respect to the LWIR beam and because of the matched segment pairs the two beams will remain antiparallel.

Both the laser beam and the HeNe No. 1 beam are reflected at a 23 degree angle by the dichroic stack (see Figure 3a). Approximately 0.05 percent of the laser beam is transmitted by the dichroic stack, propagated through a maximum of 1 mm chemical vapor deposition ZnSe (99 percent), reflected by the metallic segments (99 percent) and then another 0.05 percent is transmitted from within the ZnSe back through the dichroic and into space. Because of the index (2.4339) of CVD ZnSe at 3.8 μm the transmitted laser sample beam makes an angle of 50.9 degrees with the normal to the flat dichroic surface. The sampling efficiency is governed by the square of the dichroic transmission at 3.8 μm and is on the order of 10^{-5} percent. In excess of 99 percent of the laser radiation incident on the dichroic from within the ZnSe is again reflected, propagated through ZnSe, reflected by the segments and is again incident on the dichroic from within the ZnSe layer. If this multiple bounce beam were to be transmitted by the dichroic its intensity would be approximately the same as that of the first sampled beam (≈10^{-5} percent). For CVD ZnSe at 3.8 μm, total internal reflection occurs at an angle of 24.3 degrees. This laser beam, reflected twice by the segment, is incident on the dichroic from within the ZnSe (the second time) at an angle of 27.9 degrees and is totally reflected within the ZnSe burying layer. Hence, as shown in Figure 4, only the laser beam reflected once by the buried segment contributes to the sample and with 10^{-5} percent efficiency.

Energy losses

The high-power and low-power substrates are positioned as close together as possible to minimize space requirements and to minimize diffraction effects of the LWIR sections reflected and clipped by the different segment pairs. The minimum distance, L, between the substrates is determined so that the reflected laser beam clears the low-power substrate, as shown in Figure 5. The index dispersion of CVD ZnSe (n = 2.4173 at 8 μm and n = 2.3930 at 12 μm) causes the refracted-reflected-refracted broadband LWIR beam to be spread at the low-power substrate as shown in Figure 6. The lateral spread of the beam causes energy to be lost from the edges of each LWIR section. The approximate lateral spread is $W(\Delta n)/2n \approx$ 0.1 cm, for a beamwidth W of 20 cm. This lateral spread causes a small portion of the energy incident on the low-power substrate to be reflected by the segment adjacent to the corresponding segment of the pair. This radiation is out of phase with the coherent LWIR wavefront and is considered as lost energy. Assuming that the LWIR energy is uniformly spread among the wavelengths and across the beam width, approximately 5 percent of the

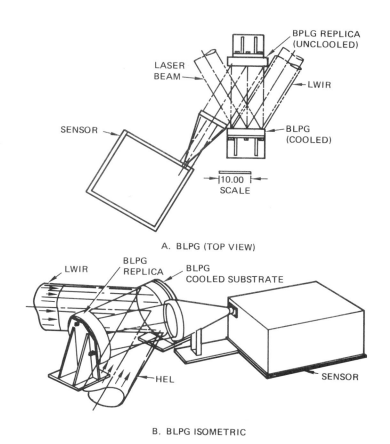

A. BLPG (TOP VIEW)

B. BLPG ISOMETRIC

Figure 2. Laser and LWIR beams for the BLPG.

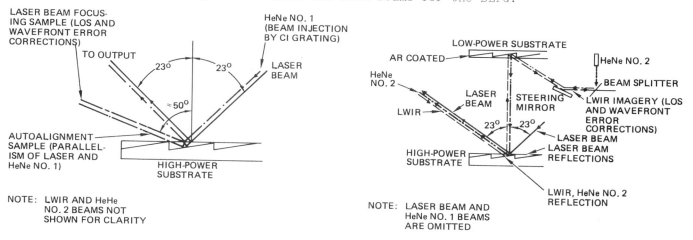

A. SAMPLED LASER AND HeNe NO. 1 BEAMS

B. LWIR AND HeNe NO. 2 BEAMS

Figure 3. Functions of the BLPG.

energy is lost for a 1 cm segment period. The end effect of the material dispersion on the output LWIR beam is to laterally shift the 8 µm radiation to one side of the beam and the 12 µm radiation to the other side of the beam (Figure 7).

During tracking the system must operate off axis. These field-of-view (FOV) requirements cause the off-axis 10 µm radiation to be tilted to one side of the on-axis 10 µm radiation

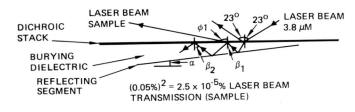

(0.05%)^2 = 2.5 × 10^{-5}% LASER BEAM TRANSMISSION (SAMPLE)

	CVDZnSe	CVDZnS
	λ = 3.8 μM	
a DEGREES	4.672	5.113
β_1 DEGREES	18.582	20.213
ϕ_1 DEGREES	50.860	51.117
β_2 DEGREES	27.926	30.438
TIR ANGLE DEGREES	24.259	26.350

Figure 4. Multiple bounces suffer TIR (total internal reflection) within the CVD material.

Figure 5. In order for the reflected laser beam to clear the low-power substrate, there is a minimum substrate separation, LMIN = W/sin α_o.

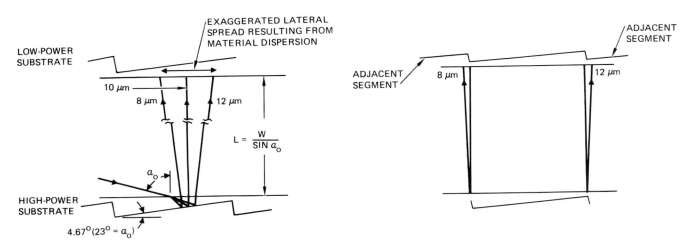

A. LATERAL SEPARATION OF THE 8μm AND 12 μm RADIATION AT THE LOW-POWER SUBSTRATE

B. BEAM DISPERSION OCCURS ON BOTH SIDES OF THE LWIR REFLECTING SEGMENT.

Figure 6. Material dispersion of a LWIR section.

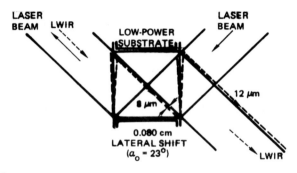

Figure 7. The effect of dispersion on the output LWIR is a lateral shift of the 8- and 12-µm radiation.

path. For a FOV, $\Delta \alpha_O$ = 1 mrad, the lateral shift of the off-axis 10 µm radiation from the on-axis path,

$$\left(\frac{W}{\tan \alpha_O}\right) \left(\Delta \alpha_O\right)$$

W = beam width at the BLP grating
α_O = angle of incidence on the BLP grating
$\Delta \alpha_O$ = FOV at the BLP grating

causes an approximate energy loss of 5 percent for a 1 cm segment period. The FOV affects the off-axis performance which does not have the tight requirements of the on-axis image.

Another potential source of energy loss and wavefront degradation is near-field diffraction of the individual LWIR sections reflected and clipped by each segment pair. As discussed below, diffracted energy loss sets a lower limit on the segment period while the wavefront aberrations due to diffraction do not significantly alter the system resolution.

As mentioned above, the BLPG functions as a segmented mirror, hence the system parameters were chosen to keep the diffraction effects small. The diffraction of a monochromatic (λ) section of the beam reflected from a single mirror segment is governed by the Fresnel number, $d^2/4\lambda L$, where d is the width of the beam section. The unfolded LWIR path is shown in Figure 9.` The segments of the high-power substrate divide the beam into sections of width d. These LWIR beam segments are then near-field propagated a distance L, clipped by the low-power substrate and then recombined into one large LWIR wavefront. Aberrations are introduced into the LWIR wavefront and some of the LWIR energy is lost as a result of the near-field diffraction over the propagation distance, L, followed by the clipping action of the low-power substrate. The fractional power loss of a single segment pair is the same as the fractional loss for the LWIR beam. The wavefront aberrations of the propagated LWIR sections are less than $+\lambda/10$ as shown in Figure 10. The energy loss as a result of diffraction is shown in Figure 11. For a period of 1 cm this diffraction loss is approximately 5 percent.

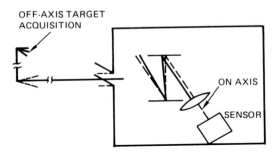

Figure 8. The FOV affects off-axis performance.

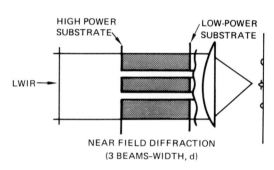

Figure 9. Schematic of the unfolded LWIR propagation path.

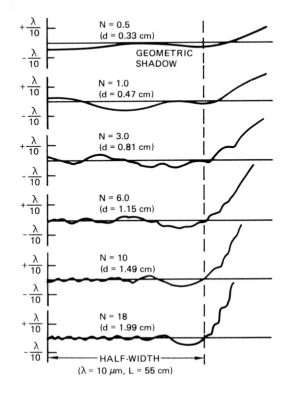

Figure 10. Propagated wavefront across slit width.

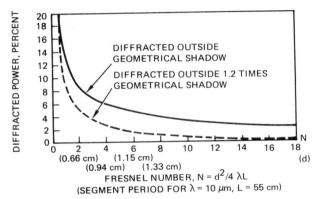

DIFFRACTED OUTSIDE
GEOMETRICAL SHADOW

DIFFRACTED OUTSIDE 1.2 TIMES
GEOMETRICAL SHADOW

FRESNEL NUMBER, $N = d^2/4\ \lambda L$
(SEGMENT PERIOD FOR $\lambda = 10\ \mu m$, $L = 55$ cm)

Figure 11. Power diffracted outside
geometrical shadow of a
near-infinite slit.

Two of the on-axis energy degradations and the
FOV degradations are summarized in Table 1. The

Table I. Energy loss due to diffraction,
dispersion and field of view (for 55 cm
BLP grating rhomb separation)

| Period, cm | Energy Loss, Percent | | Off Axis Effect (for 1 mrad FOV) |
| | On Axis | | |
	Diffraction	Dispersion	
0.2	20	25	25
0.5	12-13	10	10
1.0	5-6	5	5

approximate values assume minimum substrate separation with no reimaging. For d = 1 cm the
10-11 percent on-axis energy loss would be increased by several percent because the LWIR
transmission of the dichroic stack and burying dielectric is less than 100 percent, and the
LWIR reflection from the metallic coated segments is less than 100 percent. The resulting
signal-to-noise ratio is further degraded by thermal emission from warm mirrors in the LWIR
beam path.

Resolution

The effect of near-field diffraction on the LWIR image is considered in this section.
Each of the diffracted and clipped LWIR sections, from a given segment pair, are placed
alongside each other. The resulting wavefront is brought to a focus. The diffracted energy
loss was not considered, the tracker diffraction neglected, and a 0.4 diameter central obscuration assumed. Three cases corresponding to a uniform input (no segments), segment period of 2.08 cm, and segment period of 0.48 cm were simulated and plotted in Figure 12. For
the zeroth order diffraction patterns the image resolutions of the central peaks and the
spatial positions of the first bright rings are practically identical for all three cases.
For cases B and C there exist higher order diffraction patterns because the input waveforms
are nonuniform in both amplitude and phase. For case B the first order peak irradiance is
only 0.4 percent of its zeroth order peak, located at 0.485 mrad from the center. For case
(c), it is 3 percent and 2.083 mrad away. In either case, the peak of these ghosts are less
than the first bright ring of the zeroth order function. Thus, from the tracker imager
point of view, a pair of ideal BLPGs will not cause any problem resulting from near-field
diffraction effects.

Tolerance analysis

The segments on the low-power substrate are to be a replica of the segments on the high-
power substrate. The section of the LWIR energy reflected from a segment on the one substrate will be reflected from its corresponding segment on the other substrate. With this
scheme the entire LWIR beam will be detected by the LWIR sensor as one coherent unit keeping
the focused spot size minimal with good image resolution. If corresponding segments on the
two substrates are not replicas of one another, the spatial coherence of the LWIR beam will
be degraded as will the target resolution. In addition, wavefront quality information of
the incoming LWIR wavefront will be in error and may improperly correct the outgoing laser
beam. A finite amount of aberrations across the wavefront of the LWIR beam is tolerable. A
preliminary error budget suggests λ/10 wavefront aberrations for the spectral shared aperture component. The necessary fabrication tolerances to keep the OPD across the LWIR wavefront under 1 μm are listed in Figure 13. The tightest tolerances are for segment pairs,
i.e., a mirror segment on the high-power substrate and the corresponding mirror segment of
the low-power substrate. The alignment tolerances for the two substrates as a whole are on
the order of arc minutes which are relatively loose.

These tolerance values may be calculated by deriving analytic expressions for the optical
path difference (geometrical path difference multiplied by the index of refraction) of corresponding rays reflected by different segment pairs. One segment pair is perfectly constructed while the other segment pair has a fabrication error (segment tolerance). The
analytic expression for OPD is a function of the segment tolerance and the segment parameters (period, segment height, burying material depth, segment tilt along width and length
and the separation of the high- and low-power units). Substituting into this analytic
expression, the numerical values for the segment parameters, the OPD value may be calculated
for a given fabrication tolerance of a segment pair.

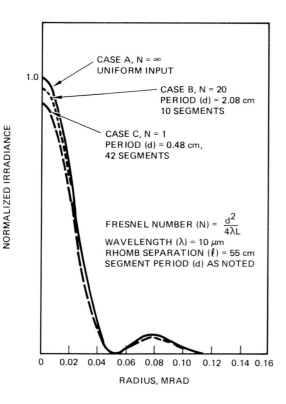

CASE A, N = ∞
UNIFORM INPUT

CASE B, N = 20
PERIOD (d) = 2.08 cm
10 SEGMENTS

CASE C, N = 1
PERIOD (d) = 0.48 cm,
42 SEGMENTS

FRESNEL NUMBER (N) = $\dfrac{d^2}{4\lambda L}$

WAVELENGTH (λ) = 10 μm
RHOMB SEPARATION (ℓ) = 55 cm
SEGMENT PERIOD (d) AS NOTED

NORMALIZED IRRADIANCE

RADIUS, MRAD

Figure 12. Point spread functions of BLP gratings for a 20 cm diameter beam.

The tolerance values of Figure 13 only consider the wavefront aberrations introduced by a single fabrication tolerance acting alone. If two fabrication errors, e.g., Δb and $\Delta \alpha$, both distort the same segment pair, the resultant OPD is very close to the sum of the individual OPDs resulting from the tolerances Δb and $\Delta \alpha$ each acting alone. With the two tolerances (Δb, $\Delta \alpha$) the exact OPD only differs from that predicted assuming Δb acting alone and $\Delta \alpha$ acting alone by one ten-millionth of a percent when the total OPD is on the order of one micron. Hence, for all practical purposes

$$OPD_{\Delta b, \Delta \alpha} = OPD_{\Delta b} + OPD_{\Delta \alpha}$$

for unmatched segment pairs. The exact algebraic equations for OPD in terms of the segment parameters and tolerances are involved but may be simplified considerably (to within 5 percent error) as follows:

$$OPD_{\Delta \alpha} \cong nd(\Delta \alpha) \cos 2\alpha$$

$$OPD_{\Delta b} \cong 2n(\Delta b).$$

Using the above equation for $\Delta \alpha$, a tilt error of $\Delta \alpha$ = 4 arcsec $\cong$20 μrad will yield an approximate phase difference slightly less than 0.5 μm between LWIR sections reflected from the unmatched segment pairs and from matched segment pairs. Hence the step OPD between a LWIR section reflected from an unmatched segment pair, tilt of $\Delta \alpha$ = 4 arcsec, and a LWIR section reflected from another unmatched segment pair, tilt of $\Delta \alpha$ = -4 arcsec, is 1 μm. This is in agreement with the specifications of Figure 13.

Thermal/structural analysis

An analysis was conducted to investigate the influence of BLPG design details on mechanical performance and to provide analytical backup for selection of optimum design parameters. The thermal/structural analysis includes the corner radii at the segment corners and analyzes the temperature, stress, and deformation characteristics as a function of ZnSe thickness. Steady-state and transient responses have been done.

Figure 14 shows the basic finite element model used to analyze the BLPG. The heavy outline indicates the bonded interface between ZnSe and the molybdenum substrates. The bondline is fixed at 0.010 inch and four thicknesses of ZnSe were analyzed. The three thickest ZnSe models are indicated by dashed lines. A one segment period was modeled using boundary interpolation to provide proper boundary conditions. The segment corner was placed in the center of the model to simplify visualization of stress and distortion near the step.

The models were subjected to a thermal load of one watt/cm^2 absorbed at the first surface. The results may be scaled to any heat loading by multiplying the results by the actual heat load in watts/cm^2.

The results are shown in Figures 15 and 16. These figures indicate the variation of peak first surface temperature, peak-to-peak surface displacement and various peak stresses as functions of ZnSe thickness. This data has been used to select a thickness for the ZnSe, with the primary objective of minimizing stress.

The peak stress is always the minimum principle stress. A design based on minimization of stress in the structure should have a ZnSe thickness above the tips of 0.025 to 0.037 inch. The peak stress will be less than 100 psi/(watt/cm^2), the peak surface temperature will range from 0.83 to 1.00°C/(watt/cm^2), and the peak-to-peak distortion varies (inversely) from 12 x 10^{-3} to 7 x 10^{-3} microns (watt/cm^2).

**DIFFERENT SUBSTRATES
DUAL SEGMENTS**

$|\beta_{A'_1} - \beta_{A_1}| \leqslant \dfrac{4 \text{ ARC SEC}}{\text{SEGMENT LENGTH, cm}}$

$|a_{A'_1} - a_{A_1}| \leqslant \dfrac{4 \text{ ARC SEC}}{\text{SEGMENT WIDTH, cm}}$

$|b_{A'_2} - b_{A_2}| \leqslant 0.1 \ \mu m$

$|d_{A'_3} - d_{A_3}| \leqslant \dfrac{2.5 \ \mu m}{\text{NUMBER OF SEGMENTS}}$

SAME SUBSTRATE, ADJACENT SEGMENTS

$|a_{A'_1} - a_{A'_2}| \leqslant 54.0 \text{ ARCSEC}$

$|\beta_{A'_1} - \beta_{A'_2}| \leqslant 48.0 \text{ ARSEC}$

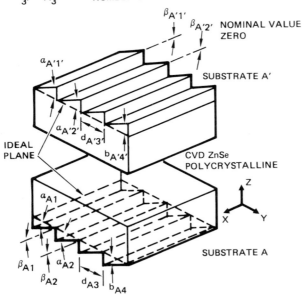

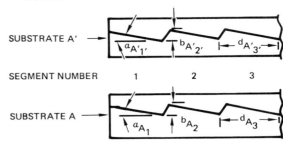

b = MAXIMUM HEIGHT (OR DEPTH) OF SEGMENT (ALONG Z AXIS), -0.0687 (34) cm

d = PERIOD OF SEGMENT (ALONG Y AXIS) - 0.8466 (67) cm

a = ANGLE THE SEGMENT WIDTH MAKES WITH THE Y AXIS - 4.6720944 DEG (4°40'19.54")

β = ANGLE THE SEGMENT LENGTH MAKES WITH THE X AXIS - 0.0 DEG

Figure 13.   Fabrication tolerances for the segmented mirror to yield 1 μ m OPD.

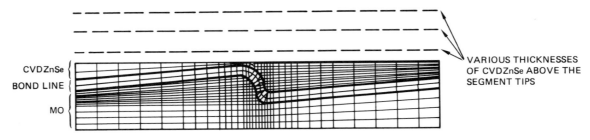

Figure 14.   Method "C" finite element model.

The stress margin of safety, MS, is defined as:

$$MS = \frac{\sigma_{ult}}{\overline{\sigma} \cdot I} - 1$$

where:

σ_{ult} = ultimate strength of material (psi)

= 7500 psi

$\overline{\sigma}$ = actual stress per absorbed intensity (psi/(watt/cm^2))

$\overline{\sigma} \cdot I$ ≡ actual stress (psi)

= 100 psi/(watt/cm^2) I

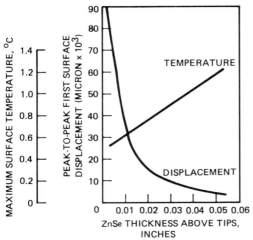

Figure 15. Temperature and displacement
versus thickness — Method
"C" BLPG.

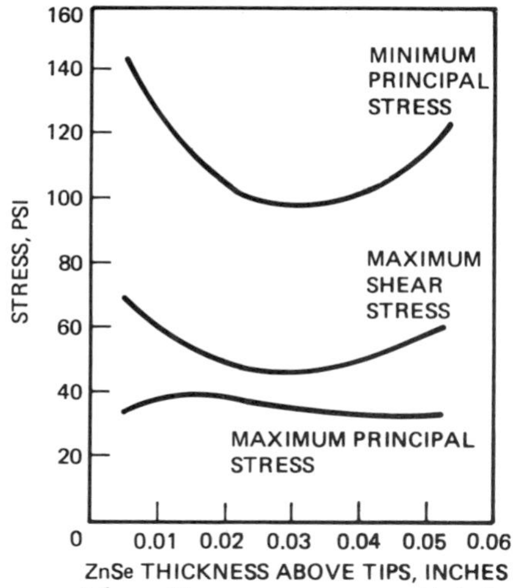

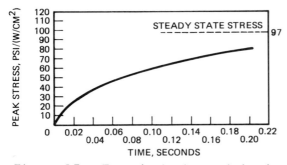

Figure 16. Stress versus thickness —
method "C" BLPG.

$I \equiv$ absorbed energy intensity (watt/cm^2)

Thus,

$$MS = \frac{7500}{100I} - 1 = \frac{75}{I} - 1$$

The margin-of-safety equals 1.0 at an absorbed energy intensity of 37.5 watts/cm^2 and becomes zero at 75 watts/cm^2.

The model constructed for the steady-state analysis was used with ZnSe thickness of 0.032 inch. The model was subjected to a thermal load of one watt/cm^2 absorbed at the first surface. It was initially expected that stresses induced in the ZnSe during the transient phase would peak above the steady-state stresses. This was found not to be true. The actual behavior is similar to any first order system subjected to a step input, as shown in Figure 17. The peak stress rises smoothly and monotonically to its steady-state value.

Figure 17. Transient stress behavior.

This analysis does not necessarily model the thermoelastic behavior of the ZnS/ThF$_4$ dichroic coating. The transient stresses in the coating stack may behave differently, but it is known that the coating has withstood significant thermal shock of this type. No analysis of this type is currently contemplated for the coating.

The transient analysis has shown that the mirror design does not experience higher stresses during the transient phase than during steady-state operation.

A steady-state thermal-structural analysis was performed on one period of the segmented high-power substrate for periods of 0.2, 0.5, 1.0 and 2.0 cm. A uniform incident flux of 30 watts/cm^2 was assumed absorbed by the substrate. For this analysis the period model was from valley to peak. The bondline was omitted and the CVD ZnSe thickness above the segment tips was assumed to be 20 microns. The thermal conductivity of CVD ZnSe is approximately one-eighth that of molybdenum and the highest temperatures appear above the regions with the thickest CVD ZnSe burying dielectric. The induced thermal stress in the CVD ZnSe and the reflecting segment distortions with respect to the bottom of the molybdenum layer were analyzed. The overall backup bending and pressure distortions (overall power) were omitted.

The breaking stress for CVD ZnSe is on the order of 5500 psi. For a safety factor of 5, it is necessary to keep the stresses at or below 1000 psi. This criterion is satisfied for a segment of 1 cm where the compression is just over 700 psi. If the period is doubled to 2 cm the maximum stress approximately doubles to 1400 psi. For smaller periods the stress is further reduced, e.g., 0.5 cm period, 400 psi. Hence the stresses which threaten the survivability of the CVD ZnSe substrate set an upper bound on the segment period, d.

It is necessary to compare the maximum out-of-plane thermal-distortion values with the fabrication tolerances necessary to hold the LWIR wavefront aberrations to 1 μm OPD. The fabrication tolerance, Δb, defined as the difference between the peak height of a segment on the high-power substrate and the peak height of its dual on the low-power substrate is analogous to the out-of-plane distortion of the high-power substrate. A Δb value of 0.2 μm is necessary for 1 μm OPD. For 1 cm period the out-of-plane segment distortion is only 0.02 μm and the dichroic stack distortion is 0.55 μm, both values well below the fabrication tolerance of Δb = 0.2 μm.

The local out-of-plane and in-plane distortions combine together to produce local tilts on the segment. The segment tilt on the high-power substrate is analogous to the fabrication tolerance, Δα, the difference between the angle along with width of a segment on the high-power substrate and the corresponding angle of its dual segment on the low-power substrate. For a 1 μm OPD, Δα is limited to 4 arcsec. For a 1 cm period the segment tilt is only 0.89 arcsec, below the Δα = 4 arcsec fabrication tolerance. In summary, for 1 cm period the aberrations introduced into the LWIR wavefront by thermal distortions of the high-power substrate are below λ/10 = 1 μm OPD. However, if the period, d, is increased significantly both the out-of-plane distortions and local segment tilts increase, degrading the quality of the LWIR wavefront.

Fabrication

The BLPG (dichroic coated buried segmented mirror) will be fabricated by bonding a metallic coated segment dielectric substrate to an appropriate molybdenum heat exchanger. The exposed dielectric surface of the hybrid substrate is then dichroic and AR coated with a 20 layer ZnS/ThF$_4$ stack to reflect the laser beam and transmit the LWIR beam.

Initial investigations consisted of bonding metallic coated CVD ZnSe flats to uncooled molybdenum substrates with three types of thermally conductive bonding materials; type I-room temperature cure hard bond, type II-medium temperature cure bond, and type III-high temperature cure flexible bond. The substrate was gold coated because the bonding material did not adhere well to bare molybdenum. Three flat substrates were fabricated with 0.75 inch x 0.75 inch aperture, 10 mil bond line and 125 mil CVD ZnSe thickness. The fringe counts before and after 185°C, 2 hour thermal cycle were recorded (Table II). For this thick CVD ZnSe layer no surface distortions were induced by the thermal cycle. The CVD ZnSe was polished to 20 mils thickness and thermal cycled (185°C, 2 hours). As seen from the last row of Table II the thermally induced fringe count was acceptable only for the high cure temperature (flexible) bond. The fringe patterns are shown in Figure 18. The low cure temperature bond was brittle and did not absorb the thermal expansion mismatch of the molybdenum and the CVD ZnSe substrates. The thickness of the CVD ZnSe on the high cure temperature bond was polished to 3 mil at which it started to crack. The CVD ZnSe thickness on the medium cure temperature bond was polished to 5 mil thickness with a figure of 1 fringe (astigmatism). After a 185°C thermal cycle there were 2 to 3 fringes with astigmatism.

Table II. Three bonds and associated thermally induced surface
distortions for CVD ZnSe (metallic overcoated) flats
bonded to molybdenum flats (0.75 inch x 0.75 inch
aperture; 10 mil bond line).

Bond	(125 mil CVD ZnSe thickness) Figure (fringes) (0.6328 μm)		(20 mil CVD ZnSe thickness) Figure (fringes)	
	Initial	185°C Thermal Cycle, 2 hours	Initial	185°C Thermal Cycle, 2 hours
low temperature cure (brittle)	3	3	3	10+ (astigmatism)
medium temperature cure	3	3	7 (astigmatism)	10 (astigmatism)
high temperature cure (flexible)	2	2	3 (astigmatism)	3 (astigmatism)

The first segmented substrates were fabricated with the high cure temperature bond. This bond was classified as having "light outgassing" (3 to 5 percent by weight) and is acceptable for space applications. The reflecting segments were ground/polished into the CVD ZnSe with a preformed oil hardened tool steel substrate and slideway. The segmented substrate has a 1 inch x 1 inch aperture with three segments, each segment has a 0.847 cm periods, 0.069 cm depth, and 4.672 degree angle. The back side of these polished CVD ZnSe segments

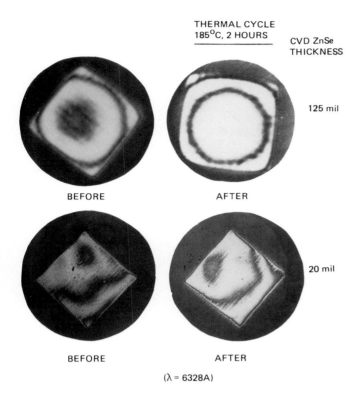

THERMAL CYCLE
185°C, 2 HOURS

CVD ZnSe
THICKNESS

125 mil

BEFORE   AFTER

20 mil

BEFORE   AFTER

(λ = 6328A)

Figure 18.   Fringes for high cure
temperature bond line
(10 mil) with flat
CVD ZnSe and
molybdenum substrate.

were sputtered with Cr-Au, for LWIR (8-12 μm) reflection, and then bonded to a machined and coated segmented molybdenum substrate. The exposed surface of the CVD ZnSe substrate was polished flat. Thermal cycling for 3 hours at 185°C did not increase the fringe count (100-125 mil CVD ZnSe thickness above the segment tips).

The CVD ZnSe was polished to 30 mil thickness above the segment tips and three fringes flatness. After 185°C thermal cycle the figure of the exposed CVD ZnSe surface distorted to a total of 7 fringes (+4). A second 185°C thermal cycle yielded a total fringe count of 6.

Two additional segmented substrates were fabricated with 250 mils CVD ZnSe thickness above the segment tips. Thermal cycling at 185°C did not increase the fringe count. After polishing the exposed CVD ZnSe surface to within 30 mils of the segment tips and 3 fringes flatness the substrate was thermal cycled at 185°C for 3 hours, which increased the surface distortion to 5 fringes. A second thermal cycle (185°C, 3 hours) did not change the 5 fringes distion on the CVD ZnSe surface (Figure 19).

Two more segmented substrates were bonded; one with a 7.5 mil bondline and one with a 5.0 mil bondline. Thermal cycling at 185°C yielded little change in the surface figure only if the CVD ZnSe was greater than 30 mils above the segment tips.

Deliverable hardware was fabricated on water cooled molybdenum substrates. One

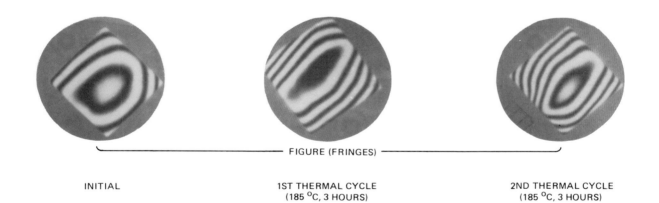

FIGURE (FRINGES)

INITIAL

1ST THERMAL CYCLE
(185 °C, 3 HOURS)

2ND THERMAL CYCLE
(185 °C, 3 HOURS)

Figure 19.   Fringes for segmented substrate.

substrate had 2 segments (4.672 degree angle, 1.270 cm period, 0.103 cm depth) and another had 6 segments (4.672 degree angle, 0.423 cm period, 0.034 cm depth) across the 1 inch x 1 inch aperture of the CVD ZnSe. Two other cooled substrates had 3 segments (4.672 degree angle, 0.847 cm period, 0.069 cm depth) across the 1 inch x 1 inch aperture; one substrate had CVD ZnSe and the other had CVD ZnS. All substrates had a 7.5 mil bond line. The two substrates with 3 segments were thermal cycled at 185°C three times and the CVD ZnSe and ZnS thicknesses polished from 250 mils down to 35-42 mils above the segment tips. The exposed surfaces were flat to within six to seven fringes at 0.6328 μm. The substrates were then vacuum baked at 175°C for 24 hours with increased surface distortion as shown in Figure 20. All of the interferograms were convex implying that the bond thickness around the perimeter of the substrate contracted relative to that in the center. This depression of the edges of the substrate relative to the center could be explained qualitatively by vacuum baking induced outgassing and resultant shrinking of the exposed bond line thickness around the edges of the substrate. Present experiments, with this and other "light outgassing" conductive bonding materials and with hermetic seals for the exposed edges of the bond, are being conducted. The substrates were then dichroic coated with a 20 layer ZnS/ThF₄ stack for laser beam reflection and an AR coat for LWIR transmission.

Acknowledgements

The authors wish to thank A. Lau for numerically calculating the effects of diffraction on energy losses and resolution.

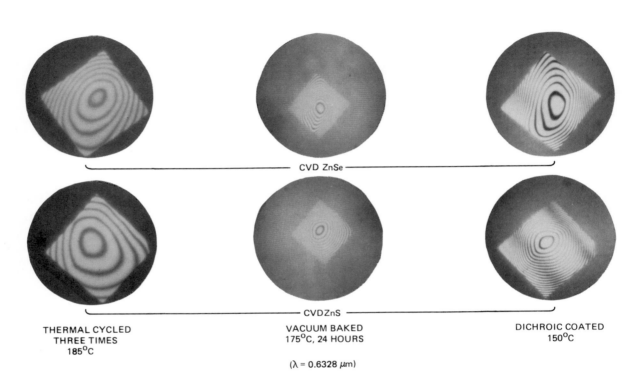

CVD ZnSe

CVDZnS

THERMAL CYCLED
THREE TIMES
185°C

VACUUM BAKED
175°C, 24 HOURS

DICHROIC COATED
150°C

(λ = 0.6328 μm)

Figure 20. Fringe patterns for water cooled segmented substrate.

Vacuum ultraviolet gratings

Bernhard W. Bach

Hyperfine Inc., 149 North Main Street, Fairport, New York 14450

Abstract

Grating efficiencies in the vacuum ultraviolet have historically ranged up to 42% absolute on plane gratings overcoated with AlMgF2 for 1216A. This has been primarily achieved with spacings of 1200 l/mm to 3600 l/mm at blaze angles of 4° and greater. Gratings with less than 1200 l/mm at blaze angles of 2° and below have shown to reflect 10 to 30%. Gratings finer than 3600 l/mm were, (untile now), only available when holographically produced. A better understanding of the factors limiting groove smoothness has made it possible to rule improved diffraction gratings for the vacuum ultraviolet. At 200 l/mm we have achieved absolute efficiency of 63% @ 1216A and with 6000 l/mm, 33 to 35% @ 1216A. These improved efficiencies have been possible on spherical and non spherical surfaces.

Introduction

For the last several years we have concentrated on improving the groove efficiency of gratings used in the vacuum ultraviolet. All the technology of optical polishing, vacuum deposition, diamond tool manufacturing, ruling engine control, replication and testing have contributed to the improved performance. The most critical technology of all is the vacuum deposition of the metal for the master coating.

Grating technology

The VUV presents new challenges to us in terms of optical tolerances. Optics tested at 6328A to 1/10 fringe are used where this 1/10 fring becomes a 1/2 wave or even a full wave. Attention to the surface finish of an optical blank becomes vey important, and diamond tool edges (1) have to be polishied to the extreme sharpness. We have viewed these tools with a SEM at 50,000x and no rounding could be seen. Ruling engines using the closed loop interferrometric control (2) have been universally adopted by the industry. Testing of VUV gratings has been undertaken by W. R. Hunter at The Naval Research Laboratory (3) and most recently by W. G. Fastie and G. H. Mount at The John Hopkins University (4). This testing gives feedback on efficiencies needed by the grating manufactures.

Ruling the groove into the deposited metal layer is a complex problem. None of the material is removed but merely compressed and pushed sideways. The diamond tool is loaded with weights until it floats at the right groove depth in the metal layer. The pressure required to accomplish this is about 4.6 million psi. The tool is then dragged through the metal layer, displacing and compressing it. I like to think of it in terms of the cold flow of materials such as nylon and teflon. It is the control of this flow with proper tool orientation, loading and choice of deposition parameters that allows the increased efficiency. To achieve a blaze angle of 2° or less, the deposited material must be compressed. Idealy a coating that is slightly spongy in composition allows vertical compression. Blaze angles larger than 2° require a slight downward and sideways movement. In addition, a wave like motion is generated that influences both the virgin material yet to be ruled and the already burnished grooves. With a single pass, (sometimes two), we try to achieve groove facet geometry and smoothness to within a few angstroms. To achieve a good groove for a plane grating all this is necessary only once; since the diamond is in the same position for the entire ruling.

The concave grating (used almost exclusively in VUV) presents a unique set of problems. As the concave grating is ruled, the diamond tool is clamped at the end of a lever which is hinged by a det of piviots attached to the diamond carriage. This allows the diamind point to follow the shape of the blank (spherical, non spherical or parabolic). As it rules the grating, the tool traces the shape of the blank with a step size equal to the grooves per mm required. Being a non-plane surface, the angular relationship between the optical surface and diamond facet varies. On some concave gratings the angular change approches 100 to 200% of the blaze angle. A multipartite grating, with the central groove in each panel blazed at the wavelength of interest, is the best solution. It is the angular change of the concave blank that demands perfect malleability from the coating for the VUV grating. There is a general feeling that the faster the deposit the better the metal burnishes. We have found the best results by adjusting the rate by a factor of 5 and operating the vacuum chamber from the mid 5 to high 7 range during evaporation.

Efficiency

In the visible, we hear absolute and relative measurements primarily in reference to a good aluminum mirror In the VUV, different coatings are used for different wavelength regions. Gold, Platium and osmium are used for short wavelengths while AlMgF2 for the longer wavelength. A more consistent way of comparing performance would be groove efficiency. As an example, a grating blazed at 304A reflecting 2% absolute reflectivity has a groove efficiency of 66% because gold reflects 3%. A grating overcoated with AlMgF2 reflecting 64% absolute has a groove efficiency of 80% since AlMgF2 reflects 78 to 80%. The best way to demonstrate improvements achieved with VUV is to list the differences in measurements of the old rulings vs the new rulings.

Old measurements

Spacing	Radius	Blaze	Blaze angle	Efficiency	Groove efficiency
3600 1/mm	4 mcc	304A	3.06°	1.8%	60%
3600 1/mm	plano	1216A	12.6°	43%	54%
313 1/mm	1.3 mcc	1400A	1.2°	11%	14%

New measurements

Spacing	Radius	Blaze	Blaze angle	Efficiency	Groove efficiency
1200 1/mm	1 mcc	304A	1.5°	2.5%	83%
2400 1/mm	.4mcc	584A	4°	11%	69%
200 1/mm	.5 mcc	1216A	.7°	61%	76%
2400 1/mm	.85 mcc	1300A	8.9°	65%	81%
313 1/mm	1.3 mcc	1400A	1.2°	55%	69%
6000 1/mm	plano	1216A	21.4°	33%	41%

Summary

   With a better understanding of the diamond tool and coating interaction, we have demonstrated significant improvement in VUV performance of gratings with less than 1200 1/mm. This has benefited the 1200 1/mm to 3600 1/mm spacings, but not with the same percentage of increase. With the 6000 1/mm grating we have been achieving a performance almost identical to the very best 3600 1/mm.

References

1. Precision Diamond Tools, J Robert Moore Co., Petersham, Mass.
2. Harrison, GR., Thompson, W.W.; Kazukonis H.; Connell, J.R. 750mm Ruling Engine Producing Large Gratings and Echelles. Journal of the Optical Society of America, Volume 62, No.6 June 1972.
3. D.J. Michels, T.L. Mikes, and W.R. Hunter, Optical Grating Evaluator: a Device for Detailed Measurement of Diffraction Grating Efficiencies in the Vacuum Ultraviolet. Applied Optics, Vol. 13, May 1974.
4. Fastie, E.G. and Mount, G.H. Study of Ultraviolet Properties of Optical Components for The Large Space Telescope System and for the International Ultraviolet Explorer, Interm Report, the John Hopkins University, Dept of Physics, Baltimore, Md.

Adjustable mosaic grating mounts

G. A. Brealey, J. M. Fletcher, W. A. Grundmann, E. H. Richardson
Herzberg Institute of Astrophysics, Dominion Astrophysical Observatory
Victoria, British Columbia, Canada V8X 3X3

Abstract

Three 30-cm mosaic gratings are being built for the Coudé spectrograph of the Canada-France-Hawaii 3.6 meter telescope. The largest mosaic is 30 by 60 cm and is composed of four 15 by 30 cm echelle gratings. The individual gratings have coarse, fine, and super-fine adjustments. The alignment can be checked, and corrected if necessary, by one person in a few minutes. An earlier version of these mosaics has been in operation at the Dominion Astrophysical Observatory for many years and has proven to be practical and popular with the astronomers.

Introduction

To increase spectral resolution without loss of efficiency using a grating of given groove spacing and blaze angle, it is necessary to enlarge the grating proportionately. For operation at the diffraction limit the enlargement is necessary to decrease the width of the diffraction pattern. However, the entrance slits of most spectrographs are set much wider than the diffraction limit when used with telescopes in which case the enlargement of the grating is required to maintain the same entrance slit width when the focal length of the camera is increased to achieve the higher dispersion. The wide slit is required for efficient spectroscopy of extended objects such as nebulae or images of stars that have been blurred by "seeing" effects in the atmosphere of the earth. Because the spectrograph is not required to operate at the diffraction limit the enlarged grating can be composed of a mosaic of several smaller gratings where the diffraction limit of the component grating is smaller than the width of the slit.

The advantages of using, say, four 15 cm gratings instead of one 30 cm grating are: (1) Few 30 cm gratings are available, particularly the exceptionally efficient ruling of 830 grooves/mm blazed for the 2nd order blue, and other rulings of even finer spacing; (2) A ruled 30 cm grating is less efficient than an identical 15 cm grating which could be replicated from the best portion of the larger grating or from the smaller master which should suffer less diamond wear; (3) Four 15 cm gratings are cheaper than one 30 cm grating even in the case of holographic gratings where there is no expected advantage in efficiency or selection of groove spacing (but holographic gratings have been less efficient than ruled gratings in most cases.)

Pioneering work in the use of an adjustable mosaic grating in a stellar spectrograph was done at Palomar Observatory which supplied detailed drawings on which the first DAO design was based. By adding a periscope, etc. to the Palomar system we made it possible for one person to check and adjust the alignment of the gratings (rather than two people, one watching the spectrum and one adjusting the gratings) in a few minutes.

In the following discussion, "rotation" refers to the axis perpendicular to the surface of the grating and this fine adjustment is used to make the individual spectra parallel and horizontal in our case. The conventional, large rotation to change spectral region is not mentioned here but would be about the axis parallel to the rulings which is vertical; in the adjustment of the mosaic components this rotation is referred to as "tilt" and is the super-fine correction done by bending the mounting using leaf springs. The other "tilt" is about the axis parallel to the gratings surface but perpendicular to the rulings, i.e. horizontal, and moves the spectra up and down, perpendicular to the direction of dispersion.

The First D.A.O. Mosaic Grating Mount

A mosaic grating mount, consisting of four gratings (16.5 x 19 x 3.5 cm), was designed and constructed at D.A.O. in 1965-1966. As shown in Figure 1 the gratings were arranged in a 2 x 2 array and the separation between segments was minimized. A centre block supports and defines the central corners of the four gratings. Each grating rests on three adjusting screws and is retained by two opposing cork-faced, spring-loaded pads at the outer edge and by a central nylon pad, in common with the adjacent grating. The three adjustment screws provided for coarse adjustment of grating tilt about both the vertical and the horizontal axis. With UNF No. 10-32 screws, adjustment sensitivity was of the order of 20 arc minutes per turn.

The rotational adjustment was accomplished by the corner block arrangement shown in Figure 1 in sectional orthographic projection. The centre block was provided with 92° cut-outs for each grating corner and the UNC No. 6-32 screws at the outer corner provided for an adjustment sensitivity of the order of 11 arc minutes per turn.

The super-fine adjustment of each grating is in the direction of dispersion (which is horizontal in our case) and is a tilt about the axis parallel to the rulings achieved by a design feature called the "Michelson Spring". The gratings are mounted on a relatively stiff, thick plate which in turn is deformed by a relatively thin, flexible plate. By further tapering the thickness of the thick plate, and by tapering the width of the thin plate a ratio of relative stiffness of about 170 was obtained. With a pitch of 1.6 mm for the deformation screw a fine adjustment sensitivity of 40 arc seconds per turn is achieved.

The "Michelson Spring" components are joined to a stand and the assembly is arranged and mounted on a standard 30 cm diameter rotary indexing table, so that the axis of rotation is in the common plane of the ruled surface of the grating.

Problems

Although this early design had all the necessary adjustments, and in the right sensitivity range, it was found difficult to align the gratings. In particular, it was noticed that adjustments to rotation affected the tilt, while coarse adjustment to the tilt affected the rotation. This inability to have one adjustment unaffected by changes to the other is due to the fact that the grating blank has to slide finite amounts, and experience frictional drag, on the support screws and pads for the rotational adjustment, and movement and drag on the corner and centre block for the tilt coarse adjustment. "Hammering" the back of the mounting support pillar with the soft side of a clenched fist settled the gratings down and after a few hammer-adjust cycles, the adjustments remained stable for long periods. (Fist hammering does not effect the alignment of the new mosaic mount.)

Unfortunately, very little attention was given to component material compatability in this early grating mounting. The thick plate was made from aluminum, the thin plate was made from tool steel and the stand was made from cast iron. These three materials had widely different coefficients of thermal expansion and it was soon found that the gratings would not stay aligned for ambient temperature changes of even less than 1° C. The "Michelson Spring" in fact had become a bi-metallic strip.

A Second Generation Mosaic Grating Mount

Improvements made to the DAO mount and other changes are incorporated in the design of the CFHT mount shown in Figure 2. The "Michelson Spring" is maintained and in this version the flexible plate and the stiff plate are of identical aluminum alloy. The stand is made of cast aluminum having an expansion coefficient almost identical to wrought aluminum. The deformation screw is an M5 x 0.8 and it is also free to articulate so as not to impose a bending moment or friction effects in either plates.

In this version each grating is mounted on a seat where it is supported by three adjusting screws and by three opposing cork-faced, spring-loaded pads. Vertical support for each grating is provided by the two lower brackets. Coarse adjustments for either tilt is accomplished as before, except that it is now independent of the rotational adjustment. Each grating seat is centrally supported on a shoulder screw which is only torqued to a degree to provide firm seating. The enlarged detailed inset illustrates the rotation adjustment using a stud with different pitch threads. One complete turn of this stud rotates the grating of the order of 3 arc minutes.

Grating Alignment Procedures

The position of the mosaic grating in relation to the slit jaws and the camera mirror is shown in Figure 3. The four gratings are first roughly aligned by visual inspection and direct physical measurements. The fine alignment is accomplished by incorporating a 4-aperture mask on a wheel, a periscope and a viewing lens in the Coude spectrograph train. The lens and the mask are removed during normal operation, but the periscope is left in place because it falls in the shadow of the plateholder. Three positions of the mask permit illumination of a choice of two gratings, one of which is always the same grating selected as the reference standard. The illumination is switched between the two gratings by a manually operated sliding shutter. The fourth position illuminates all four gratings.

During alignment the photographic plateholder is removed. With the assistance of the mask and by systematically comparing the jumping of the image of a mercury line from one grating with the three others, it becomes a relatively simple procedure to fine tune the alignment of the mosaic grating. See Figure 4. It normally takes one man about 5 minutes to complete this task.

The alignment mask for the CFHT does not use a wheel but instead has all shutters stacked in line and operated electrically allowing oscillation of illumination. Provision has been made for the viewing of the mercury line by television.

Current Status

The mosaic grating mount has now given many years of satisfactory service at the DAO. For the Canada-France-Hawaii telescope three further mosaic grating mounts for the Coudé spectrograph have been built at D.A.O. The gratings for these were two sets of four blanks 22 x 16.5 x 3.5 cm and one Echelle set of four blanks 32 x 16.5 x 5 cm. The first two mountings have been assembled and tested. The Echelle grating is expected to be completed by August, 1980.

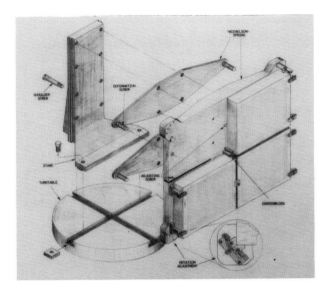

Figure 1. First version of a mosaic grating mount.

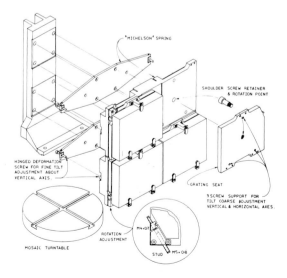

Figure 2. Improved version of a mosaic grating mount.

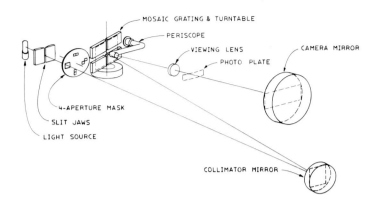

Figure 3. Coudé spectrograph component layout.

Figure 4. Mosaic grating under alignment test.

PERIODIC STRUCTURES, GRATINGS, MOIRÉ PATTERNS AND DIFFRACTION PHENOMENA

Volume 240

SESSION 6

FABRICATION TECHNIQUES, APPLICATIONS II

Session Chairman
William R. Hunter
Naval Research Laboratory

Tunable picosecond transient grating method to determine the diffusion constants of excitons and excitonic molecules

Yoshinobu Aoyagi, Yusabro Segawa, Tetsuya Baba, Susumu Namba

The Institute of Physical and Chemical Research, Wako-shi, Saitama 351, Japan

Abstract

We made a first success to observed a diffusion constant of excitonic molecules in CuCl by a tunable picosecond transient grating method. This method is proved to be a powerful method to examine the spatial diffusion of uncharged particles like excitons and excitonic molecules in solid states.

The examination of the diffusion of excitons and excitonic molecules in a crystal has been very difficult so far because of the lack of the direct method to measure the spatical behavior of those particles. In this paper we report the observation of the diffusion constant of the excitonic molecules in CuCl by the tunable picosecond transient grating method.

Figure 1 shows a block diagram of our high power tunable picosecond laser system. In this system two dye oscillators and two dye amplifiers pumped by the third harmonics of modelocked Nd-YAG laser were used to get high power tunable picosecond sources which oscillate at two independent wavelengths.

Figure 2 shows the schematic diagram of the tunable picosecond transient grating method. A laser for the excitation of the sample with a wavelength of the resonant two photon absorption of the excitonic molecules (3891 A) was splits into two beams by a half mirror and they were crossed each other in the sample with an angle Θ to produce a density grating of the excitonic molecules in the crystal. The probe light (3900 A) was diffracted by this grating. Since the decay time of the diffracted light T is given as

$$\frac{1}{T} = \frac{1}{\tau} + \frac{4\pi^2}{\Lambda^2} D$$

by the analysis of the differential equation of the diffusion, we can estimate the diffusion constant D and the lifetime τ by observing the decay time T for the various grating periods Λ as shown in the fig. 3. The slope and the cross point to the y-axis in the figure give the diffusion constant and the lifetime of the excitonic molecules at 1.8°K and are estimated to be 45 cm^2/sec and 200 psec, respectively. The lifetime observed in this method is in good agreement with the value observed by the decay time of the luminescence of the excitonic molecules in CuCl[1]. The mean free

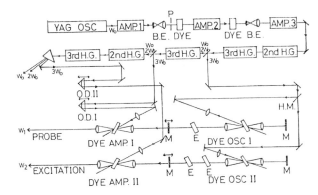

Fig.1. Block diagram of our high power tunable picosecond laser system. B.E., Beam expander, 2nd H. G., 2nd harmonics generator, 3rd H.G., 3rd harmonics generator, J. M., half mirror, M., mirror, E., etalon, O.D. I, O.D. II., optical delay line.

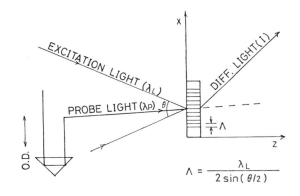

Fig.2. Schematic diagram of a tunable picosecond transient grating method.

path and the mean free time of the excitonic
molecules are estimated to be 0.4 μm and
40 psec from the observed diffusion constant,
respectively.   This mean free time is larger
than the calculated values, i.e. 10 psec for
the molecule-molecule collision and 20 psec
for the molecule-phonon collision which are
estimated by a simple gas model and emitting
rate of LA-phonon[2] respectively.   This
discrepancy is now under studying.

The tunable transient grating method is
a powerful method to determine the diffusion
constant of uncharged perticles like
excitons and excitonic molecules.   We can
determine the diffusion constant of the
excitonic molecules in CuCl by this method,
clearly.

References

1. Segawa. Y., Aoyagi. Y., Baba. T., and Namba.
   S., "Luminescence and raman processes under
   resonant two photon excitation in CuCl",
   J. Luminescence 18/19,262 (1978).
2. Ojima. M., Kushida. T., Tanaka. Y. and
   Shionoya. S., "Picosecond time analysis of
   excitonic molecule luminescence spectra in
   CuCl", J. Phys. Soc. Japan 44, 1294 (1978).

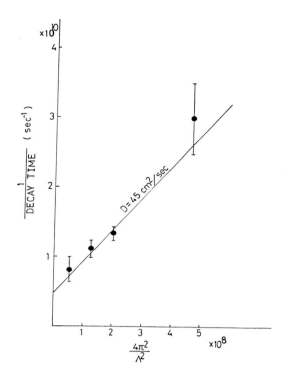

Fig. 3. $1/T$ vs. $4\pi^2/\Lambda^2$ plots for excitonic
molecules in CuCl.

Hole grating beam sampler—a versatile high energy laser (HEL) diagnostic tool

James E. Harvey, Marion L. Scott

University of Dayton Research Institute, AFWL/ARAO, Kirtland AFB, New Mexico 87117

Abstract

The hole grating beam sampler consists of a uniform array of small circular holes in a highly polished mirror. When used as a turning flat in an HEL optical train, it produces a multiplicity of angularly separated low-power replicas of the HEL beam in both the transmitted and reflected directions. It thus provides the opportunity to perform many simultaneous diagnostic tests such as power monitoring, bore-sight alignment, jitter sensing, wavefront sensing, power-in-the-bucket measurements, and spectral purity measurements in real time during HEL operation on a non-interference basis with the intended mission. The principle of the hole grating beam sampler will be discussed in detail and several relationships useful to the HEL systems engineer will be derived. The capabilities and limitations of this versatile HEL diagnostic tool will then be discussed in the context of several specific applications.

Introduction

Certain high-energy laser (HEL) applications require the simultaneous measurement and/or control of various laser beam characteristics in real time. Power monitoring, bore-sight alignment, jitter sensing, wavefront sensing, power-in-the-bucket measurements, and spectral purity are a few of the diagnostic tests routinely required by HEL system engineers. The hole grating beam sampler, consisting of a uniform array of small circular holes in a highly polished mirror used as a turning flat in the HEL optical train, has been suggested by several investigators for some of these functions.[1-3] Since the hole grating produces a multiplicity of angularly separated low-power replicas of the HEL beam, it provides the opportunity in principle, to perform many separate tests simultaneously in real time during HEL operation on a noninterference basis with the intended mission. The hole grating beam sampler therefore has the potential of being a very valuable and versatile diagnostic tool for many applications requiring advanced beam-control techniques.

A detailed analysis of both the transmitted and reflected beams produced by the hole grating beam sampler results in several mathematical relationships useful to the HEL systems engineer. The capabilities and limitations of this versatile HEL diagnostic tool will then be disucssed in the context of several specific applications.

Hole grating beam sampler analysis

A typical geometrical configuration using a hole grating beam sampler is illustrated in Figure 1.

Transmitted beam

If we now assume an HEL beam with complex amplitude distribution $U_O(\hat{x},\hat{y})$ incident upon the hole grating, then the transmitted wave field can be written as

$$U_T(\hat{x},\hat{y}) = U_O(\hat{x},\hat{y}) \, t(\hat{x},\hat{y}) \tag{1}$$

where $t(\hat{x},\hat{y})$ is the complex amplitude transmittance of the hole grating.

Assuming that the Fourier transform of this function exists (and Bracewell[4] emphasizes that physical possibility is a valid sufficient condition), we can write

$$A_T(\alpha,\beta) = \int\!\!\int_{-\infty}^{\infty} U_T(\hat{x},\hat{y}) e^{-i2\pi(\alpha\hat{x} + \beta\hat{y})} \, d\hat{x} d\hat{y} \tag{2}$$

$$U_T(\hat{x},\hat{y}) = \int\!\!\int_{-\infty}^{\infty} A_T(\alpha,\beta) e^{i2\pi(\alpha\hat{x} + \beta\hat{y})} \, d\alpha d\beta \; . \tag{3}$$

Note that we have used a scaled coordinate system where

$\hat{x} = x/\lambda$, $\hat{y} = y/\lambda$, $\hat{z} = z/\lambda$, and $\hat{r}^2 = \hat{x}^2 + \hat{y}^2 + \hat{z}^2$.

The reciprocal variables in Fourier transform space

$$\alpha = \hat{x}/\hat{r}, \quad \beta = \hat{y}/\hat{r}, \quad \text{and} \quad \gamma = \sqrt{1-(\alpha^2 + \beta^2)} = \sqrt{1-\rho^2} \tag{4}$$

are then the <u>direction cosines</u> of the propagating light. The kernel of Eq. (3) is precisely the mathematical expression for a <u>plane wave</u> propagating with direction cosines α and β. We have thus decomposed U_T into a superposition of plane waves whose amplitudes are given by A_T.

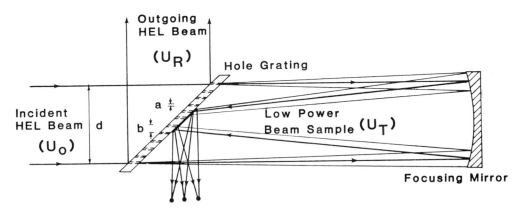

Figure 1. Geometrical configuration for obtaining low-power beam sample.

Consider the hole grating illustrated in Figure 2, which consists of a rectangular array of circular holes in a mirror of reflectance R. The complex amplitude transmittance of the hole grating can be written in symbolic notation as

$$t(\hat{x}, \hat{y}) = \mathrm{cyl}\left(\frac{\hat{r}}{\hat{a}}\right) * \frac{1}{\hat{b}^2} \mathrm{comb}\left(\frac{\hat{x}}{\hat{b}}, \frac{\hat{y}}{\hat{b}}\right) . \tag{5}$$

The symbolic representation of mathematical functions and operations used throughout this paper demonstrates the ease and convenience with which certain mathematical manipulations can be performed when dealing with Fourier analysis or linear systems theory. These techniques are becoming widely used by electrical engineers and optical scientists as evidenced by the recent emergence of several good text books that develop and use this "Fourier short-hand" notation.[4-6] A brief review defining those functions and operations used in this paper is provided in the Appendix.

The angular spectrum of plane waves emitted by the hole grating beam sampler is thus obtained by substituting Eqs. (1) and (5) into Eq. (2) and applying the convolution theorem of Fourier transform theory to yield

$$A_T(\alpha, \beta) = A_0(\alpha, \beta) * [T \, \mathrm{somb}(\hat{a}\rho)\hat{b}^2 \, \mathrm{comb}(\hat{b}\alpha, \hat{b}\beta)] \tag{6}$$

where

$$A_0(\alpha, \beta) = \iint_{-\infty}^{\infty} U_0(\hat{x}, \hat{y}) e^{-i2\pi(\alpha\hat{x} + \beta\hat{y})} d\hat{x}d\hat{y}$$

is the Fourier transform of the complex amplitude distribution incident upon the hole grating, and

$$T \equiv \frac{\pi}{4} \frac{\hat{a}^2}{\hat{b}^2} = \underbrace{\left(\frac{\pi}{4} a^2 \right)}_{\substack{\text{Area} \\ \text{of} \\ \text{Hole}}} \underbrace{\left(\frac{1}{b^2} \right)}_{\substack{\text{Density} \\ \text{of Holes}}} = \frac{\text{Transmitted Power}}{\text{Incident Power}} = \frac{P_T}{P_O} \tag{7}$$

is the fraction of the mirror surface made up of holes, or the transmittance of the hole grating. From Figure 3 it is apparent that if the value of the envelope function, $\text{somb}(\hat{a}\rho)$, does not change significantly over the width of A_O, then Eq. (6) can be re-written as

$$A_T(\alpha, \beta) = T \, \text{somb}(\hat{a}\rho) [A_O(\alpha, \beta) \ast \hat{b}^2 \, \text{comb}(\hat{b}\alpha, \hat{b}\beta)] \tag{8}$$

or due to the replicating property of the comb function

$$A_T(\alpha, \beta) = T \, \text{somb}(\hat{a}\rho) \sum_{-\infty}^{\infty} \sum_{-\infty}^{\infty} A_O(\alpha - n/\hat{b}, \ \beta - m/\hat{b}) \ . \tag{9}$$

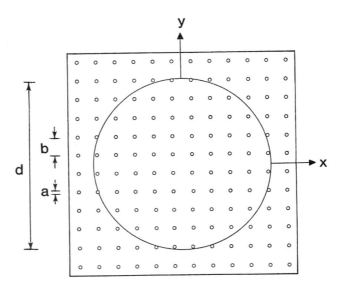

Figure 2. Hole grating beam sampler.

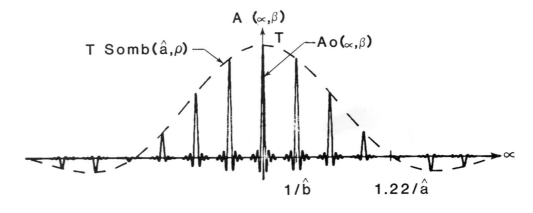

Figure 3. Plane wave spectrum transmitted by the hole grating.

If we make the more stringent restriction that the width of A_O is small compared to the separation, $1/\hat{b}$, then there is negligible overlap between the individual functions being summed and we can write the intensity as

$$I_T(\alpha,\beta) = \left| A_T(\alpha,\beta) \right|^2 = T^2 \text{somb}^2(\hat{a}\rho) \sum_{-\infty}^{\infty} \sum_{-\infty}^{\infty} A_O^2(\alpha-n/\hat{b},\ \beta-m/\hat{b}) \tag{10}$$

(i.e., there are no cross terms produced by squaring the summation). We can now write

$$I_T(\alpha,\beta) = T^2[\text{somb}^2(\hat{a}\rho)\hat{b}^2\text{comb}(\hat{b}\alpha,\hat{b}\beta)] * A_O^2(\alpha,\beta)\ . \tag{11}$$

But this is just a two-dimensional rectangular array of attenuated replicas of the incident HEL beam. Furthermore, the envelope function controling the relative intensity of these diffracted orders is merely the Fourier transform of one of the small holes making up the hole grating. Figure 4 is an actual photograph of the transmitted diffraction patterns obtained while testing a particular hole grating with visible light and clearly illustrates this two-dimensional array as well as the associated envelope function.

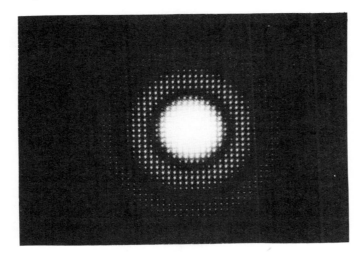

Figure 4. Transmitted diffraction pattern illustrating both the two-dimensional array of diffracted orders and the associated envelope function.

The intensity distribution of the m,nth diffracted order is obtained from Eq. (11) and given by

$$I_{T_{m,n}}(\alpha,\beta) = T^2\text{somb}^2(\hat{a}\sqrt{m^2+n^2}/\hat{b})[\delta(\alpha-m/\hat{b},\beta-n/\hat{b}) * A_O^2(\alpha,\beta)] \tag{12}$$

and the radiant power contained in the m,nth diffracted order is obtained by integrating over direction cosine space

$$P_{T_{m,m}} = \int\int_{-1}^{1} I_{T_{m,n}}(\alpha,\beta)\,d\alpha d\beta = T^2\text{somb}^2(\hat{a}\sqrt{m^2+n^2}/\hat{b})\int\int_{-1}^{1} A_O^2(\alpha-n/\hat{b},\beta-m/b)\,d\alpha d\beta \tag{13}$$

or

$$P_{T_{m,n}} = T^2\text{somb}^2(\hat{a}\sqrt{m^2+n^2}/\hat{b})P_O \tag{14}$$

where P_O is the total power incident upon the hole grating. This result is completely independent of the amplitude or phase variations in the incident beam so long as the width of its angular spectrum is small compared to $1/\hat{b}$. Note that the power contained in the central, undiffracted order is given simply by

$$P_{T_{o,o}} = T^2 P_o \quad . \tag{15}$$

Furthermore, it has been shown that the central order of the transmitted diffraction pattern remains a true replica of the incident HEL beam even in the presence of small random hole-positioning errors[2] and atmospheric turbulence.[3]

Reflected beam

A similar analysis of the outgoing HEL beam results in the following expression for the complex amplitude distribution reflected from the hole grating beam sampler

$$U_R(\hat{x},\hat{y}) = U_o{}'(\hat{x},\hat{y}) \sqrt{R}[1-cyl(\frac{\hat{r}}{\hat{a}}) * \frac{1}{\hat{b}^2} comb(\frac{\hat{x}}{\hat{b}}, \frac{\hat{y}}{\hat{b}})] \tag{16}$$

where R is the reflectance of the hole grating mirror and

$$U_o{}'(\hat{x},\hat{y}) = U_o(\hat{x},\hat{y}) \exp[i2\pi\hat{W}_s(\hat{x},\hat{y})] \tag{17}$$

contains both amplitude and phase variations of the incident beam plus any additional phase variations, $\hat{W}_s$, introduced by the hole grating mirror itself.

The plane wave spectrum of the outgoing HEL beam is given by the Fourier transform of Eq. (16)

$$A_R(\alpha,\beta) = A_o{}'(\alpha,\beta) * \sqrt{R}[\delta(\alpha,\beta) - T \, somb(\hat{a}\rho)\hat{b}^2 \, comb(\hat{b}\alpha,\hat{b}\beta)] \tag{18}$$

and illustrated in Figure 5.

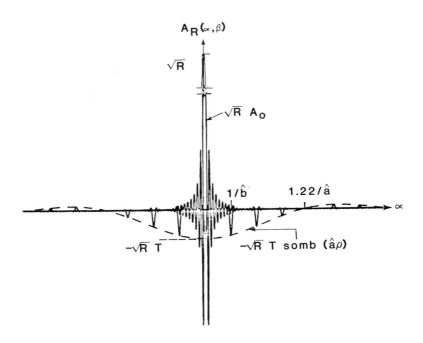

Figure 5. Plane wave spectrum reflected by the hole grating.

Applying the same restrictions as before on the width and separation of these individual diffracted orders, and ignoring subtle interference effects that will be discussed later, we can write the intensity distribution of the outgoing HEL beam reflected from the hole grating mirror as

$$I_R(\alpha,\beta) = R[(1-2T)\delta(\alpha,\beta) + T^2 somb^2(\hat{a}\rho)\hat{b}^2 comb(\hat{b}\alpha,\hat{b}\beta)] * A_O^{'2}(\alpha,\beta) \ . \tag{19}$$

Note that care was taken to include the nonzero cross-term between the specularly reflected HEL beam and central diffracted order. With the exception of the central order, which obviously contains most of the power in the incident beam, we see that the outgoing beam exhibits diffracted orders very similar to the transmitted beam. In fact for a perfect hole grating mirror surface (i.e., $R = 1$, $A_O^{'}(\alpha,\beta) = A_O(\alpha,\beta)$), the outgoing diffracted orders and transmitted diffracted orders are precisely the same in magnitude, shape, and angular separation. This result should come as no surprise since it could have been obtained without the preceding mathematical treatment by the mere application of Babinet's principle.[7]

The radiant power contained in the m,nth diffracted order in the outgoing beam is thus given by

$$P_{R_{(m,n)\neq o}} = RT^2 \ somb^2(\hat{a}\sqrt{m^2+n^2}/\hat{b})P_O \tag{20}$$

and the power contained in the central HEL (undiffracted order) beam is obtained from Eq. (19)

$$P_{R_{o,o}} = R(1-T)^2 P_O \ . \tag{21}$$

Therefore, a hole grating with $a \approx 1$ mm and $b \approx 6$ mm and a reflectance of 0.99 will deliver approximately 95% of the incident power to the target while producing a multiplicity of low-power replicas of the high-power beam in both the transmitted and reflected directions. These low-power beam samples provide the opportunity to perform a variety of diagnostic tests or controls functions simultaneously on a noninterference basis with the intended mission.

HEL diagnostic applications

There are three distinct applications where the hole grating can potentially play a valuable role in HEL diagnostics: (1) as a beam attenuator for laboratory diagnostics, (2) as a beam sampler for real-time field tests, and (3) as a diagnostic tool for local-loop beam control.

Beam attenuator for laboratory diagnostics

The problem of determining the relevant beam characteristics of an HEL device is complicated because the flux densities are too great to be handled by conventional detectors. Furthermore, most attenuators or beamsplitters are unreliable in terms of maintaining uniform attenuation, or they exhibit distortion due to the high incident flux loads.

A transmission hole grating attenuator for analysis of high power laser beam quality was first reported by Marquet.[1] In addition to providing an attenuated beam amenable to measurement by conventional infrared imaging techniques, he illustrates how a lucite burn pattern (Figure 6) can provide a qualitative determination of the flux density distribution at the focus in terms of the threshold level required for burning lucite.

Figure 6. Lucite burn pattern.

Beam sampler for real-time field tests

The hole grating beam attenuator has also been used to evaluate propagation effects such as thermal blooming by comparing spot size at a down-range target for successive shots with the hole grating first in the beam then removed to allow the high-power beam propagate to the target. The results of such tests always suffer from uncertainties due to shot-to-shot variations in the laser beam characteristics. Figure 7 illustrates that real-time diagnostics in the target plane can be performed on the reflected beam samples during field tests designed to study propagation effects or target effects. However, the reflected central diffracted order (HEL beam) is three to four orders of magnitude more powerful than the adjacent low-power beam samples. The nth diffraction ring of the HEL beam can thus interfere either destructively or constructively with the nearest diffracted orders and cause a significant variation in their magnitude (Figure 8). For reasonable values of beam quality, the second and higher reflected orders will be unaffected by this interference phenomena and can be safely used for target plane diagnostics.

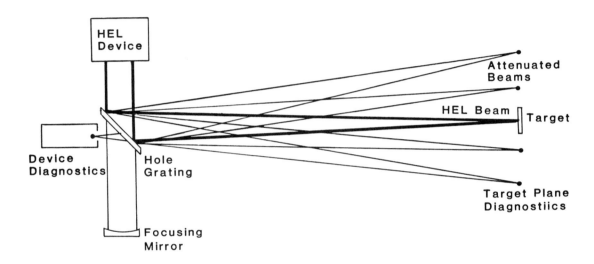

Figure 7. Beam sampler for real-time field tests.

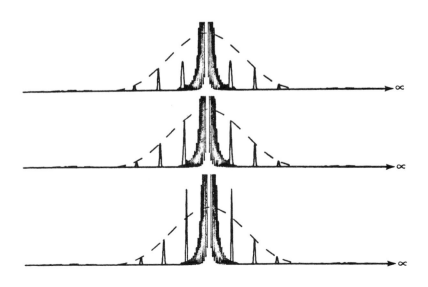

Figure 8. Interference effects upon reflected diffracted orders.

It should be emphasized that this configuration requires a high quality surface on the hole grating mirror and may require cooling to control thermal distortion for continuous wave or long pulse devices.

Diagnostic tool for local-loop beam control

A beam control system will in general consist of four main components: (1) a wavefront sampler that provides one or more low-power replicas of the high-power beam, (2) a wavefront sensor that measures the required wavefront parameters in real time, (3) a control system that accepts and processes the measured data, and (4) a wavefront corrector driven by the control system that implements the required correction, thereby improving the beam quality.

The low-power beam samples produced by the hole grating can be used to perform various local-loop beam control functions such as jitter control and wavefront error compensation (Figure 9). One of the diffracted orders can be directed onto a jitter sensor that drives a tilt mirror through a closed-loop control system to eliminate the beam jitter. Similarly, a real-time wavefront sensor, such as a shearing interferometer, can be used to drive a deformable mirror to compensate wavefront errors inherent in the HEL device or produced by imperfections in the optical train.

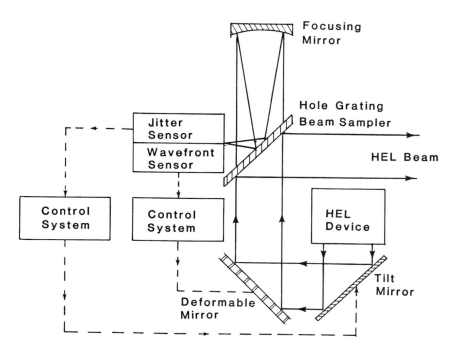

Figure 9. Local-loop beam control.

Specific diagnostic tests

Since a multiplicity of low-power replicas of the high-power beam are produced by the hole grating beam sampler, a variety of diagnostic tests can be performed simultaneously in real time during field tests or other HEL missions. The following are some specific diagnostic tests that can be readily performed.

Total power monitoring (from the laser or on the target) can be performed by merely measuring all of the power in an arbitrary diffracted order and applying Eq. (21) and/or Eq. (14). The moderate amount of dispersion exhibited by low diffracted orders will not affect this diagnostic test.

Similarly, power-in-the-bucket measurements can be performed (for various sized buckets) by placing the appropriate size aperture in front of a detector centered upon a given diffracted order. A very low diffracted order should be used for this test to minimize errors due to disperson.

Both static bore-sight alignment errors and beam jitter can be measured and compensated with appropriate sensors measuring small displacements of their respective beam samples and driving autoalignment systems or jitter compensation systems through the proper use of feedback control systems. The sensitivity of these tests will be degraded somewhat by the dispersion exhibited in the higher diffracted orders.

Wavefront sensing should probably be performed upon the central diffracted order since it has been shown that only this order is unaffected by the presence of small random hole-positioning errors as well as small random phase errors.[2] Furthermore, the central order will not exhibit any aberrations inherent to the diffraction process.[8] Either a real-time shearing interferometer or a real-time Hartmann wavefront sensor can be used.

Finally, the substantial dispersion exhibited by the higher diffracted orders can be used to our advantage to measure the spectral purity of the laser output or to observe and study mode-hopping during laser operation. Figure 10 illustrates the dispersion of the various diffracted orders for a particular hole grating design when there are three discrete wavelengths present. Note that for the low diffracted orders the spectral components within a given order overlap, and for the high diffracted orders the spectral component between adjacent orders overlap; however, for the intermediate diffracted orders the spectral components are completely separated and thus allow relative spectral power measurements to be made. Similarly, mode hopping can be readily observed by monitoring abrupt displacements of a given diffracted order. The sensitivity of these measurements is enhanced for the higher diffracted orders.

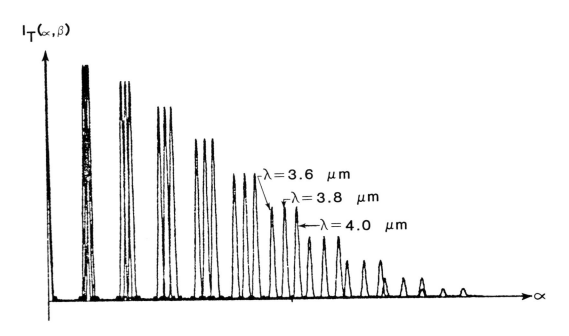

Figure 10. Spectral purity measurements (mode hopping).

Summary

The hole grating beam sampler consists of a uniform array of small circular holes in a highly polished mirror. When used as a turning flat in an HEL optical train, it produces a multiplicity of angularly separated low-power replicas of the HEL beam in both the transmitted and reflected directions. Fourier optical techniques have been used to derive engineering equations useful to the HEL systems engineer in the design and analysis of hole grating beam samplers for a variety of diagnostic applications. It is then suggested that the reflected diffracted orders as well as the transmitted diffracted orders are available for performing a variety of diagnostic tests simultaneously in real time during HEL field tests. Potential pit-falls due to interference effects when doing diagnostics on the reflected beam samples are presented. Finally, after discussing a variety of specific diagnostic tests that can readily be performed upon the low power beam samples produced by the hole grating, we show that the dispersion exhibited by the higher diffracted orders can be used to make spectral purity measurements or to observe and study mode-hopping in the laser device.

Appendix

The symbolic representation of mathematical functions and operations used throughout this paper demonstrates the ease and convenience with which certain mathematical manipulations can be performed when dealing with Fourier analysis or linear systems theory. A brief review defining those functions and operations used in this paper is provided here. Note that we have used a scaled coordinate system where.

$$\hat{x} = x/\lambda, \ \hat{y} = y/\lambda, \ \hat{z} = z/\lambda, \ \text{and} \ \hat{r}^2 = \hat{x}^2 + \hat{y}^2 + \hat{z}^2 \ .$$

The reciprocal variables in Fourier transform space

$$\alpha = \hat{x}/\hat{r}, \ \beta = \hat{y}/\hat{r}, \ \text{and} \ \gamma = \sqrt{1-(\alpha^2+\beta^2)} = \sqrt{1-\rho^2}$$

are then the <u>direction cosines</u> of the propagating light.

Symbolic representation of mathematic functions

Dirac delta function $= \delta(\hat{x}-\hat{a}) \equiv \begin{cases} 0 & \hat{x} \neq \hat{a} \\ \\ \int_{-\infty}^{\infty} \delta(\hat{x}-\hat{a})\,d\hat{x} = 1 \end{cases}$

Replicating function $= \dfrac{1}{|\hat{a}|} \text{comb}\left(\dfrac{\hat{x}}{\hat{a}}\right) \equiv \displaystyle\sum_{n=-\infty}^{\infty} \delta(\hat{x}-n\hat{a})$

Cylinder function $= \text{cyl}\left(\dfrac{\hat{r}}{\hat{a}}\right) = \begin{cases} 1, & 0\leq\hat{r} < \hat{a}/2 \\ 1/2 & \hat{r} = \hat{a}/2 \\ 0, & \hat{r} > \hat{a}/2 \end{cases}$

Sombrero function $= \text{somb}\left(\dfrac{\hat{r}}{\hat{a}}\right) = \dfrac{2J_1\left(\frac{\pi\hat{r}}{\hat{a}}\right)}{\frac{\pi\hat{r}}{\hat{a}}}$

Convolution operation

$$f(\hat{x}) * g(\hat{x}) = \int_{-\infty}^{\infty} f(u)\,g(\hat{x}-u)\,du$$

Fourier transform pairs

Definition:
$$F(\alpha) = \int_{-\infty}^{\infty} f(\hat{x}) e^{-i2\pi\alpha\hat{x}} d\hat{x}$$

$$f(\hat{x}) = \int_{-\infty}^{\infty} F(\alpha) e^{i2\pi\alpha\hat{x}} d\alpha$$

Examples:
$$f(\hat{x}) \longrightarrow F(\alpha)$$
$$\delta(\hat{x}) \longleftrightarrow 1$$
$$\text{comb}(\hat{x}) \longleftrightarrow \text{comb}(\alpha)$$
$$\text{cyl}(\hat{r}) \longleftrightarrow \frac{\pi}{4} \text{somb}(\rho)$$

Similarity theorem

Given: $f(\hat{x}) \longleftrightarrow F(\alpha)$, then $f(\hat{x}/\hat{a}) \longleftrightarrow |\hat{a}| F(\hat{a}\alpha)$

Convolution theorem

Given: $f(\hat{x}) \longleftrightarrow F(\alpha)$

$g(\hat{x}) \longleftrightarrow G(\alpha)$,

then: $f(\hat{x}) * g(\hat{x}) \longleftrightarrow F(\alpha)G(\alpha)$

Acknowledgments

This work was supported by the Air Force Weapons Laboratory (AFWL), Air Force Systems Command, Kirtland Air Force Base, NM, under contract F29601-79-C-0027.

References

1. Marquet, L. C., "Transmission Diffraction Grating Attenuator for Analysis of High-power Laser Beam Quality," App. Opt. Ltrs, Vol. 10, p. 960. 1971.

2. Sziklas, E. A., Appendix J, "Theory of the Diffraction-Grating Attenuator" (1973) (see J. E. Harvey for further details).

3. Hogge, C. B. and Butts, R. R., "Hole Grating Samplers for High Power Laser Beam Diagnostics," AFWL-TR-78-15 (1978).

4. Bracewell, R. N., The Fourier Transform and Its Application, McGraw-Hill Co., 1965.

5. Goodman, J. W., Introduction to Fourier Optics, McGraw-Hill, 1968.

6. Gaskill, J. D., Linear Systems, Fourier Transforms, and Optics, Wiley, 1978.

7. Born, M. and Wolf, E., Principles of Optics, Pergamon Press, 1964.

8. Harvey, J. E. and Shack, R. V., "Aberrations of Diffracted Wave Fields," Appl. Opt., Vol. 17, p. 3003. 1978.

Scattering from high efficiency diffraction gratings

G. J. Dunning, M. L. Minden

Hughes Research Laboratories, 3011 Malibu Canyon Road, Malibu, California 90265

Abstract

Scatter from high efficiency, reflective diffraction gratings has been characterized. Three major types of scatter are identified: (1) random, which occurs over 2π s.r., (2) band or ghost scatter which occurs in the plane of incidence and (3) structured scatter which is a symmetrical pattern repeated about each order. Measurements were taken at 10.6 μm on gratings made by ruling, ion-etching, holography, or a combination of these techniques. We have found that the characteristic scatter from these high efficiency gratings depends strongly on the way the gratings are fabricated.

Introduction

Scatter from diffraction gratings has not been defined and measured as extensively as other properties of gratings. However the use of diffraction gratings as optical components in increasingly complex systems has made the understanding and minimization of scatter increasingly important. Diffraction gratings are now being used in high power laser systems as beam splitters, beam samplers and aperture sharing components. [1,2] These gratings are typically designed to have high efficiencies and low losses.

Earlier measurements by Verrill [3,4] at visible wavelengths and the statistical analysis of Sharpe and Irish [5] based on Scaler diffraction theory have contributed substantially toward characterizing grating scatter. However, scatter theory and visible measurements cannot be scaled to the very high efficiency gratings presently being incorporated in high power systems. Such gratings are typically used in littrow or near-littrow configuration with a small period to wavelength ratio ($0.6 < d/x < 1$) and deep grooves ($h/x \sim 0.25$). In this regime scalar theory is invalid. i.e., the polarization of the incident beam strongly effects the scattered energy.

This study addresses the scattered energy from very high efficiency gratings designed for CO_2 laser wavelengths. The scatter has been characterized both qualitatively and quantitatively [6] for gratings fabricated by various techniques.

A representative cross section of fabrication techniques were chosen in order to extract as much information as possible from a limited number of samples. The grating grooves were shaped either by a ruling engine or by ion etching through a holographically-exposed or contact-printed mask.

Fabrication

Each grating was designed to have diffraction efficiency of at least 90% [7,8] at 10.6 μm in the TM ($E\perp$ to grooves) polarization when used in the near-littrow configuration. The substrates were 1.5-in. cooper discs, polished flat to $\lambda/4$. Most of the substrates were subsequently Kanigen coated, to aid in gold deposition, and then repolished.

The two ruled gratings that we used were ruled by Hyperfine, Inc. Sample RB37 is a conventional blazed grating (37° blaze) with a triangular groove profile. Sample RR1 was ruled with a "rectangular" groove profile of width 1.4 μm and depth 2.4 μm. According to the vendor, the groove is probably wider than 1.4 μm at the top, giving a trapezoidal structure comparable to the contact-printed, ion-etched gratings. Both gratings are ruled into greater than 3 μm of evaporated gold, with a groove period of 6.67 μm.

The ion-etched gratings were made at Hughes Research Laboratory using sputtered argon ions to etch into gold through a photoresist mask. [9] The gold on some samples was sputter deposited; on others it was evaporated. The photoresist masks were exposed either holographically or by contact printing.

The holographic exposures were made with a HeCd laser at 4416A. An antireflection (AR) coating [10] on the gold was required before the photoresist was spun on. This coating helped to minimize the standing waves formed by multiple reflections within the photoresist and greatly enhanced the edge definition of the groove walls. After the gratings were etched and the photoresist mask was dissolved, a light etch was required to remove the remaining AR coat from the top of the groove. The period of these gratings was 6.45 μm.

The contact-printed patterns were exposed with multi-line UV light (a medium-pressure mercury lamp) through a ruled aluminum-on-quartz master pattern. This pattern was also ruled by Hyperfine. One of the contact-printed gratings was given an AR coating in the same sequence as the holographic gratings.

Overcoats of evaporated gold, silver, and silver/thorium fluoride were also added to some of the gratings. In Figure 1, SEM photographs show the final groove profile. Photographs were taken of the grating grooves and of the groove cross-section at magnifications which best illustrate surface structure.

HOLOGRAPHICALLY EXPOSED GRATING

CONTACT—PRINTED GRATING

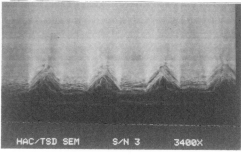
END VIEW OF HOLOGRAPHIC GRATING

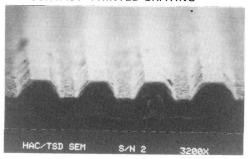
END VIEW OF CONTACT—PRINTED GRATING

Figure 1. SEM photographs of ion etched gratings

Scattering patterns

To describe qualitatively the spatial distribution of scattered energy, we photographed the pattern of light surrounding two diffracted orders at 6328A. The patterns for each of the gratings are shown in Figure 2. To emphasize the surrounding scatter, we photographed two orders that were near the blaze direction yet of relatively low efficiency. We were able to identify three major types of scatter: random, band, and structured.

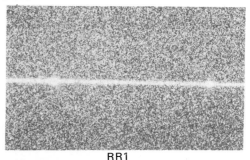
RR1

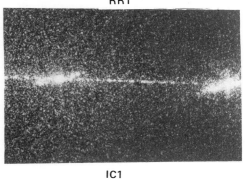
IC1

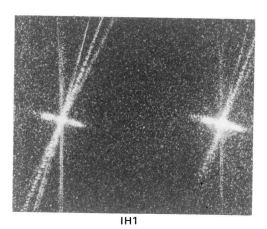
IH1

Figure 2. Scatter patterns photographed at 6328 Å to emphasize variations in scatter produced by ruled, contact printed and holographic gratings.

Random scatter, also called diffuse scatter, is distributed into 2π S.r. It is caused by surface roughness and random deviations from a perfect groove shape. The grooves in a ruled grating are typically much rougher than in an ion-etched grating (see Figure 3).

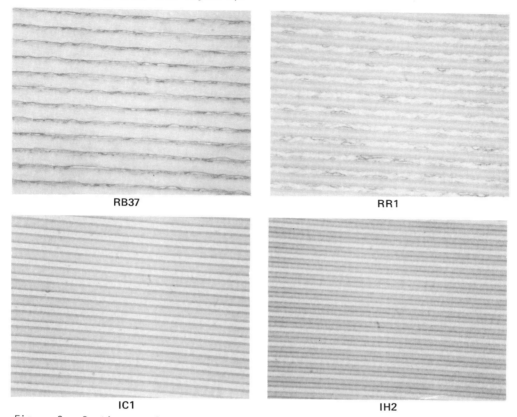

RB37

RR1

IC1

IH2

Figure 3. Grating samples photographed through a visible microscope: 1000X

Band or ghost scatter, also called "grass", is defined as light that is scattered in the plane of incidence. Ghosts are spurious orders found in this plane that are caused by periodic errors in the ruling of a ruled grating. Spurious orders were a serious problem until the advent of interferometrically controlled ruling engines. Although gratings ruled on the new engines do not have such clearly defined ghost orders, they still exhibit a lower intensity band of light in that plane. This is because the control is not perfect [11]. Errors are introduced because it is the diamond carriage rather than the groove position which is monitored. In addition, the correction for a sensed error may be applied only to a subsequent groove therefore this feedback loop eliminates accumulated errors but not random ones. This band shows up clearly in the visible scattering photographs of the ruled samples as well as in the ion-etched gratings contact printed from a ruled master pattern. The holographically exposed gratings have no distinctive band. However, by defining band scatter to include any plane-of-incidence scatter we can use it to compare the performance of each of the gratings in any practical system with in-plane detectors.

Structured scatter is less well defined. We identify it as a symmetrical pattern that is repeated about each order. The pattern varies in orientation and magnitude as the beam is scanned across the grating surface, but its symmetry about the order does not vary. Structured scatter in the holographically exposed gratings appeared as a flare extending to either side of the order. This flare was not as strong in the contact-printed gratings. However, these gratings also exhibited bands of light parallel to the plane of incidence. No structural scatter was observed in the ruled gratings. We therefore concluded that the scatter was caused by a slight but systematic modulation of the groove pattern in the photoresist exposure. This may be due to stray reflections or multiple images in the exposure setup (for example, an imperfect AR coating on the master pattern used in contract printing) or due to index or depth variations in the photoresist. In either case, the groove modulations caused by the imperfections will act as a hologram upon playback.

Measurements at 10.6 μm

Experiments at 10.6 μm were made both to determine what types of scattering measurements are meaningful for gratings, and to estimate the magnitude of the different types of scatter. This work was performed in an optical scattering measurement facility which was designed for measuring scatter from highly polished mirrors. Only minor modifications were needed for making measurements on gratings. The measurements were made for both TE (E // to grooves) and TM (E ⊥ to grooves) waves.

The measurement of band scatter was the most straightforward of the three. A converging beam is incident onto a grating whose grooves have been carefully aligned to the vertical. The beam is then diffracted horizontally. Since these gratings have no optical power (the grooves are straight lines), both the reflected (n = 0) and diffracted (n = -1) beams focus approximately at the same distance from the grating. At this distance, band scatter forms a semicircle focus of light, with the orders appearing as points along the arc.

At the focal distance (31.5 cm from the grating), we placed a HgCdTe detector with a 2 x 2 mm aperture and scanned between the two orders at 5° intervals. The results are shown in Figure 4. The band scatter varies by over three orders of magnitude. As expected, the holographic gratings, which have no distinct band of on-axis scatter at 6328Å, are also the best at 10.6 μm. Similarly, the ruled grating with the rectangular groove profile (RR1), which showed the strongest band at 6328Å, is the strongest scatterer at 10.6 μm.

In Table 1 (end of text), order-of-magnitude intensities are listed for the band scatter of each grating in a 50° arc midway between the n = 0 and n = -1 orders. Since the detector was larger than the width of the band, we have effectively integrated over the scatter in the vertical direction. The scattering intensities relative to the incident power are therefore listed in radians^{-1}. We confirmed that the on-axis band scatter retained the polarization of the incident beam. This may be useful in eliminating noise from systems which use these gratings.

Radiation out of the plane of incidence was collected by a spherical mirror which was scanned between the n = o and n = -1 orders. This mirror was set on a scanning boom with the grating at its center of curvature (see Figure 5). A pyroelectric detector was set on the boom, as close to the grating as possible, and the scattered light collected by the spherical mirror converged onto the detector. The angle of scatter intercepted by the mirror was 0.1 sr. We measured the collected scatter from seven horizontally adjacent areas, with a slight overlap between them. The first and last of the seven were centered about the n = 0 and n = -1 orders. These orders were focused onto a small (0.001 sr) rejecting mirror and deflected into a beam dump.

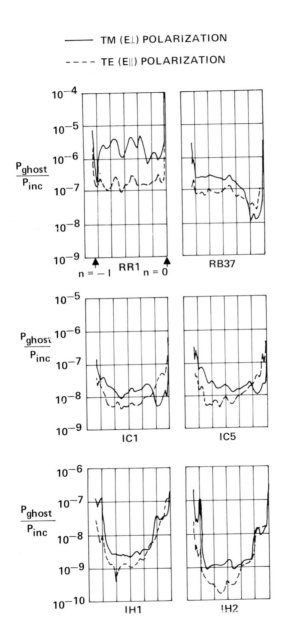

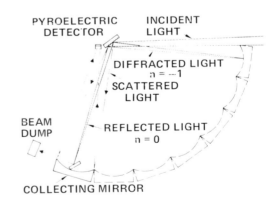

Figure 5. Method of measuring out of plane random scatter in 0.1 sr. A 0.001 sr rejecting mirror in front of the collecting mirror is used to deflect the n=o and n=-1 orders.

Figure 4. Interorder band or ghost scatter, measured at 10.6μm into 6.3mrad for ruled, contact printed and holographic gratings.

The results of the measurements are shown in Figure 6. Again, the holographic gratings are best, followed by the contact-printed gratings, and last by the ruled gratings. From these measurements, we can extract information about both order scatter and random scatter.

To estimate the intensity of the random scatter between the orders, we averaged over the three central readings of collected energy for each grating and polarization. From these values, we subtracted an average value for the band scatter (based on the in-plane measurements), which was intercepted by the collecting mirror along with the random scatter. We then divided by the solid angle of the collecting mirror (0.1 sr). The resulting numbers, listed in Table 1, thus give the average intensity of randomly scattered light between the two orders, measured within 10° above and below the plane of incidence.

Order scatter is the dominant component of the collected scatter about the $n = 0$ and $n = -1$ orders in ion-etched gratings. Unfortunately, we were unable to define a consistent figure of merit for this phenomenon because the measured values changed significantly as the incident beam impinged on different portions of the gratings.

The symmetry of this type of scatter dictates that, at any one position of the grating, the shape of the scattering pattern should be the same about both the $n = -1$ and $n = 0$ orders. Our experiments suggest that the magnitude of this scatter is also comparable about orders of similar intensity. This can be seen in all the ion-etched gratings. The value of scatter about the $n = 0$ order in the TE polarization is invariably similar in magnitude to the $n = -1$ order of the TM polarization. These are the two orders that carry most of the incident energy in their respective polarizations. The two values remain comparable to each other even when the absolute magnitude varies widely.

Summary

The characteristic scatter from high-efficiency gratings at 10.6 μm depends strongly on the way the grating is made. Holographically exposed, ion-etched gratings were better than two orders of magnitude lower in scatter than ruled gratings when measured in the plane of incidence halfway between the reflected and diffracted orders. This occurred because the in-plane band of high-intensity scatter (ghost band) characteristic of ruled gratings is absent in holographic gratings. Off axis, the holographic gratings were still almost two orders of lower in scatter than the ruled gratings. Symmetrical scattering patterns about the orders of the ion-etched gratings (both contact printed and holographic) are caused by uneven photoresist deposition and stray reflections during exposure.

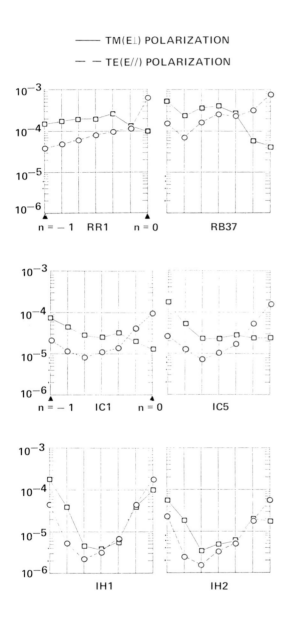

——— TM(E⊥) POLARIZATION

— — TE(E//) POLARIZATION

Figure 6. Out of plane random scatter at 10.6μm for ruled, contact printed and holographic gratings.

TABLE 1. Comparison Chart of Grating Fabrication Techniques and Properties.

	Sample	period	gold deposition	A/R coated before exposure?	groove depth	over coat	Interorder ghost scatter relative to P_{inc} (rad^{-1}) $E\perp$	$E\parallel$	Interorder random scatter relative to P_{inc} ($sterrad^{-1}$) $E\perp$	$E\parallel$
Contact Printed & Ion-Etched	IC1	6.67μm	sput	no	2.3-2.4μm	none	2×10^{-6}	1×10^{-6}	3×10^{-4}	1×10^{-4}
	IC2	6.67μm	evap	no	2.3-2.4μm	none	6×10^{-6}	1×10^{-6}	1×10^{-4}	2×10^{-4}
	IC3	6.67μm	sput	no	2.3-2.4μm	Ag/AgThF$_4$	2×10^{-5}	5×10^{-6}	2×10^{-3}	5×10^{-4}
	IC4	6.67μm	sput	no	2.3-2.4μm	evap Au	5×10^{-6}	1×10^{-6}	3×10^{-4}	2×10^{-4}
	IC5	6.67μm	evap	yes	2.3-2.4μm	none	3×10^{-6}	1×10^{-6}	2×10^{-4}	1×10^{-4}
	IC6	6.67μm	evap	no	2.3-2.4μm	none	1×10^{-5}	1×10^{-6}	5×10^{-4}	9×10^{-5}
Holographic Exposure & Ion-Etched	IH1	6.45μm	evap	yes	2.3-2.4μm	AgThF$_4$	5×10^{-7}	2×10^{-7}	4×10^{-5}	4×10^{-5}
	IH2	6.45μm	sput	yes	2.3-2.4μm	none	2×10^{-7}	6×10^{-8}	5×10^{-5}	3×10^{-5}
Ruled	RR1	6.67μm	evap		2.4μm	none	3×10^{-4}	3×10^{-5}	1×10^{-3}	7×10^{-4}
	RR37	6.67μm	evap		37° blaze	none	2×10^{-5}	1×10^{-5}	3×10^{-3}	2×10^{-3}

Acknowledgements

The authors are indebted to the following people for their contributions to this study: H. Garvin and K. Robinson for the fabrication of the ion etched gratings, A. Au and R. Mullen for the exposure of the holographic gratings, B. Bach of Hyperfine, Inc. for supplying the ruled gratings and R. Loveridge and C. Stevens for their assistance in adapting the optical scattering facility for use with diffraction gratings. We would especially like to thank Gary Janney for helpful discussions throughout this program.

References

1. Janney, G., "Ring Resinator Gas Dynamic Laser with Diffraction Grating Coupling and Axial Mode Selection," presented at IEDM, Washington, D.C., Oct. 1971.
2. Janney, G., Garvin, H., Close, D., "Nonconventional Diffraction Grating Coupling and Axial Mode Selection," presented at Sixth Winter Colloquium on Quantum Electronics," Steamboat Springs Co., Feb. 1976.
3. Sharpe, M., Irish, D., Optica Acta 1978, Vol. 25, No. 9, pg. 861-893.
4. Verrill, J. F., Optica Acta 1970, Vol. 17, No. 7, pg. 747.
5. Verrill, J. G., Optica Acta 1978, Vol. 25, No. 10, pg. 531-547.
6. Minden, M., Dunning, G., Hughes Research Report No. 528, Oct. 1979.
7. Loewen, E., Neviere, M., Maystre, D., Applied Optics, Oct. 1977, Vol. 16, No. 10.
8. Kalhor, H., Neureuther, A., Journal of Optical Society of America, Jan. 1971, Vol. 61, No. 1.
9. Garvin, H. et al., Applied Optics 1973, Vol. 12, No. 3, pg. 455.
10. Garvey, J. et al., Rome Air Development Center Report, RADC-TR-78-275, Vol. I, March 1979, "Holographic Grating Study".
11. Stroke, G., Handbuch Der Physick, Vol. 25, 1967, pp. 687-730.

Polarization-dependent phase shift in high-efficiency gratings

M. L. Minden, G. J. Dunning

Hughes Research Laboratories, 3011 Malibu Canyon Road, Malibu, California 90265

Abstract

Light diffracted from a high efficiency, reflective grating may undergo a phase shift which depends strongly on the incident polarization. If the incident wave includes both TE (E // groove) and TM (E⊥ groove) polarization, then the diffracted orders may be elliptically polarized. This effect is of special interest in curved-line (holographic) gratings in which the eccentricity of the ellipse may vary across the diffracted wavefront. In this paper the existence of a substantial phase shift between the two polarization is demonstrated in gratings with a period-to-wavelength ratio less than one. Measurements are made at 10.6 μm on both ruled and ion-etched gratings. Comparison with theory shows that the measured effect is due to groove shape rather than to finite conductivity of the substrate.

I. Introduction

Studies of high efficiency gratings have normally been concerned with the distribution of diffracted energy; for example, the polarization sensitivity of the diffraction efficiency[1], the effect of dielectric coatings,[2] or the characterization of scattering and absorption losses.[3] Phase shifts in the diffracted waves have been of little concern since gratings were ruled in straight lines and were examined with only one polarization component in the incident wave. A wave polarized with its electric field either perpendicular to the grooves (TM wave) or parallel to the grooves (TE waves) will always be linearly polarized upon diffraction. However, if both polarization components are present in the incident wave, a phase shift on the grating of one component relative to the other will cause the diffracted light to be elliptically polarized.

The advent of high efficiency gratings with curved grooves, such as those generated holographically or by reduction of computer-generated patterns, introduces the possibility that spatial variations in groove orientation will be translated into spatial variations in polarization state in the diffracted wave.

In this study we measured the polarization-dependent phase shift (PDPS) generated by a set of high efficiency, linear gratings. Our purpose was to determine whether the phase shift was large enough to be of practical interest and to compare the measured values to those predicted by a computer code. The codes could then be used to predict regimes in which the expected phase shift is significant.

II. Experiment

The high efficiency gratings used in this experiment were either ruled or ion-etched into gold. A diffraction efficiency of at least 90% in the TM polarization was required of each grating, measured at 10.6 μm in the near-littrow configuration of Figure 1. Grating frequency was 150 lines/mm.

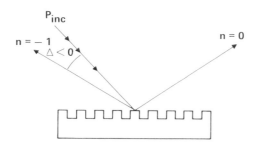

Figure 1. Beam geometry in near-littrow configuration. Δ ≈ -15°

Ellipsometry was used to analyze the polarization of the diffracted light from various gratings. The natural coordinate system to work with is the one defined with one axis parallel to the grating groove direction and the other axis perpendicular. Any plane waves can then be decomposed into these two orthogonal components. These two components may have arbitrary amplitudes and phase with respect to each other, and may thus be defined by angles Δ and α, where

$$\vec{E} = \left[\hat{x} + \hat{y} \ \tan\alpha e^{-i\Delta} \right] E_0$$

Elliptically polarized light can be completely characterized by the angle ψ that the major axis makes with the axis perpendicular to the grooves and by the ratio of the length a of the semimajor axis to the length b of the semiminor axis, where $a/b \equiv \tan \chi$ (see Figure 2).

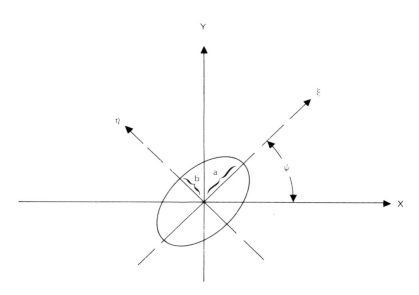

Figure 2. Orientation of polarization ellipse. The grating grooves define the x-y coordinate system. The major and minor axes of the ellipse define the η - ξ coordinate system, and ψ is the rotation angle between the two systems.

From Born and Wolf,[4] the two pairs of angles are related by

$$\tan 2\psi = \tan 2\alpha \ \cos\Delta$$
$$\sin 2\chi = \sin 2d \ \sin\Delta$$

which can be solved for the phase difference, where $\tan \chi > 0$ and ψ is restricted so that $0 < \psi < \pi$.

$$\Delta = \cos^{-1} \left[\frac{\cos 2\chi}{\left(1 + \dfrac{\sin^2 2\chi}{\tan^2 2\psi} \right)^{\frac{1}{2}}} \right] . \qquad (1)$$

The following experiment was used to determine ψ, χ, δ, and α for each grating. The experimental arrangement is shown in Figure 3. The output from a line stabilized CO_2 laser was passed through a $\lambda/2$ plate and then a polarizer prior to the grating. This combination ensured the linearity of the polarization and allowed the E vector to be rotated with respect to the grating grooves. The diffracted beam of elliptically polarized light then passed through a $\lambda/4$ plate and a second polarizer. The $\lambda/4$ plate was rotated until the emergent light was linearly polarized (i.e., until a major or minor axis of the ellipse was parallel to a principal axis of the $\lambda/4$ plate). The second polarizer was then rotated until the emergent light was extinguished. The angle the principal axis of the $\lambda/4$ plate makes with respect to the grating grooves is the angle ψ. χ, defined previously, is found by the angle the second polarizer makes with respect to the axes of the ellipse. Eq. 1 is used to compute the phase difference Δ.

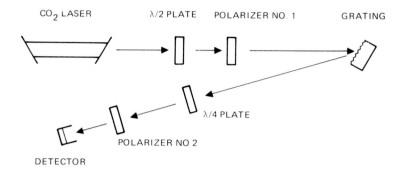

CO$_2$ LASER   λ/2 PLATE   POLARIZER NO. 1   GRATING

λ/4 PLATE

POLARIZER NO 2

DETECTOR

Figure 3.   Schematic of grating ellipsometry experiments

The experimental results are tabulated in Table 1 (end of text). Values ranged from a low phase shift of 15° for a blazed (triangular profile) grating to a high of 55° for an ion-etched sample which had been over-coated.

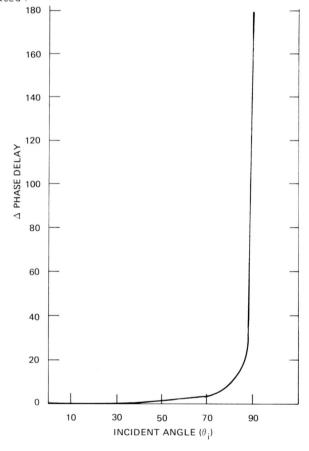

Figure 4.   Relative phase delay between TE/TM
polarization components on
reflection from a gold surface at
10.6 μm

III.   Discussion

Polarized light reflecting from an unmodulated surface of finite conductivity will exhibit a phase shift predicted by the Fresnel reflection coefficients. The coefficients depend on the optical constants of the material, the angle of incidence, and the polarization of the incident light.

When 10.6-μm radiation is reflected from a gold surface, Figure 4 shows that the phase difference between the TE and a TM wave is less than 3° for an incident angle of less than 60°. The phase delay Δ can be found as a function of the incident angle θ$_i$ by solving the following simultaneous equations from Born and Wolf.[4]

$$n = - \frac{\sin \theta_i \, \tan \theta_i \, \cos 2\psi}{1 + \sin 2\psi \, \cos \Delta}$$

$$K = \quad \tan 2\psi \, \sin \Delta \quad . \tag{2}$$

Some indication that the surface optical constants were affecting the high PDPS grating measurements was given by the large phase shift in sample IC4. The poor quality of the surface in this sample was indicated by its low reflectivity. Absorption measurements at 10.6μ showed IC4 had higher absorption than other samples. The samples had a grating fabricated on only half of the surface. Absorption measurements on both halves allowed us to distinguish between absorption due to the grating structure or absorption due to surface preparation. The flat surface of IC4 showed an absorptivity of 1.8% compared to the 1.0% typical of other samples.

The importance of groove shape, however, was also indicated clearly by the measurements. In particular, samples RR1 (ruled, rectangular groove profile) and RB37 (ruled, blazed at 37°) were ruled on identical substrates but were over 25° apart in measured phase difference.

At this point, however, we were unable to determine whether the large phase shifts observed in the uncoated gratings were due to the grating structure or whether the surface modulation was simply enhancing the effect of the finite conductivity and the fresnel coefficients. We wanted to know whether an ideal structure, one with infinite conductivity and perfectly smooth walls, would show the same effect, therefore we used computer modeling.

IV. Computer model

The computer code we used to model the PDPS effect was a modified version of a code developed originally by Kalhor and Neureuther[5] for grating efficiency. The grating profile is modelled by a set of contiguous straight line segments and we assumed infinite conductivity in the substrate.

We initially chose ideal groove profiles as shown in Figure 5. The ruled grating RR1 we assumed to have slightly narrower grooves than the ion etched gratings based on the description given by the vendor. For sample RB37 we assumed the apex angle to be 90°.

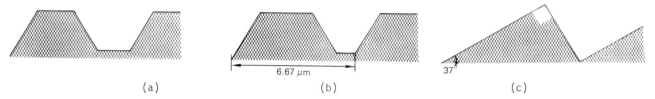

(a)   (b)   (c)

Figure 5. Assumed groove profiles of (a) ion-etched, contact printed, (b) ruled "rectangular" and (c) ruled, blazed gratings.

TABLE 2. Polarization dependent phase shift (PDPS) measured values versus predicted values for the groove profiles of figure (5).

Sample Number	Measured Phase Shift (Δ)	Code Prediction
IC1	35.2°	
IC2	32.5°	30.6°
IC5	24.0°	
RR1	45.6°	40.2°
RB37	16.7°	13°

Good agreement is seen in Table 2 between the predictions of the code and measurement of the uncoated gratings. Even better agreement was found where accurate profiles could be determined. Each set of gratings fabricated by ion etching was accompanied by a witness sample on a silicon substrate. Photographs of these samples under SEM magnification (figure 6) were then used to determine the widths of ridge tops and groove bottoms. Calibration of groove depth was made to within 0.1 µm. In figure 7 we show calculations of relative phase shift Δ vs. groove depth for the three witness samples. Superimposed are measured values for the uncoated samples IC1, IC2, IC5 and IH2, which fall very nicely on the appropriate curves. Since the code assumed infinite conductivity, the agreement found between theory and experiment must indicate that the large phase shifts we observed are due to the grating structure itself, rather than to an enhancement of the finite surface conductivity.

This is not difficult to explain physically. Consider an ideal grating with narrow, rectangular grooves. The fields within the grooves will be of the same form as those within an open rectangular waveguide. If the grooves are narrower than $\lambda/2$, they can be penetrated by light in the TM (E∥) polarization, which can support a zero-order mode, but not by light in the TE(E⊥) polarization, which cannot. In effect, the two polarizations will perceive the "surface" of the grating to be at different levels, and the optical phase on diffraction will be different.

The lower phase shift of the blazed grating is consistent with this physical explanation. This structure is the most open, and the least like a rectangular waveguide, of the gratings listed. The two polarizations can perceive nearly the same structure.

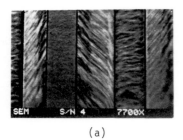

(a)

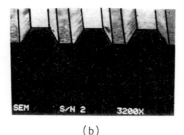

(b)

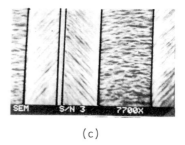

(c)

Figure 6 (above). SEM photographs of grating witness samples. Lines were drawn to facilitate measurement of ridge top and groove bottom. Corresponding samples are (a) IC1 through IC3 (b) IC4 and IC5 (c) IH1 and IH2.

Figure 7 (right). PDPS predicted from witness sample groove shapes and corresponding measurements of the uncoated gratings. The brackets indicate uncertainty in the experimentally measured groove depth.

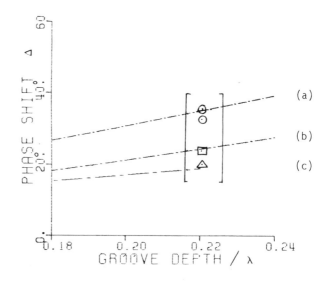

For similar reasons we found that the PDPS diminished quickly with larger periods, $d/\lambda > 1$. A calculation of the relative phase shift Δ is shown in figure 8 for blazed gratings in the littrow configuration. At each point on the curve, the blaze angle is assumed equal to the littrow angle in order to guarantee high efficiency.

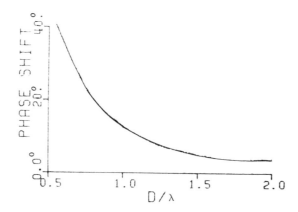

Figure 8. Calculated PDPS for high efficiency gratings blazed at the littrow angle. Phase shift rises sharply as n=-1 order approaches cutoff.

V. Conclusion

We have seen that a large phase shift can exist between the TE and TM polarization components diffracted from a high efficiency grating when the groove period is less than a wavelength. The phase shift is strongly affected by groove profile and overcoating, and it can be accurately modelled by existing diffraction efficiency codes. This suggests that, in the future, ellipsometric measurements of PDPS could be used in conjunction with computer codes and efficiency measurements to assess accurate groove and/or coating profiles.

Table 1. Summary of measured samples

	Sample	period	gold deposition	A/R coated before exposure?	groove depth	over coat	polarization-dependent phase shift
Contact Printed & Ion-Etched	IC1	6.67μm	sput	no	2.3-2.4μm	none	35.2°
	IC2	6.67μm	evap	no	2.3-2.4μm	none	32.5°
	IC3	6.67μm	sput	no	2.3-2.4μm	Ag	47.3°
	IC4	6.67μm	sput	no	2.3-2.4μm	evap Au	53.7°
	IC5	6.67μm	evap	yes	2.3-2.4μm	none	24.0°
Holographic Exposure & Ion-Etched	IH1	6.45μm	evap	yes	2.3-2.4μm	$AgThF_4$	25.2°
	IH2	6.45μm	sput	yes	2.3-2.4μm	none	20.2°
Ruled	RR1	6.67μm	evap		2.4μm	none	45.6°
	RB37	6.67μm	evap		37° blaze	none	16.7°

Acknowledgements

We would like to express our appreciation of the many people who contributed to the fabrication of the grating samples: A. Au, B. Back, H. Garvin, R. Mullen, K. Robinson and E. Rudisill. We would also like to thank A. Neureuther of UC Berkley for access to the grating diffraction code and to D. Fink for technical discussions.

References

1. Loewen, E. G., Neviere, M. and Maystre, D., "Grating Efficiency Theory as it Applies to Blazed and Holographic Gratings," Applied Optics, V. 16, No. 10, Oct. 1977, p. 271.
2. Maystre, D., "A New General Theory for Dielectric Coated Gratings," JOSA, V. 68, No. 4, April 1978, p. 490.
3. M. Minden, G. J. Dunning, "High Performance Diffraction Gratings," Research Report 528, October 1979, Hughes Research Laboratories.
4. Born, M. and Wolf, E., Principles of Optics, 4th ed., Pergamon Press.
5. Kalhor, H. A. and Neureuther, A. R., "Numerical Method for the Analysis of Diffraction Gratings," JOSA, V. 61, No. 1, Jan. 1971, p. 43.

Increase in Raman excitation of surface polaritons with surface roughness explained in terms of Wood anomalies of gratings

Raymond Reinisch

ENS-IEG—Laboratoire de Génie Physique—ERA 836, Domaine Universitaire
B.P. n° 46-38 402 St. Martin d'Hères, France

Michel Nevière

Laboratoire d'Optique Electromagnétique, E.R.A. n° 597
Faculté des Sciences et Techniques de St-Jérôme, 13 397 Marseille Cédex 4, France

Abstract

In a recent publication Ushioda et al. (Phys. Rev. B. 19, 8, (1979), 4012) (henceforth referred to as I) reported an experiment in which Raman excitation of surface polaritons was performed using several rough surfaces. The authors of I observed an increase, with surface roughness, of the Raman scattering intensity, but were unable to give a theoretical explanation of this surprising effect. We suggest an explanation which is based on the similarity existing between Raman excitation of surface polaritons of frequency ω along a grating and grating generation of surface polaritons using an incident E.M. wave of the same frequency ω. We determine the surface polariton wave along the grating and find a strong enhancement of its intensity when increasing the groove depth. Thus there is a good qualitative agreement between our results and those published in I. This allows us to show that the phenomenon observed by the authors of I is related to the existence of Wood anomalies of gratings.

Introduction

In a recent publication (1) (henceforth referred to as I) Ushioda et al. reported an experiment in which Raman excitation of surface polaritons was performed using several rough surfaces. The authors obtained the following main results when the surface roughness was increased :

1) A dispersion curve which is slightly depressed below the theoretical one based on flat surface assumption.
2) A surface polariton linewidht which broadens.
3) An increase in Raman scattering intensity

The first two points are well accounted for by the existing theories, but the authors of I wer unable to give a theoretical explanation concerning point 3.

The purpose of this paper is to suggest an explanation to this last point i.e. to the increase of the observed Raman scattering intensity with surface roughness.

Surface polaritons may be excited, along a flat or rough surface, by various experimental methods which can crudely be classified as follow :
In the first one, in order to obtain a resonant interaction, a non linear phenomenon is used, such as spontaneous (2) or stimulated (3) Raman scattering where the frequency of the surface wave differs from those of the incident E.M. fields.
In the second kind of experiments, a linear coupling takes place between an incident radiation field at a given frequency and the surface polariton wave at the same frequency : this is achieved by using attenuated total reflection (4) or grating coupling (5).

From a mathematical point of view the difference between stimulated Raman excitation of surface polaritons of frequency ω and linear excitation of surface polaritons (using an incident E.M. wave of frequency ω)along a rough surface lies in the fact that in the first case one deals with a source term in Maxwell equations i.e the non linear polarization at ω which is diffracted by the rough surface, whereas, in the second one, the incident E.M. wave, also diffracted by this surface, does not act as a source term in Maxwell equations but is taken into account via the boundary conditions.

Therefore, both types of surface polariton excitations along a rough surface, the non-linear one and the linear one, although rather different at first glance, are in fact similar in the sense that both deal with the excitation of surface polaritons along a given surface and the conclusions relative to the interaction of the surface wave with the surface along which it travels remain the same : the dependance of the dispersion curve, linewidth and surface polariton intensity with surface roughness is independant of the way of excitation.

Thus, instead of studying the experiment described in I, i.e. Raman excitation of surface polaritons taking into account the random surface roughness, we turn our attention to the more simple following problem : considering surface polariton generation by a grating coupler we ask the following question : given an incident E.M. field, what is the effect of the groove depth of the grating on the surface polariton intensity ?

Theory

Surface plasmon anomalies

Several authors reported theoretical results, derived in the case of linear excitation of surface polaritons and valid in the case of small depth, concerning the effect of surface roughness on the attenuation length (6), dispersion curve and damping (7), complex shift Δk in the wavevector (8) of surface polaritons. On the other hand, investigating a somewhat different subject, namely the existence of total absorption of light by a metallic grating, Maystre and Nevière (9) (henceforth referred to as II) relate this phenomenon to "surface plasmon anomalies". They found the following key results :
- the intensity of the specularly reflected wave exhibits a dip for a given frequency and incidence angle of the incident wave.
- the depth of this dip increases with the groove depth : there even exists a critical value of the groove depth for which the intensity of the specularly reflected wave is zero (i.e. the incident E.M. wave is completely absorbed by the metallic grating). If the groove depth is further increased the absorption peak then becomes broader and weaker.
- the position of the dip only slightly depends on the groove depth.
All these results have been accounted for theoretically in II and checked experimentally (10). As shown in II, this phenomenon is closely related to the so-called Wood anomalies due to the excitation of surface plasmons along the metallic grating : a resonant interaction takes place whose efficiency not only involves the values of the longitudinal component of the wavevector and frequency of the diffracted E.M. field but also the groove depth of the grating. Since, for low groove depths, the absorption of the incident beam increases with groove depth so does the intensity of the surface plasmons.

We shall show that this result, obtained in the case of surface plasmons, remains valid for surface polaritons, and allows one to understand point 3. But let us first recall the underlying mathematics of the theory developped in II.

Outline of the theory of II

The method used in II is a differential one[9, 11] in which a Fourier decomposition of the E.M. field is done in the three regions (x : transverse coordinate, h : groove depth) : $x > \frac{h}{2}$ (region 1), $x < - \frac{h}{2}$ (region 2), $- \frac{h}{2} < x < \frac{h}{2}$ (region 3).

In regions 1 and 2, the Fourier coefficients of the E.M. field have a simple dependance on x. They are equal to the product of constant coefficients, respectively B_n and T_n, by an exponential function, the form of which shows that in regions 1 and 2 the field can be described by a superposition of plane waves. On the contrary, in region 3, the Fourier coefficients cannot be expressed in an analytical way. They appear to be the solution of an infinite set of linear second order differential equations which is an homogeneous one since, in II, one deals with a kind of experiment where a linear coupling takes place between and incident E.M. wave and the surface plasmons. The boundary conditions written at $x = \pm \frac{h}{2}$ linked with this infinite set allows the determination of the E.M field everywhere in space, i.e in regions 1, 2, 3. This is achieved on a computer using an algorithm described in ref. 11. One then gets the results summarized in above.

It is also shown in II that one can avoid tedious computer calculations using the following simplified expression for the Fourier coefficient B_o of the specularly reflected wave :

$$B_o = B_o (\alpha_o , h) = r(\alpha_o) \frac{\alpha_o - \alpha^z (h)}{\alpha_o - \alpha^P (h)}$$

where

$\alpha_o = \sin \theta,$ (the grating is illuminated under incidence θ)

$r(\alpha_o)$: reflection coefficient in the case where h = 0

By making an analytic continuation of α_o in the complex plane, the authors of II determine, for different groove depths, the locus, in the complex α_o plane, of the zero $\alpha^z(h)$, and the pole, $\alpha^P(h)$ of $B_o (\alpha_o, h)$.

It is pointed out that for low values of the modulation $\frac{h}{d}$ (d : grating periodicity) :

i) $Re\,[\alpha^Z(h)] \;\simeq\; Re\,[\alpha^P(h)]$ $\hspace{6cm}$ (1 a)

$\hspace{2.5cm}$ Re [...] : real part of [...]

ii) and consequently the minimum reflected intensity, $|B_o|^2_m$, is accurately derived from :

$$|B_o|^2_m \;\propto\; \frac{J_m\,[\alpha^Z(h)]}{J_m\,[\alpha^P(h)]}^2 \hspace{5cm} (1\ b)$$

$\hspace{2.5cm}$ J_m (...) : imaginary part of (...)

$\hspace{1cm}$ Thus the smaller will be the ratio $(J_m\,[\alpha^Z(h)]/J_m\,[\alpha^P(h)])^2$ the greater will be the maximum of the transmitted intensity.

$\hspace{1cm}$ For more details the interested reader will refer to references (9 - 10 - 11) and to the references cited there in.

Application to surface polaritons

$\hspace{1cm}$ The system under consideration, invariant along the y coordinate, is depicted in Fig. 1 : an E.M. field of frequency ω is incident (under incidence θ) on a GaP grating (GaP being the active material used in I) of groove depth h and spatial periodicity d.
The frequency of the incident wave is choosen in such a way that ω belongs to the absorption band of GaP.

$\hspace{1cm}$ The surface of I is a rough random one. Thus, its Fourier spectrum presents every spatial frequencies. It means that the groove spacing d can take any values.

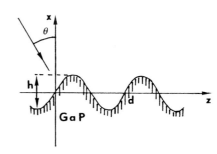

Figure 1 : The excitation geometry

$\hspace{1cm}$ As will be now shown the range of validity of d is however limited (between d_{min} and d_{max}) since one is interested in an efficient excitation of surface polaritons for which one has to fulfill :

$$K_n = Re\,[\beta_s\,(\omega,\ h)] \hspace{5cm} (2\ a)$$

K_n, $\beta_s(\omega, h)$: longitudinal component, for a grating of groove depth h, of the wave-vector of respectively the nth diffracted order, and the surface polariton wave.
The value of θ derived from 2a being such that :

$$-1 < \sin\theta < +1 \hspace{5cm} (2\ b)$$

Notice that no incidence θ leads to a resonant excitation of surface polaritons when equation 2b is not fulfilled.

$\hspace{1cm}$ The quantity K_n of equation 2a is given by :

$$K_n = n\,\frac{2\pi}{d} + k_1\sin\theta \hspace{5cm} (3\ a)$$

where :

$\hspace{1cm}$ $\dfrac{k_1^2 c^2}{\omega^2} = \varepsilon_1$ $\hspace{2cm}$ ε_1 being the permittivity of the outside medium, equal to 1 whithout loss of generality

As shown in (9) the pole $\alpha^P(\omega, h)$ is related to $\beta_s(\omega, h)$ by :

$\hspace{2cm}$ $\alpha^P(\omega,\ h) = \dfrac{\beta_s(\omega,\ h)}{2\pi/\lambda}$ $\hspace{1.5cm}$ $(\omega = \dfrac{2\pi c}{\lambda})$

equations 2a and 3a lead to :

$$Re\,[\alpha^P(\omega,\ h)] = n\,\frac{\lambda}{d} + \sin\theta \hspace{5cm} (3\ b)$$

$\hspace{1cm}$ It is worth noting that the condition $\sin\theta + \frac{\lambda}{d} = Re\,[\alpha^P(\omega, h)]$ together with $-1 < \sin\theta < +1$ can be fulfilled in an infinite number of ways. Indeed, let us call d_1 the possible values of d which correspond to n = 1. As soon as, for a given θ, d_1 is determined, the values $2d_1$, $3d_1$, ..., pd_1, ... also verify equations 2b and 3b since they may be associated to integers 2, 3, ... p, ... Among this infinite possible values of d, the smallest ones d1, give the highest modulation h/d1. As it will be shown hereafter, in the low modulation range, the highestmodulation gives the strongest resonant interaction. Thus, we focus in the following, on the n = 1 diffracted order for which we calculate d_{min} and d_{max}. Therefore, the n = 1 diffracted order, of longitudinal wavevector component $K_1 = Re \cdot [\beta_S (\omega, h)]$, is the most efficiently excited. It is then consistent to consider that this diffracted order represents the surface polariton field whose amplitude for $x < - \frac{h}{2}$ is T_1. According to equations 2b and 3b we get :

$$d_{max} = \frac{\lambda}{Re \ [\sqrt{\varepsilon(\omega)}] - 1} \tag{4 a}$$

$$d_{min} = \frac{\lambda}{Re \ [\sqrt{\varepsilon(\omega)}] + 1} \tag{4 b}$$

Where one has :

$$\frac{\beta_S^2 c^2}{\omega^2} = \varepsilon \ (\omega) = \frac{\varepsilon_1 \ \varepsilon_2 \ (\omega)}{\varepsilon_1 + \varepsilon_2 (\omega)} \tag{5 a}$$

$\varepsilon_2(\omega)$ being the permittivity of GaP :

$$\varepsilon_2(\omega) = \varepsilon_\infty + \frac{S \ \omega_0^2}{\omega_0^2 - \omega^2 + j\omega\Gamma} \tag{5 b}$$

ω belonging to the absorption band of GaP we have : $Re \ [\varepsilon_2(\omega)] < 0$
ε_∞, ω_0, Γ, S being respectively the high frequency dielectric constant, transverse frequency, damping constant and oscillator strengh of the optical mode of GaP, the numerical values of which are (12) :

$$\varepsilon_\infty = 9.09, \quad \omega_0 = 367.3 \ cm^{-1}, \ \Gamma = 0.0035 \ \omega_0, \ S = 1.8$$

The transmitted intensity $|T_1|^2$ is a function of ω, θ, d and h. In the range d_{min}, d_{max}, for a given incidence θ, it exists a value of ω for which $|T_1|^2$ is maximum. In the following we are mainly interested in the behavior of the peak value and corresponding peak position of the transmitted intensity $|T_1|^2$ as a function of d and h. This leads us, according to the formalism developped in II, to the study of the quantity $[J_m (\alpha^z)/J_m(\alpha^P)]^2$. This is done :
   - for the two groove depths h used in I and several grating periodicities
   - for a given grating periodicity and different groove depths greater than 0.3 μm.

Constant groove depth

Starting from equations 1, we plotted on fig. 2 (for a groove depth h = 0.3 μm) the ratio $[J_m (\alpha^P)/J_m(\alpha^P)]^2$ versus ω for several values of d in the case of surface polaritons propagating along a GaP grating. Also reported are some values of θ derived from equation 3b, for some couples d and ω.

We may draw the following conclusions :

   - large values of d lead to rather large values of $(J_m(\alpha^z)/ J_m(\alpha^P))^2$ i.e. to low values of the maximum of the surface polariton intensity $|T_1|^2$.

   - for each value of ω, there exists a specific value of d, d_M, for which the maximum surface polariton intensity is itself maximum : for the considered groove depth d_M corresponds to the maximum maximorum of $|T_1|^2$. The knowledge of the pole $\alpha^P (\omega, h)$ linked with equation 3b then allows us to get the corresponding value θ_M of θ.

   - using fig. 2, we are able to get, for a given ω, the corresponding value of d_M. For example for $\omega = 393 \ cm^{-1}$, $d_M = 12 \ \mu m$ and $\theta_M = 72.9°$ (point 1), point 2 refers to $\omega = 397 \ cm^{-1}$, $d = 12 \ \mu m \neq d_M$ and $\theta = 43.8° \neq \theta_M$ leading to a weaker resonance.

   - Obviously θ_M and d_M are frequency dependent.

These results have been checked making a direct computer calculation, like those described in II, of $|T_1|^2$ for the two groove depths used in fig. 2 of I. The curves of the surface polariton intensity $|T_1|^2$ versus ω are reported on figures 3, 4, 5 Curves A refer to h = 0,3 μm and curves B to h = 0,05 μm.
$|T_1|^2$ is the square modulus of the complex amplitude of the n = 1 diffracted order which, in the limit case of lossless grating, corresponds to an evanescent wave. Therefore it is not concerned by the energy balanced criterion and has no reason to be less than unity.

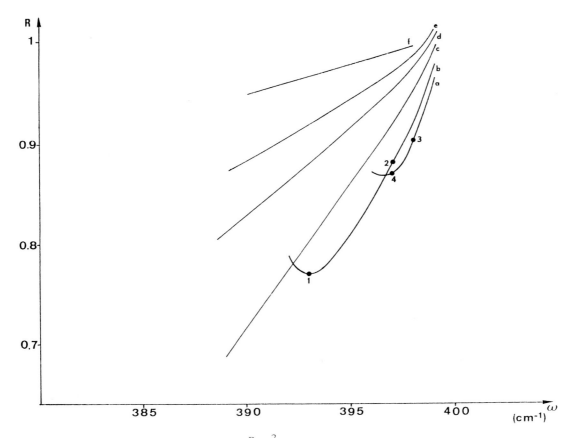

Figure 2 :   R = $(J_m (\alpha^z)/J_m (\alpha^P))^2$ versus ω. Material GaP. Groove depth h = 0.3μm
a : d = 11 μm ; b : d = 12 μm ; c : d = 14 μm ; d : d = 17 μm ; e : d = 20 μm ;
f : d = 30 μm. Point 1 : θ = 72.9°, ω = 393 cm^{-1} ; Point 2 : θ = 43.8°,
ω = 397 cm^{-1} ; Point 3 : θ = 44.6°, ω = 398 cm^{-1} ; Point 4 : θ = 62°, ω = 397 cm^{-1}

An inspection of figures 3, 4, 5 shows that the following features appear :
    - In the range of frequency we studied the resonant excitation of surface polaritons is the strongest for ω = 393 cm^{-1}, θ = θ_M = 72.9° and d = d_M = 12 μm (fig. 3).
    - Comparison of figures 4 and 5 shows that this resonance is stronger for ω = 397cm^{-1} d = 12 μm, θ = 43.8° than for ω = 398 cm^{-1}, d = 11 μm, θ = 44.6°.
    All these results are in agreement with those reported on fig. 2.
    In paper I, for a given couple (ω,θ), the total measured intensity results from the contribution of all the spatial FOURIER components of the rough random surface. Since each of them leads to a scattered intensity which increases with the groove depth h (figures 3, 4, 5 and others non published) we thus can deduce that the total intensity will still depend on h according to figures 3, 4, 5. Therefore, considering only the h dependence, anyone of figures 3, 4, 5 is in qualitative agreement with fig. 2 of I. As in fig. 2 of I :
    - The intensity of the excited surface polariton (which of course, behaves like the Raman scattering intensity) increases with h.
    - The position of the peak intensity of the surface polaritons very slightly depends on h. We, therefore, expect that in the h range considered in I the surface polariton dispersion curve is rather insensitive to surface roughness (see fig. 2 and 4 of I).
    This dispersion curve has been reported in fig. 6 (for h = 0.3 μm) as the locus of the maximum maximorum of the surface polariton intensity $|T_1|^2$ in a {ω, Re [$\beta_s(\omega, h)$] } diagram. Also reported is the usual surface polariton dispersion curve based on flat surface assumption and derived from equation 5a by plotting ω versus Re [$\beta_s(\omega, h = 0)$]. Both curves are superimposed. Thus, for the groove depths considered in I, the surface polariton dispersion curve only slightly depends on h. This is in agreement with figures 2 and 4 of I.

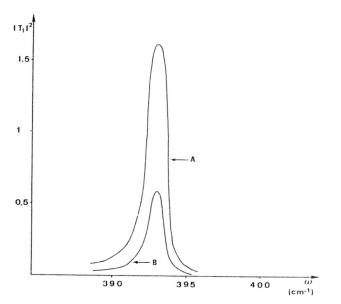

Figure 3 : Surface polariton intensity for two groove depths. A : h = 0.3 μm ; B : h = 0.05 μm. Vertical axis : curve A : $|T_1|^2$; curve B : 10 x $|T_1|^2$. Groove spacing d = 12 μm. Incidence θ = 72.9°. Material GaP

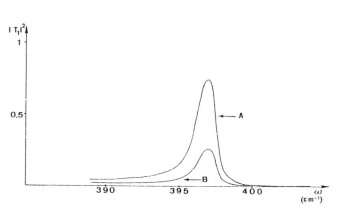

Figure 4 : Surface polariton intensity for two groove depths. A : h = 0.3 μm ; B : h = 0.05 μm. Vertical axis : curve A : $|T_1|^2$ curve B : 10 x $|T_1|^2$. Groove spacing d = 12 μm. Incidence θ = 43.8°. Material GaP.

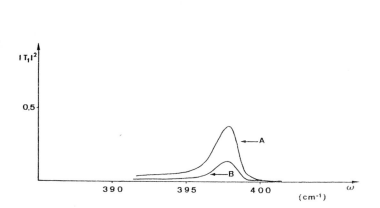

Figure 5 : Surface polariton intensity for two groove depths. A : h = 0.3 μm ; B : h = 0.05 μm. Vertical axis : curve A : $|T_1|^2$; curve B : 10 x $|T_1|^2$. Groove spacing d = 11 μm. Incidence θ = 44.6°. Material GaP.

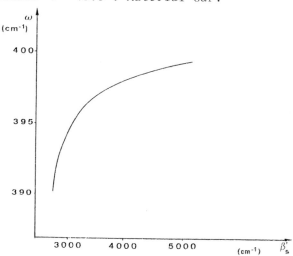

Figure 6 : Surface polariton dispersion of the maximum maximorum of the surface polariton intensity $|T_1|^2$ in a $\{\omega, \beta_s'\}$ diagram (β_s' = Re [β_s (ω, h)]) for a groove depth h = 0.3 μm. Also reported is the usual surface polariton dispersion curve based on flat surface assumption. Both curves are superimposed. Material GaP.

The following is worth noting : at first glance Raman excitation of surface polaritons along a rough random surface should not depend on the angle ψ between the incident laser beam and the STOKES one, since an infinite number of spatial periodicities d are in principle possible which would allow us to associate to different values of the surface polariton frequency ω, the same value of ψ by simply choosing different values of d. But this does not occur as can be seen on fig. 4 of I. Therefore, it is the spatial Fourier component $d_M(\omega)$ which mainly determines the peak position of the Raman scattering intensity leading to a one to one correspondance between the choosen frequency ω and the value ψ_M (ω) of ψ corresponding to $d_M(\omega)$. This explains why, although using a random surface, Ushioda et al. observed a ψ dependent scattering process.

Constant grating periodicity
d = 17 μm

We study, for several values of h greater than 0.3 μm, the behavior of $|T_1|^2$. The corresponding results are reported on fig. 7 for h = 0.9 μm, 1,3 μm and 2.4 μm. It is seen that :
- the maximum of $|T_1|^2$ first increases when increasing the groove depth h and then decreases when h is further increased beyond h = h_{opt}. In our case, for d = 17 μm, the maximum maximorum of $|T_1|^2$ is obtained for h_{opt} = 1.3 μm (which no longer stricly corresponds to the minimum of $[J_m(\alpha^z)/J_m(\alpha^P)]^2$ since for these high values of the ratio $\frac{h}{d}$ $Re(\alpha^P) \neq Re(\alpha^z)$)

- The surface polariton line-widht broadens as h is increased beyond 0.3 μm. Other results, obtained for different values of h, and not reported in this paper, have allowed us to draw the curve of figure 8, which clearly shows the increase of width at half height of $|T_1|^2$ as a function of the groove depth.
- The peak position of $|T_1|^2$ shifts when h is increased above 0.3 μm, thus leading to a surface polariton dispersion curve which is depressed below the one based on flat surface assumption.

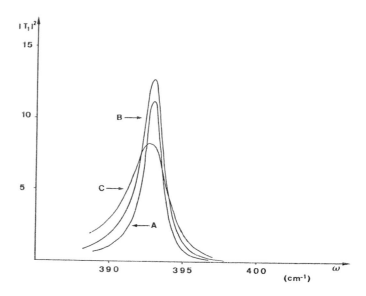

Figure 7 : Surface polariton intensity for the grating periodicity d = 17 μm and several groove depths : curve A : h = 0.9 μm ; curve B : h = h_{opt} = 1.3 μm ; curve C : h = 2.4 μm. Incidence θ = 19.875°. Material GaP.

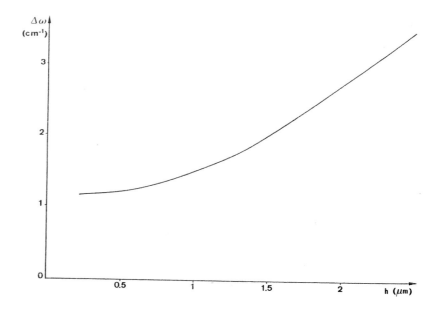

Figure 8 : Evolution of the linewidth Δω of $|T_1|^2$ as a function of the groove depth h. θ = 19.875° , d = 17 μm. Material GaP.

Conclusion

We have suggested an explanatation which well accounts for the increase of Raman scattering intensity with surface roughness. The simple calculation presented here does not allow a quantitative comparison between our results and the corresponding ones of I since this requires to take into account the fact that :
        a) Raman excitation of surface polaritons results from a nonlinear interaction.
        b) the surface used by the authors of I is a rough random surface.

Point a) will be dealt with by making the new theory of diffraction by a nonlinear grating. Works are now in progress to developp such a theory.
The effects related to the random aspect of the surface used in I are impossible to evaluate since the Fourier spectrum of this surface is not known. Moreover, as noticed in I, this Fourier spectrum depends on the part of the surface striked by the incident laser beam.

We see that the phenomenon observed in I is closely related to the so-called Wood anomalies : in the experiment described in I, or in grating generation of surface polaritons, we deal with two coupled resonant systems, the surface polariton wave and the incident E.M. one, which are coupled by the rough surface of I, or by the grating, and, therefore, it is not suprising to find that the strength of this coupling also depends on the groove depth. Extending to surface polaritons results derived in II in the case of surface plasmons shows that it is possible to find , for a given grating periodicity, a value of h, h_{opt}, for which the peak Raman scattering intensity is maximum, thus optimizing the coupling described above. The unexpected result is that h_{opt} is different from zero.

Acknowledgments

This research has been made possible by the CNRS GRECO Micro-Ondes.

References

1. Ushioda S., Aziza A., Valdez J.B., Mattei G. Phys. Rev. B19, 8, p. 4012, 1979.
2. Evans D.J., Ushioda S., Mc Mullen J.D., Phys. Rev. Lett. 31, p. 369, 1973
3. De Martini F., Giuliani G., Mataloni P., Palange E., Shen Y.R., Phys. Rev. Lett. 37, p. 440, 1976.
4. Otto A., Z. Phys. 216, p. 398, 1968
    Barker A.S., Phys. Rev. Lett. 28, p. 892, 1972
5. Teng. Y.Y., Stern E., Phys. Rev. Lett. 19, p. 511, 1967
    Richtie R.H., Arakawa E.T., Cowan J.J., Hamm R.H., Phys. Rev. Lett. 21, p. 1530, 1968
6. Mills D.L., Phys. Rev. B12, p. 4036, 1975
    Mills D.L., Maradudin A.A., Phys. Rev. B12, p. 2493, 1975
    Mills D.L., Phys. Rev. B15, p. 3097, 1977; J. of Appl. Phys. 48, p. 2918, 1977
7. Maradudin A.A., Zierau W., Phys. Rev. B14, p. 484, 1976
8. Kroger E., Kretschann E., Phys. Status Solidi B76, p. 515, 1976
9. Maystre D., Nevière M.; J. Optics (Paris) 8, p. 165, 1977 and references cited there in
10. Hutley M.C., Maystre D. Optics Comm. 19, p. 431, 1976
11. Petit R. Nouv. Rev. d'Opt. 6, p. 129, 1975
    Nevière M., Vincent P., Petit R.; Nouv. Rev. d'Opt. 5, p. 65, 1974
12. Marschall N., Fischer B., Phys. Rev. Lett. 28, p. 811, 1972
    Faust W.L., Henry C.H., Phys. Rev. Lett. 25, p. 1265, 1966

Experimental investigation in microwave range of diffraction by classical and crossed gratings

Patrick Bliek, Roger Deleuil

Départment de Radioélectricité, Université de Provence
Rue Henri Poincaré, 13397 Marseille Cédex 4, France

Abstract

The behaviour of classical gratings (echelette, lamellar and sinusoïdal) is correctly described by present electromagnetic theories by means of which new interesting profiles and more elaborate Fourier gratings can be obtained. Confrontation of theory and experiment is critical in the optical field because the groove spacing is less than one micrometer and the profile is not always regular. On the other hand, it is possible to work in the microwave range (26.5 to 40 GHz) with isomorphic-gratings having periods of several millimeter and perfectly ruled profiles. Our experimental equipment consists of a tridimensional 3 m radius goniometer, placed in an anechoïc chamber. The antennae for emission and reception are parabolic D = 60 λ diameter mirrors. The measurements are obtained using a phase-amplitude receiver, a ratiometer and a chart-recorder made by "Scientific-Atlanta". The system of data acquisition and analysis is obtained with a "Hewlett-Packard 9825" minicomputer. The relative accuracy of the grating efficiency measurement is better than 3%. We have already studied echelette, lamellar and Fourier gratings and, recently, we have studied crossed-gratings (round-holed grids and planar distributions of pyramids or half prolate spheroïds - both intended for solar captation). Agreement between theory and experiment is excellent and this apparatus is very powerful for diffraction or diffusion effect measurements - this is very important when the grating models are not treated theoretically.

Introduction

Development of space optics has lead to a new interest in diffraction grating studies. The gratings used often have a period that is close to the wavelengths incident on the grating. For values of the wavelength to period ratio close to one the efficiency of gratings is much influenced by polarization of incident light. This means that a rigorous electromagnetic theory has to be applied to solve (numerically) the problem of grating efficiency determination.

With the help of these theories attempts are made to find profiles giving a blaze effect simultaneously for the two polarization cases. It has primarily been a question of optimizing classical profiles (the echelette but also the lamellar profile). Nowadays, with new techniques for making Fourier gratings, very high groove densities can be obtained at low cost, thus arousing new interest in the sinusoïdal grating and the more elaborate Fourier grating (holoharmonic grating).

New interesting fields of application have been presented: using gratings in cavity resonaters in lasers, or as grating couplers in integrated optics and even for recording color images by embossable surface relief structures. Two-dimensional gratings, or crossed-gratings, up to now only used as inductive or capacitive grids for microwave filtering, are now studied as potential solar absorbers. Attempts are made to find the shape and dimensions of the grids which allow both high absorption for wavelengths in the solar emission spectrum and low emission for wavelengths in the infrared. In the same way any regular two-dimensional structure (half spheres, half prolate spheroïds, pyramids) distributed periodically on a plane can be considered as a crossed grating.

Rough surfaces to be used as solar absorbers must have high selectivity, i.e. absorptivity close to one or low reflectivity. Such a performance could be obtained by carefully selecting the structure and its distribution throughout the surface of a dissipative material.

Experimental studies of diffraction gratings in the microwave region have already been published by several authors[1-19].

Interesting comparisons between theoretical and experimental studies were undertaken during the 1950's[3,4] and later by Palmer et al.[9,10,11], Wirgin and Deleuil[15] and Einstein, Juliano and Pine[16]. More recently Ebbeson, Heath and Jull[17,18,19] have successfully compared their theoretical results concerning "comb gratings" with measurements made at 35 GHz.

The aim of this paper is to summarize the major experimental results obtained, in our laboratory, on one-and two-dimensional grating models studied in the microwave region. The microwave goniometer, constructed by one of the authors[12] to study classical gratings[14,15], has been improved so that, radar targets[20] and different models of rough surfaces[21,22] can be studied. More recently this apparatus was used to study crossed gratings[23,24] and Fourier grating[25].

The study of diffraction gratings with microwave radiation is very interesting. Thus, grating models that have a perfectly known geometry and a very smooth surface, can be utilized. This is important for echelette gratings[26] and very important for lamellar and comb gratings[17,18], which are very sensitive to small variations of the profile. This remark can be compared to the result which has recently been found with metallic Fourier gratings[25].

Use of microwave radiation is likewise advantageous in that one works with perfectly conducting models and thereby difficulties which occur in the optical region are diminished (finite conductivity, surface roughness, oxide layers and plasmon anomalies)[19,27]. Nevertheless, use of microwave radiation demands caution and we would like to make some important remarks. It is imperative to use parabolic reflectors to obtain an incident wave as plane as possible and to have a relatively large "quiet volume" so that gratings consisting of a sufficient number of grooves can be studied[3,11,15,19].

Experimental apparatus

The experimental setup consists of a threedimensional goniometer working in the Ka band (26.5 to 40 GHz). With this apparatus, a linearly polarized plane harmonic electromagnetic wave can be directed to a target, arbitrarily orientated, and the directivity diagram (bistatic scattering pattern) can be measured.

The equipment can be used for radar target scattering studies as well as for studies of diffraction from classical gratings and from crossed gratings. The most fundamental components of the setup are (Fig. 1, 2 and 3).

1. Target support. This support can be rotated around the vertical axis Z'Z. In the cartesian system OXYZ the orientation of the target is defined relative to a characteristic direction (in this case this direction is that of the normal to the plane of the grating), and by the elevation angle γ and the azimuthal angle α.

2. An azimuthal arm, with radius $\simeq$ 3 m, which can be rotated around the vertical axis Z'Z. At its end is placed a parabolic reflector A1, with a diameter of 53 cm. The principal axis of A1, defined by the angle ϕ to the X'X axis, passes through the point O.

3. A parabolic reflector A2, identical to A1, carried by an elevation arm. A2 can be moved along a half circle in the Z'Z-X'X plane. The principal axis of A2, defined by the angle θ, also goes through the point O.

4. "Scientific-Atlanta" electronic measurement equipment, consisting of: a source model n° 2150 which covers the Ka band, a lock-in phase-amplitude receiver model n° 1753 with a dynamic of 100dB, a ratiometer model n° 1823-20, a chart recorder model n° 1523 and a digital positioner programmer n° 2011.

5. A desk computer HP 9825 with plotter permitting reading and processing of measurements simultaneously.

6. A hemispherical anechoïc chamber of 6 meters diameter covered with absorbing material (Eccosorb CV3). The primary sources of the parabolic reflectors A1 and A2 are rectangular horns whose orientation can be changed to have any polarization of the incident wave and to analyse the polarization of the diffracted wave.

Measurements

The "quiet volume" is about 30 λ long, the length of the grating is 26 cm and the width is 30 cm. The classical gratings are made of duraluminium and have Ng grooves, where Ng is between 20 and 50. The number of diffracting elements for a crossed grating is Ng $\times$ Ng.

The power diffracted into the order (p,q), i.e. P(p,q) is obtained by integrating the directivity pattern $r(\theta,\phi)$ over a solid angle $\Omega(p,q)$ centered on the direction of the diffracted order (p,q). The diagram $r(\theta,\phi)$ is measured as a function of ϕ for several values of θ, giving:

$$P(p,q) = \int_{\Omega_{p,q}} P(\theta,\phi)d\Omega = P_M \int_{\phi_{p,q}} d\phi \int_{\theta_{p,q}} r(\theta,\phi)\sin\theta \ d\theta \tag{1}$$

The increment for ϕ and θ is of the order of 0.1 degree and 0.5 degree respectively.

To measure the total incident power P(i), incident on the grating, we use a reference metal plane. This metal plane has the same dimensions as the grating being measured and the same angle of incidence is used. The specularly reflected power from this reference plane is measured in the same way as described above.

In the case of classical (one-dimensional) gratings studied in non-oblique mountings the diffracted power in the order (p,0) is obtained by just one integration with respect to ϕ. The efficiency E(p,q) of a grating in a given order (p,q) is the ratio between the power P(p,q) diffracted into this order and the incident power P(i). The efficiency is measured for the two fundamental polarization cases, $E_{||}$ and $E_\perp$, corresponding to whether the incident electric field is parallel or perpendicular to the plane of incidence. The apparatus permits measurements with less than three percent relative error.

We have studied, experimentally, several different plane metallic gratings having profiles which are of interest for optical and infra-red spectroscopy[28],[29]. Below, we give a brief summary of the most important results we have found with echelette and lamellar gratings. These specific gratings have also been the subject of theoretical results.

Figures 4a and 4b show echelette gratings used in the Littrow mount. The efficiency curves, + 1 order, plotted for the two polarizations show a very good agreement between our measurements and the theoretical results of Petit[30]. The efficiency curves reveal the great interest of this profile as an important blaze effect can clearly be seen. The $E_{||}$ curve is especially good - efficiency is better than 0.9 over 1/2 octave. However, as the maximum of the two curves, $E_\perp$ and $E_{||}$, are displaced the diffracted light will be polarized.

To understand the performance of the echelette gratings better we have studied the effects of non-perfect groove profiles (deformations and surface roughness) on efficiency. We have studied a profile which is close to the one described above and used under the same conditions (Littrow mount, first order). In Figures 5a and 5b it can clearly be seen that presence of deformations influences efficiency. A decrease of efficiency and a shift of the blaze wavelength towards shorter wavelengths can be noticed.

Figures 6a-b and 7a-b show two lamellar gratings which have been studied by Wirgin and Deleuil[15],[31]. In $E_{||}$ polarization the gratings have a very pronounced blaze effect for a broad wavelength interval. However, continuous exploration of + 1 order was not feasible in the domain of interest because of the geometry of the apparatus which made measurements impossible about the back-scatter direction. The zeroth order only could be measured. The $E_{||}$ curves, which exhibit a filtering property (Fig. 6a, 7a), are very different from the $E_\perp$ curves (Fig. 6b, 7b), hence these gratings strongly polarize the incident field.

Excellent agreement between theory and experiment is observed and this is particularly true for the first profile (Fig. 6a, 6b). For the second profile the comparison is less satisfactory (Fig. 7a, 7b). The discrepancy is probably due to minor imperfections of the groove profile.

Furthermore, we have shown that lamellar gratings are sensitive to small variations of the depth and width of the grooves. Figure 8 shows the variation of the efficiency in the zeroth order of a lamellar grating as a function of the angle of incidence. The differences between the theoretical and experimental curves are larger than the experimental errors. New calculations were made where the parameters of the profile (period, depth and width of the grooves) and the wavelength were varied one by one. It was only the variation of the depth or the width which led to theoretical efficiencies close to our measurements. The variations were of the order of 3 or 4 percent[14].

The same narrow tolerances of the profile and the smoothness of the surface have been found for another type of grating. This latter is composed of parallel dielectric cylindrical rods, having a diameter D = 10.5 ± 0.5 mm, a groove spacing d = 30.2 mm and dielectric constant ε = 2.70 + i 0.01. We have measured the efficiencies of the transmitted -1, 0 and + 1 orders as a function of the angle of incidence at λ = 11.12 mm. The results obtained have been compared with the calculations done by Cerutti-Maori[32]. In Figures 9a, b and c one can see that the agreement between theory and experiment is not always satisfactory. The general behaviour is conserved and the anomalies are found at the same angle of incidence (this is especially true for the zeroth order) but quantitatively the discrepancies can be large. This is due essentially to the fact that the diameter of the rods was not constant and the rods were not rigorously aligned parallel. These remarks can be compared with those of Kerr and Palmer[33] in a similar study.

We conclude this summary, which so far has dealt with known results for classical gratings, with a short note on work at our laboratory. We are at present studying sinusoïdal gratings, not the simple ones but the more complicated Fourier gratings[34] (see "Optimization of Fourier gratings" by M. Breidne and D. Maystre in this Proceedings). These gratings can exhibit very interesting blaze effects for U.V. spectroscopy. This is the subject of a coming publication.

At present two kinds of crossed-gratings, which can be used for solar captation, are studied by our laboratory. The first type is a metallic grid with circular holes, which acts as a frequency filter and the second type is a rough regular surface which is composed of infinitely metallic pyramids and which can act as a solar absorber.

Inductive and capacitive grids have long been used as filters (high pass, low pass and notch filters) in the field of microwave engineering[35,36]. More recently, their application in the field of solar energy has been advocated, since there one requires a device which can selectively absorb visible radiation and reject long wavelength re-emission[37,38,39]. For this second application it is particularly important to know the spectral behaviour of the grid in the region of diffraction anomalies where the wavelength and period are of comparable order. Thus, it is of interest to have accurate theoretical models of grid behaviour and to confirm these models with careful experimental measurements[23]. As we will see, the behaviour in the region of diffraction anomalies is critically dependent on the grid parameters, and so it is in this region that the most stringent tests of both the theoretical treatment and the experimental technique can be made.

The phenomena to be studied here (filtering and resonance effects) extend over a wavelength range which exceeds the usable Ka band. Thus we have been forced to construct three isomorphic grids whose characteristics are given in table (1). It is then possible to vary the ratio of the wavelength to the grating period (λ/d) between 0.8 and 2.2.

Table 1

Characteristics of the three isomorphic grids used in the measurements

Number of the grid	1	2	3
Groove spacing (mm)	5	7.5	10
Diameter of the holes (mm)	4.5	6.75	9
Thickness of the metal (mm)	2	3	4
Number of holes	3120	1360	780

In Figure 10 a theoretical curve of the energy transmission by the isomorphic grids of table 1 is compared with the experimental measurement at normal incidence. The agreement is excellent at all save one of the points. It should be noted that this latter is the first of the five points obtained with grid 1. This is the thinnest of the three grids, and the only one with noticeable departures from planarity due to manufacture. We attribute the discrepancy to the effect of the non-planarity on the shape of the wavefront of the transmitted zeroth order.

It will be noted that the experimental points are in good agreement with the theoretical predictions concerning the Wood and resonance anomalies. The first of these occur at any wavelength at which a diffracted order ceases to propagate. For normally-incident light, Wood anomalies occur at normalized wavelengths

$$\lambda(p,q)/d = 1/\sqrt{p^2 + q^2} \ , \tag{2}$$

where, p and q, define the (p,q) diffracted order.

At $\lambda/d = 1.0$ the four orders $(\pm1,0)$ and $(0,\pm1)$ cease to propagate. Below this wavelength they can carry reflected energy away from the top surface of the grid. It is the energy carried by these waves which is responsible for the abrupt drop in the transmitted energy curve at $\lambda/d = 1.0$.

The resonance anomaly peaks at $\lambda/d \simeq 1.25$ and corresponds to the resonant frequency of the equivalent circuit of the grid[35]. The filtering region of the grid occurs on the long-wavelength side of this anomaly.

Figures 11a-b show the variation of the transmitted $(0,0)$ order efficiency with angle of incidence i at a fixed normalized wavelength of $\lambda/d = 1.12$ for grid 2. Although the energy properties of the grid are polarization independent at normal incidence $(i = 0)$, the effects of polarization become apparent in a remarkably rapid manner as one moves off axis. It is astonishing that a variation of only $6°$ can transform a structure from being polarization independent to one being a good polarizer. This strong polarizance is caused by the presence of the $(0,-1)$ Wood anomaly $(i = 6.5°)$.

In the case of $E_\perp$ polarization (Fig. 11a), the agreement between theory and experiment is almost perfect. The Wood anomaly is evident only as a change in the gradient of the curve. However, for $E_\parallel$ polarization (Fig. 11b), the experiment shows the anomaly as being an abrupt loss of energy from the transmitted spectrum, this energy plunging by 80% in $2°$. In the narrow angular band $(6.2° < i < 6.8°)$ numerical convergence is particularly difficult to achieve because of the very rapid fluctuation in the distribution of the transmitted energy carried by the various modes in the grid apertures.

The pyramidal crossed grating, studied experimentally, is composed of infinitely conducting pyramids with a quadratic base. The pyramids have been constructed by ruling consecutively, in two orthogonal directions, two echelette gratings having the same groove spacing $d = 8$ mm, symmetric profile and an apex angle of $90°$ (the depth-to-groove spacing ratio is equal to 0.5).

Figure 12 shows the variations of the efficiency in the order $(0,0)$ as a function of the angle of incidence, in the two principal polarization cases, and with $\lambda/d = 1.375$. For angles of incidence less than the limit value only the order $(0,0)$ propagates and the efficiency in the two polarizations $E_\parallel$ and $E_\perp$ are identical and equal to unity. For values greater than the limit value the $E_\perp$ efficiency remains close to one, but the efficiency in $E_\parallel$ falls rapidly to 0.1 and stays below 0.2. Thus, the structure studied behaves like a high performance polarizer. This interesting feature of pyramidal crossed gratings will be the subject of a systematic study concentrated on the case of a high h/d ratio. In this latter case the pyramidal crossed grating can be thought of as a model of dendritic structure. Such structures are of great interest for solar captation[24].

Theoretical studies of pyramidal crossed gratings have been made, which were limited to small h/d ratio[40]. Recently, however, the model used for our measurements has been treated theoretically and the first results are in very good agreement with our measurements[41].

Conclusion

The study of diffraction, in the microwave region, of models of gratings usually used in the optical region is very fruitful. This procedure allows relatively easy manufacture of a variety of different profiles and in particular the actual grating profile of a manufactured grating can be known. It is thus possible to make a valid comparison between a rigorous theory and a precise and reliable experiment.

In addition to known results on classical gratings (echelette, lamellar,...) we have also presented results, not yet published, concerning crossed gratings, for which the field of application seems to be very large (microwave filters and various rough surfaces for solar captation).

References

1. Kodis, R.D., J. Appl. Phys., Vol. 23, p. 249, 1952.
2. Row, R.V., J. Appl. Phys., Vol. 24, p. 1448, 1953.
3. Meecham, W.C., and Peters, C.W., J. Appl. Phys., Vol. 28, p. 216, 1957.
4. Rohrbaugh, J.H., Pine, C., Zoellner, W.G., and Hatcher, R.D., J.O.S.A., Vol. 48, p. 710, 1958.
4. Decker, M.T., J. Res. Natn. Bur. Stand. D, Vol. 63, p. 87, 1959.
6. Sueta, T., J. Inst. elect. commun. Engrs. Japan, Vol. 42, p. 677, 1959.

7. Aksenov, V.I., Radiotekh. Elektron., Vol. 5, p. 782, 1960.
8. Stroke, G.W., Rev. Opt., Vol. 39, p. 291, 1960.
9. Palmer, C.H., Jr., in Symposium on Quasi-Optics, Polytechnic Press, Brooklyn, N.Y., 1964.
10. Palmer, C.H., Jr., and Evering, F.C., Jr., J.O.S.A., Vol. 54, p. 844, 1964.
11. Palmer, C.H., Jr., Evering, F.C., Jr., and Nelson, F.M., Appl. Opt., Vol. 4, p. 1271, 1965.
12. Deleuil, R., C.r. hebd. Séanc. Acad. Sci. Paris, Série B, Vol. 262, p. 1676, 1966.
13. Evering, F.C., Jr., Appl. Optics, Vol. 5, p. 1313, 1966.
14. Deleuil, R., Optica Acta, Vol. 16, p. 23, 1969.
15. Wirgin, A., and Deleuil, R., J.O.S.A., Vol. 59, p. 1348, 1969.
16. Einstein, F., Juliano, R., Jr., and Pine, C., J.O.S.A., Vol. 61, p. 621, 1970.
17. Ebbeson, G.R., J.O.S.A., Vol. 66, p. 1363, 1976.
18. Jull, E.V., Heath, J.W., and Ebbeson, G.R., J.O.S.A., Vol. 67, p. 557, 1977.
19. Heath, J.W., and Jull, E.V., J.O.S.A., Vol. 68, p. 1211, 1978.
20. Deleuil, R., Rapport final du contrat D.R.M.E. N° 565/73, Dec. 1975.
21. Deleuil, R., Rapport final du contrat D.R.M.E. N° 677/68, Dec. 1971.
22. Deleuil, R., Rapport final de l'A.T.P. du C.N.R.S. N°3295, Dec. 1979.
23. Bliek, P., Botten, L.C., Deleuil, R., Mc Phedran, R.C., and Maystre, D., 8th. European Microwave Conference, Paris, Sept. 4-8, 1978.
24. Bliek, P., Deleuil, R., Papini, F., and Papini, M., Rev. Phys. Appl., to be published.
25. Bliek, P., and Deleuil, R., Journées Nationales de Microondes, Lille, June 26-28,1979.
26. Deleuil, R., and Varnier, F., Rev. Phys. Appl., Vol. 6, p. 195, 1971.
27. Roumiguières, J.L., Maystre, D., and Petit, R., J.O.S.A., Vol. 66, p. 772, 1976.
28. Madden, R.P., and Strong, J., "Concepts of Classical Optics", Freeman and Co., San Francisco, 1958.
29. Stroke, G.W., "Diffraction gratings", Handbuch der Physik, Vol. XXXIX, Springer-Verlag, Berlin, 1967.
30. Petit, R., Optica Acta, Vol. 14, p. 301, 1967.
31. Wirgin, A., Nouv. Rev. d'Optique appliquée, Vol. 1, p. 161, 1970.
32. Cerutti-Maori, G., Thèse de 3ème Cycle, Marseille, 1971.
33. Kerr, D.W., and Palmer, C.H., J.O.S.A., Vol. 61, p. 450, 1971.
34. Breidne, M., and Maystre, D., SPIE 24 th. Annual Technical Symposium, 1980.
35. Ulrich, R., Infrared Phys., Vol. 7, p. 37, 1967.
36. Chen, C.C., I.E.E.E. Trans. M.T.T., Vol. 18, p. 627, 1970.
37. Horwitz, C.M., Opt. Commun., Vol. 11, p. 210, 1974.
38. Fan, J.C., Bachner, F.J., and Murphy, R.A., Appl. Phys. Letters, Vol. 28, p. 440, 1976.
39. McPhedran, R.C., and Maystre, D., Appl. Phys., Vol. 14, p. 1, 1977.
40. Maystre, D., Nevière, M., Vincent, P., Derrick, G., and Mc Phedran, R., Proceedings of ICO 11 Conference, Madrid, Spain, 1978.
41. Maystre, D., Mc Phedran, R., and Neviere, M., Private Communication.

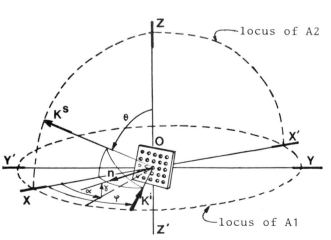

Figure 1. Schematic representation of the device.

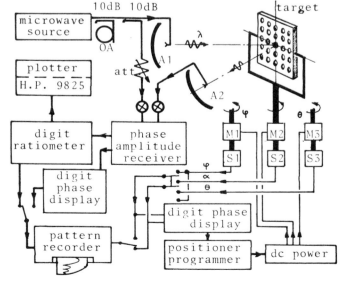

Figure 2. Synopsis of the microwave system.

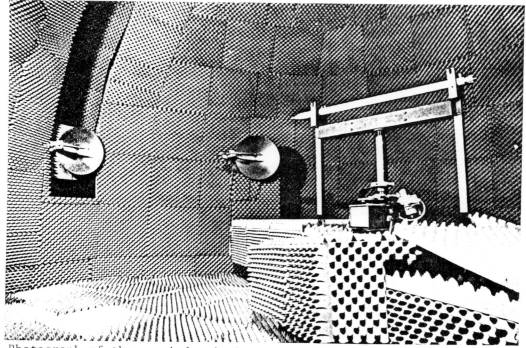

Figure 3. Photograph of the anechoic chamber. Here, the scatterer is a radar target.

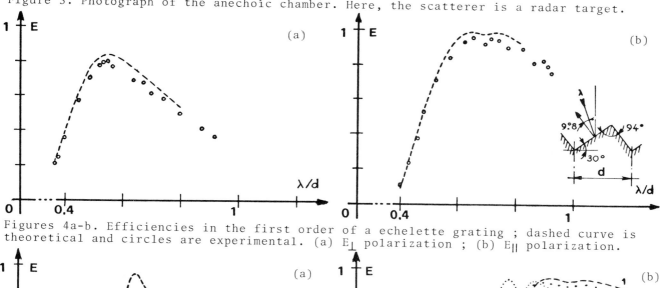

Figures 4a-b. Efficiencies in the first order of a echelette grating ; dashed curve is theoretical and circles are experimental. (a) $E_\perp$ polarization ; (b) $E_\parallel$ polarization.

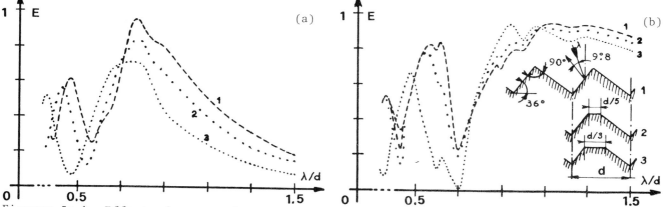

Figures 5a-b. Effect of non perfect groove profiles on the efficiency of echelette grating; experimental efficiency in the first order for the two polarization states $E_\perp$ (a) and $E_\parallel$ (b)

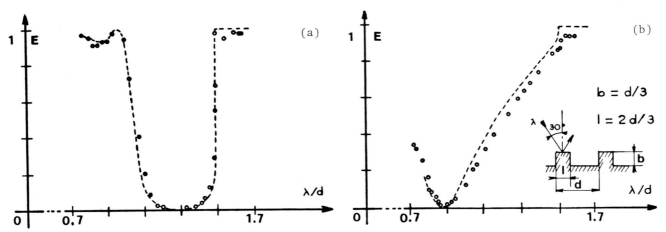

Figures 6a-b. First lamellar profile. Efficiencies in the zeroth order. Dashed curve is theoretical and circles are experimental. (a) $E_{||}$ polarization ; (b) $E_{\perp}$ polarization.

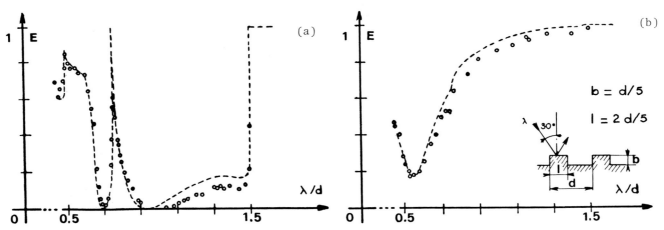

Figures 7a-b. Second lamellar profile. Efficiencies in the zeroth order. Dashed curve is theoretical and circles are experimental. (a) $E_{||}$ polarization ; (b) $E_{\perp}$ polarization.

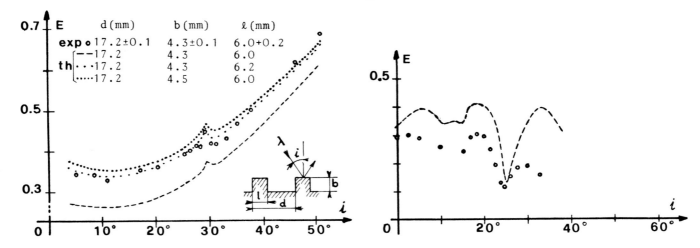

Figure 8. Effect of small variations of groove profile on the efficiency of lamellar grating. Efficiency in $E_{||}$ case as a function of incidence ; circles are experimental and dashed and dotted curves are theoretical.

Figure 9a. Transmittance as a function of incidence ; grating of parallel dielectric cylindrical rods ; dashed curve is theoretical and circles are experimental. -1 order.

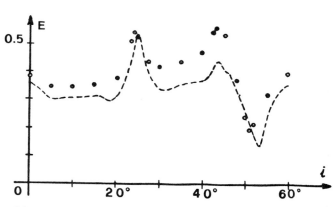

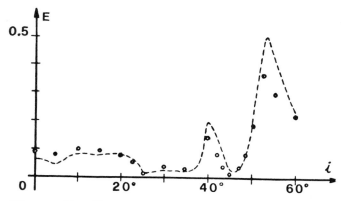

Figure 9b. The same as Figure 9a, but in the zeroth order.

Figure 9c. The same as Figure 9a, but in the -1 order.

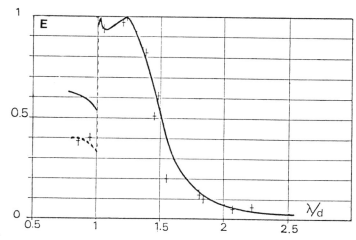

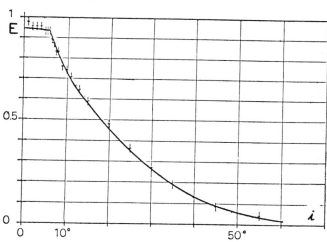

Figure 10. Efficiency transmitted in the zeroth order vs λ/d. Solid curve is theoretical and crosses are experimental.

Figure 11a. Efficiency transmitted in zeroth order by round-holed inductive grid as a function of angle of incidence in $E_\perp$ polarization. Solid curve is theoretical and crosses are experimental.

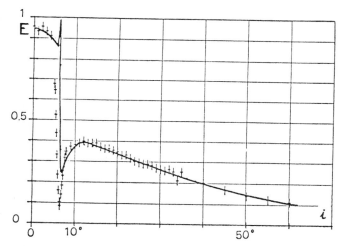

Figure 11b. The same as Figure 11a, but in $E_{||}$ polarization.

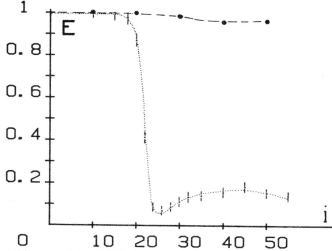

Figure 12. Efficiency in the zeroth order as a function of the incidence with a pyramidal crossed grating. Dashed and dotted curves are $E_\perp$ and $E_{||}$ polarization respectively.

Panel discussion on grating technology

Panel Members

Chang H. Chi (CC), Hughes Aircraft Company, Moderator

Hugh L. Garvin (HG), Hughes Research Laboratories

Erwin G. Loewen (EL), Bausch and Lomb, Incorporated

William R. Hunter (WH), Naval Research Laboratory

Jeremy M. Lerner (JL), Instruments SA, Incorporated

Michael C. Hutley (MH), National Physical Laboratory, England

CC: I would like to ask the panel four questions of general interest, and since our time is rather limited, I would like to ask you to try to limit your comments to three or four minutes.

The questions are the following:

1. How would you characterize the progress of grating technology over the past 5 years, particularly in the area of theoretical analysis, experimental applications, and diagnostics?

2. What are the areas that need innovative ideas and technical breakthroughs. This question is addressed more for the benefit of younger people. Suppose we have, for example, a Ph.D. student wanting to do a thesis; in what areas would you like to suggest they put their study time in?

3. What are the major problems in the grating technology community? Do they include funding, work force (are we educating enough talent in the Universities?), industrial secrecy, government assistance, international cooperation, patent protection, or any other items?

4. What actions can you suggest to promote the welfare of the grating technology community? Has it been worthwhile to come to this conference? Are there some things that we can change? What recommendations do you feel we should make?

These are the questions that I feel will be of general interest. Any volunteers from the panel to address any of the questions?

EL: Starting with your first question. I think that the two areas that have really shown significant change in the last five, or more accurately, the last ten years have been first, the development and then the putting to use, for the first time, accurate diffraction theory, as it applies to just about any kind of grating, and we've had several contributions to that at this very meeting. Some things that used to be mysterious, since Fraunhofer's day in 1820, are no longer mysterious, and the benefits of that are just tremendous. It shows what you can do and what you can't do, the ways you can optimize, the way you can't. It's just tremendous. And the other change, of course, is in the area of interference of photoresist or of holographic gratings—whatever term you want to use—because that has given us a brand new tool with which to attack on several fronts, and again some of those have been discussed earlier this week and I guess there are some left to be discussed tommorrow. I think it is safe to say that what's happened in those two areas in the last ten years is equivalent to the progress in the previous forty years.

CC: I certainly agree. The fact that we have fast digital computers and the fact that we can look into the interaction of the grooves and the waves, like Professor Neviere and his colleagues have done, really have improved the standing of this art.

MH: Well, to carry on from that, I am a simple-minded soul and I am one of those who says I'll do the experiments, and let somebody else who understands it do the theory. I am extremely impressed by the way things have gone. There was a time when there were experimental results which you couldn't explain, and then we got experimental results which tied up with the theory. The theory was able to explain the experiment. And now we have gone past that stage to a stage in which the theory predicts new results. The total absorption of light by a grating is one of those examples. But for my personal simple-minded point of view, I would rather hanker after the simple thought of light being reflected from a facet, something like that. I am now at a loss and I don't really understand the physics of what's going on. For example, I would like it very much if you could tell me what aspect it is of a groove profile that causes an anomaly to take place. I've got examples at home of gratings that are very similar in groove profile, but yet the anomalies are dramatically different. Nobody really knows until he either makes it or does the full calculation, whether a given groove profile is going to give him an anomaly or not. What I would like is something which tells me the physical picture. Something I could feel that I could handle. So I list that under what areas need innovative ideas.

WH: I approach this problem from the point of view of a person who was recently associated with the space science effort, rocket spectroscopy, and the vacuum ultraviolet, and who is presently into the synchrotron radiation business, and it seems to me that in the area of theory things have gone ahead very well, as Erwin pointed out. In applications—well they just have to wait. The different applications that come, you can't really mention that. Where I feel things are falling down, and in my particular field, is in experimentation. We like to measure the characteristics of the gratings in the vacuum ultraviolet to see what they are doing, and the apparatus that I set up at the Naval Research Laboratory is capable of doing so with nonpolarized light. However, I think it would be very interesting to measure these things in polarized light, and I am thinking particularly of a synchrotron light source. There doesn't seem to be any interest in setting one up. I have approached a number of people about a computer controlled grating scanning device that would enable gratings to be measured in both polarizations and there doesn't seem to be any response at all. At least money is not forthcoming, although everyone is very sympathetic.

MH: Don't you think that will happen when somebody actually runs into trouble because they haven't gotten it?

WH: Not necessarily, they will just forget about it and do something else.

JL: From the standpoint of Instruments S.A. and Jobin-Yvon, I think that one of the most startling things that has happened over the last five years is the development of diffraction gratings that act as complete spectroscopic systems in their own right. And I believe it is very likely that this technology is going to increase the use of diffraction gratings to reimage on flat focal planes. I think it is going to become a fundamental necessity as payloads in space programs become a little bit more critical, and the use of mirrors, lenses, and other optical components, becomes cumbersome. The use of aspheric gratings, whether or not they be aberration-corrected is going to become a standard product in synchrotron radiation facilities. I believe that this can also be extrapolated into space programs for monitoring UV sources. As far as applications using ruled gratings are concerned; it's easy to feel, especially at a meeting like this, that all the future is going into holographic and ion-etched gratings. Ruled gratings are certainly going to become more and more sophisticated, and the use of ruled gratings in waveguide laser, lidar, and dye laser applications will become extremely significant. We have seen this trend over the years. Then there is the concept of making one grating, or one blank with many gratings upon it perhaps as an analytical tool. We have seen at least one paper from colleagues in Sweden in which that was done. I think that this is going to become more and more commonplace.

HG: I fully agree that this is an exciting time, in that we have tools available to use right now with which we can do some things that haven't been done before. They are exciting things to do in terms of making lenses that are sitting on a flat surface and so on, and yet they are scalable to making very real hardware capable of handling large area coherent beams. We now have available to us a phase uniformity, or phase control, across large areas that can be done by the older techniques. But by some of the new techniques this may be done more easily or with a certain amount of control that wasn't previously available. I think it is particularly interesting right now to look to the applications of these techniques and see where the greatest interest will really arise. The immediate applications, from military systems or high energy laser systems, give impetus for looking at specific areas, but I think the most exciting part is going to be in other areas which are of interest to every day life and perhaps of more interest to people than just the military applications.

CC: And now, I would like to ask if there are any questions or comments from the audience?

J. Cowan, Polaroid Corporation: Well, I was thinking about a particular application, but you pretty well covered the spectroscopic military application. I think there is another application that might be very valuable, and that is the idea of using gratings as solar absorbers. We've looked at this in the microwave region, but actually we have to look at this in the visible and infrared regions. There is much work to be done making cross-grating structures. Such things could potentially be very valuable, particularly in the energy field, and we all know how important that is these days.

MH: I just happen to have with me what we were talking about. It's being written up in the literature, but not very well. We call it the "moth eye." It's called the "moth eye," because the eye of the night-flying moth is covered with something rather equivalent to a crossed grating. Because it is so fine that the light doesn't resolve it, it has the appearance of a graded diffractive index so that the Fresnel reflection is reduced. If you put it on a dielectric surface, the transmission is increased. If you put it on a metallic surface, the transmission is increased, but it just gets absorbed. So what you have is something that will absorb—a metallic surface which will absorb—visible radiation. But because the height of the pimples is so small it isn't noticed by the infrared radiation, it has a low emissivity. I've got one of these here; it has about 30 megapimples per square millimeter on it, and you can see it is quite black. That's electroformed nickel. We would like somebody to...

Audience: There is a use for electroforming then?

MH: Oh, there is a use for it. We would like to see that on a roller for example, turning it out by the acre on plastic sheet. If you could get it cheap enough I'm sure it would be useful.

Audience: Thank you for saying that. I didn't expect it.

CC: Anybody else like to comment on this?

Monica Minden, Hughes Research Labs.: Over the last three days I have felt that there is a crying need for someone to group together all the different areas of applications and analyses that people have talked about here, because I've seen that we have spectroscopists and we have theorists on diffraction efficiency and we have all these different applications, but we don't speak the same language. You know, you can walk in and out of each other's talks with absolutely no comprehension, if the person speaking was from one of the different areas, and I think there's a crying need for somebody to pull all these different areas together, get a common notation between them and show in which places these areas overlap.

CC: I agree. Let's take a vote and ask Professor Neviere and Dr. Mike · Hutley to write a nice book for us. These are typical things that could be covered so that all these different areas could be explained in the coherent mathematics of an understandable form. That is a very good thought, and I feel the same way.

MH: I wonder if it's even worth calling a meeting so that we would all get together and agree on the terminology. If we all knew what we meant by blazed, for example, it would be nice. Thereafter, everyone

would agree to abide by this terminology. If we all agree what we meant by the minus one order, it might help. I've got various other pet things that I won't go into now, but one of them is...well, my talk tomorrow is entitled "Interference Gratings" and that's because I feel strongly that there is nothing holographic about holographic gratings. I wonder if there is any sense in having a meeting to discuss all the terminology. There is quite a lot of it which is a bit loose.

CC: Well, I think past experience shows that nobody can tell people what something should be called. Rather, these things tend to be decided by the force of habit. Someone uses a certain nomenclature, then his student says the same thing. After many years it is called that way by habit; so I suppose the same thing will happen in this case. The holographic grating, for example, is called a non-linear. Yet, there is nothing non-linear about it even when the period is not a constant.

I would like to add one more area that needs more work. It is the grating technology for high power applications. In the low power applications, the absorption of power, scatter and grating substrate distortions are not of paramount importance. However, in the high power applications, a fraction of one percent absorption, for instance, is a significant power when the input beam has a huge power level. This absorption of power causes the grating substrate to distort, which in turn results in the change in the grating period as well as the degradation of the optical flatness. As you know, the grating characteristics change significantly when the change is on the order of a wavelength. The same is also true for the scatter by the grating in the high power applications. These are a few of the obvious technical issues we need to consider in the high power applications. Other issues include coating design, survivability, fabrication techniques because the elements tend to be dimensionally large in high power applications, and diagnostic techniques. Having recognized these problems, people begin to think of the solutions, such as better fabrication control, compensation techniques which negate the effects of these undesirable distortions, and better analysis and understanding of the phenomena. Any other comments?

Gilmore Dunning, Hughes Research Labs.: Branching off on your absorption, I was wondering if anybody is familiar with any work that has been done in terms of absorption as a function of whether the grating surface has been annealed or not? In some of our absorption studies we found that because of the groove profile there are sharp edges. This enhanced the absorption, and we were wondering if anybody had looked at annealing, or laser annealing. We also found that when the gratings were used over long periods of time that there was laser annealing on the surface and the absorption was reduced as a function of time. And, I just was wondering if anybody had studied this in a more systematic manner than our off-hand...

JL: Was this not on the subject of the 1976 paper by Brian Newnan from Los Alamos in the CLEOS show? He did a study on groove profiles taking ruled and holographic gratings and studied them simply from the point-of-view of damage thresholds. I think he postulated at that time that a ruled grating was a very sharp peak and was subject to localized burning at points where there may be a burr due to the ruling technique. That creates a nucleation point, if you like, that expands, and so on. The holographic gratings which had a nice round top, or the ones he was looking at, had damage thresholds ranging from 2 to 5 times that of the ruled gratings.

Monica Minden: I think those were made on a low efficiency beam sampler. The problem with the types of gratings that Gil and I have been looking at are somewhat different. Just as an interesting point, one of the things we found is that our ruled gratings were actually lower absorbing than some of the ion-etched ones. We have no idea why. We don't know if this is a temporary effect that time will change, or what it is. I don't think that one study that was made, as far as I know, can be considered a definitive one.

JL: I agree.

HG: I think in discussions I've had with Brian, that he had actually used it as a technique looking for the relaxation of the grating, and its effective change in efficiency, and looking at the deformation, and using that as a technique for measuring. I think we are getting to a point where considerable work is required in the materials aspect. Our experiences have been with the materials that we are most familiar with because they are immediately at hand. But you have to get to the area where you look for better materials, better reflectivity, better durability, better adherence to substrates, and over-coatings of materials that can tolerate a buildup of thick coatings and subsequent polishings, and over-coating of those. I think that we are sort of coasting along; we haven't come to that stone wall we are going to run into one of these days. Until we get better materials for these particular applications, we are in bad trouble. And, I think that's the kind of thing we have to look downstream to, particularly in some of these applications.

JL: In addition, as far as laser gratings are concerned, while it's generally agreed that everyone has to maintain very high efficiency, the problems of diffracted wavefront errors have become just as important. Sometimes when you correct one, you create problems with the other. I believe that more work should be done simply on laser gratings, which will perform well in efficiency, and also extremely well in diffracted wavefronts.

Audience: I'd like to echo Mike Hutley's comments about knowing what's happening at the grating surfaces. I would like to ask the French contingent, because their work is so good, if they can begin to tell us what's happening in some of these corners. One knows that there has to be some very large currents because of discontinuity. We would love to know what's really happening. In an experiment we conducted many years ago—mostly because we got into trouble with a Wood's anomaly with our instrument—we found that we could take the same grating and take the negative of it, and we got rid of our problem. We didn't get rid of the Wood's anomaly, we just changed the shape. The truth is that the groove profile at the corner was different, depending on whether you had the positive or negative. And I would still like to know what is going on.

Michel Neviere, University of Marseille: We have no physical explanation, but the theory indeed predicts, for example, that the bottom of a groove of a blazed grating has no influence on the efficiency curve while the top of the groove has a very important influence. It is for this reason that when you take the negative of a groove you may suppress an anomaly, because an anomaly may be caused by the top of a groove, and if you take the negative of the profile then that sharp corner may disappear and it may be at the bottom, where it has no influence. We have no physical model, no simple explanation at hand to explain the behavior of grating in terms of what the origin of the anomaly is. We are able to predict the evolution, we are able to find a mathematical model, complex poles and so on, but what is the origin of this complex pole? We know that the complex pole is related to the existence of a surface wave but we don't know why. For example, when we increase groove depth the pole goes to the upper half of the complex plane, which means that the anomaly may be reduced or under control, but we don't have any real physical explanation. I hope that before too long we will find some explanation. For the moment we cannot answer the question.

Tom Theis, IBM: I am a research physicist, so I take yet another view of the problem of terminology in the field of gratings, etc., and I would like to address these remarks to Dr. Hutley, and also to Professor Neviere. When we talk about a Wood's anomaly, we refer to it as a plasma wave, and a view that has been fairly productive in a great deal of research that has been done on surface plasma waves and thin metallic films is to consider the grating as being a mechanism whereby the wave is coupled out to the external radiation field. The simple view that I have of why that pole moves around, why the anomaly shifts when you make the groove deeper, is that you are increasing the coupling to the external field, you're essentially damping an harmonic oscillator. It explains the direction of the frequency shift. Another thing that has been discussed in some of the talks here is the shifts of the anomalies when you put a dielectric coating on the grating. Well, if you just look at the velocity of propagation of the surface wave without the grating, you will find that it changes quite a bit because you essentially screen the electric fields associated with that wave when you put a dielectric on the metal surface. I'm not sure that these physical ideas have precise analysis in the very beautiful theory that has been presented here by the French group, but I think that it would definitely be worthwhile looking into these sort of things.

Michel Neviere: I'm afraid that the simple model will predict that if you increase the groove depth of a grating, you will increase the strength of the anomaly. I'm afraid of that, and that is false. You increase the strength of an anomaly at the beginning and after that anomaly disappears. In the time that we have tried to explain the behavior of gratings in terms of harmonic oscillators or surface impedance and so on, we found some qualitative analogies, but never quantitative agreement. That is my answer. Maybe it is useful to get a feeling, but these simple models have their limits, you see.

Tom Theis: Certainly everything I said would only apply to the case of fairly shallow gradings and you certainly do need elaborate mathematical formulism. I think you can get some simple feeling with the direction of these trends, and so forth.

MH: One of the things you can do is this: You have a surface plasma, you know what it's wave vector is, you know the conditions under which you can couple energy into it, and then you can say as the groove gets deeper so the surface plasma has to go further, and so there is a change in the effective wave vector. That pushes you in the right way, it doesn't get you all the way, and so there is a limit to how far you can use even that sort of modification to the theory.

Tom Tice: I would just like to point out that in the example I gave a moment ago, both gratings have exactly the same depth, but the Wood's anomalies are very different. So you are begging the question when you say work with the depth.

MH: You're certainly right.

Audience: Which depth are we talking about? Is it the depth of the material? Is it the conductivity of the material? You know, that is what we would like to know.

MH: Sometimes with very similar gratings, you'll get at the Rayleigh wavelength, which is really a bit away from where the surface plasmon is happening. Sometimes you will get a steep cut, sometimes you will just get a sudden discontinuity—a shoulder. Why?

Jim Cowan, Polaroid Corporation: We worked out with Richie, when Richie was working on the surface plasmon theory. We tried to apply this to gratings, and it was a very beautiful theory. You had plasmon-photon coupling and surface waves and so forth and so on, and it wound up so that if you got anything over a groove depth of 700 or 800 angstroms or for a groove spacing of about one micron that the theory started to break down. So, again, it is true at least as far as these theories are concerned, that you really have to work with shallow grooves.

CC: How about the physical parameter constants? Do we know the physical constants when we plug in these formulas? Very often I've found that the index of the refraction, the loss constant, for instance, varies over a considerable range, depending on how the material is fabricated. For instance, the refractive index and interference pattern are different when the optical coating is performed at a different temperature. I remember when we were talking about some design—I think it was zinc selenide—there all kinds of constants from the different handbooks, and a person who happens to prefer certain handbooks, has a set of constants he prefers. When we make a negative from a positive grating, for instance, are we sure that the material constants are the same between the negative and positive even though they are made of the same material? Remembering that these constants vary depending on the processes, temperature, and the pressures it goes through. I was wondering if these things could be contributing to the difficulties.

MH: Are you suggesting that the grating itself has read a different handbook?

CC: Any other thoughts?

Bernhard Bach, Hyperfine: When talking about the anomaly, it is possible that in the ruling process, as you make these deep grooves, the leading burr and the trailing burr join just below the upper sharp edge of the groove. Depending on the coating, there will be a residual remnant of oxide in aluminum, and there will be a discontinuity of the surface, or the smoothness, and depending on the generation of the grating, that is possibly what is influencing the reversal of the anomaly. If you look at height graphs you will see a definite roughness on the upper portion of the grooves and depending on the coating, also on the lower portions of the grooves. The lower portion is much more tied together than the roughness of the upper part of the grooves.

EL: That sounds quite reasonable to me. I think that one facetious solution to all of this problem is to declare a moratorium on reflection gratings in order to start using transmission gratings, and the anomalies will then disappear in very short order.

Bernhard Bach: That would be a little difficult in certain spectral regions.

EL: Well, that's your tough luck.

MH: Can you make a blazed 1,200 grooves per millimeter grating in transmission in the visible?

EL: Oh, yes. You can make transmission gratings up to 1,800 per millimeter. It might not have the last word in high efficiency, but it will work.

Jim Cowan: One aspect that might be interesting to look at, in some future work, since we are relating these anomalies to surface waves, and since it does depend on a finite propagation distance along the surface, is looking at the possibility of putting selective dielectric coatings just on the facets, or just on the short side of the grooves, so that you essentially are preventing the surface wave from propagating.

EL: That sort of thing has been tried, and it doesn't work nearly as well as your physical intuition would lead you to expect. There is a Japanese patent on switching metals, for example, and people also try, with some success, to make replica gratings that have metal only on the blazed face. It all helps, but it doesn't completely suppress the problem.

Jim Cowan: I guess the question is, why doesn't that work? Perhaps the dielectric gap is not sufficiently great to really stop the waves.

CC: Well, perhaps we should ask the next question. Is the tempo of the activities in this industry sufficient? Are we getting enough work force in the industry? What is the current status of our activities? Are we going very strong? Are we educating enough young people? Is there interest in the schools in this area, or is it just older people doing the same thing for 10 or 20 years? Anybody have some comment on that?

Jim Cowan: I know one good way of getting young people interested in gratings, and that is not to show them gratings but to show them holograms. They always get interested in looking at these images, and you can explain later that this is actually a diffraction grating.

JL: I don't think that the majority of universities do anywhere near enough to teach grating theory. In fact, it is really quite grim the number of people who find themselves at one point or another with the requirement to design an instrument. The amount they know about gratings is really sometimes pitiful.

WH: But you can say this about optics in general, can't you?

JL: Yes, you can, and it is very sad.

EL: As more and more nuclear physicists go unemployed I think there will be a gradual renaissance and interest in optics. That has even taken place at an institution like MIT. They used to spend a great deal of their time sneering at optics, but no more.

CC: Are there any problems that we see that need to be rectified. Are we too secretive among companies believing that we should not divulge our industrial proprietary ideas? Are we getting enough cooperation among the international scientists? Any on patent problems?

EL: By funding, I presume you mean the only source that counts and that is the U.S. government. I think it's safe to say that the U.S. government for the last 5 or 10 years has done very little in the way of encouraging the grating R&D. They've certainly funded development efforts in high energy laser gratings, in the work that you were describing this morning, but I don't think any significant effort has been

funded at any university. On the other hand, if there were people in the government who were anxious to provide such funds, where is the university with the capacity to accept them in a really useful way? One university tried. I saw the research reports and they really regretted that they ever bothered.

JL: It's very difficult to wander into the business of making gratings. It's extremely expensive and there aren't too many people who know how to do it well. There would be a tremendous initial lag if someone just took a good ruling engine and put it in a large number of scientific institutions. There would still be a considerable period of time before that engine, under the operation of one of the faculty, was actually going to be producing good quality gratings. From the commercial standpoint, yes, there is secrecy because diffraction gratings do not have a commercial market and there will always be a lag between what is going to be the next generation and when it's actually first discussed.

Wayne McKinney, Bausch & Lomb: If I could get parochial for a minute, I think there is a fundamental difference on the two sides of the Atlantic in the way the government does business with business. There are certain areas of technology that, if you have to make a dollar, you won't do, but if you have a government subsidy you are a lot more likely to do. I think we should either quit playing the game in certain instances or we should get some government support. I mean you can't compete with some of these various organizations who are competing under different rules. It's just impossible.

WH: Of course, different governments have different operating philosophies regarding the support they give to their industrial infrastructure, and in the United States it has certainly changed. There was a long period when it was unfashionable to support industrial development; there was supposed to be the progress of industry by itself. Then came the 1960's when money was being thrown in all directions, particularly by NASA in the interest of speeding up progress. NASA had a strong mandate to support increased development by anybody in industry that was willing to accept some money and could show, with reasonable clarity that the project on hand was going to contribute to the space effort, just about any part of it. So for more than a decade the scene changed. And then as soon as we had a man on the moon, the urgency was gone and this is the situation that we find ourselves in today. Now it's only when one can demonstrate another important link to a defense effort, and it has to be fairly well and directly proven, that some kind of support can be expected. Other countries have maintained much steadier policies in supporting their instrumentation industry. But the likelihood of that changing on the basis of anything we can say here today, I think, is very small.

CC: My opinion is that the government is made up of a group of people just like us; they have their own opinion and their own judgment. I think the first thing to do is to find the application area. It looks like we have a tremendous pool of knowledge and backlog of experience. If we can find a good application in a new technology area using grating technologies, then there will be a justifiable reason to claim the support money. We are expanding in UV, we're expanding in IR, and in power levels. But if we can just leave the state where we worry about gratings only, and combine the grating technology with some other disciplines, we may see the birth of a new application area. The optical circuits technology is an example. I would now like to ask you the last question, as a way to summarize this conference. What actions would we suggest to promote whatever we did here and in the future? Any thoughts about this conference, any thoughts about what could be improved? What went wrong here, or any comments on this?

JL: I have a comment to the previous question. Don't you feel that many areas of grating developments are in fact taking place under government sponsorship at this moment. I am sure that the U.S. Army, for example, is funding a little research at least on new diffraction gratings.

CC: Well, of course, there are...

JL: I believe this is government funding for the development of diffractions gratings.

CC: Of course, but they are piecemeal and fragmental—some funding here and there. But, I feel it is unorganized. Somebody wants to do something and gets funded if he manages to sound convincing. What we are talking about, however, is a sustained effort, not just a six month or one year study contract. One reason why there has been significant progress in the high power application areas is that the funding has been steady and the work has been a sustained effort. Also, the high energy laser application is an example of finding the application area of grating technology. In it, we combine the grating technology with a multilayer coating technology and make them work in a high power environment. This means that one has to learn more than the grating parameters. Then you see you can do something more than just what gratings can do.

CC: I appreciate all of your comments and thank you, the panel members. With this, we will end the panel discussion.

Thank you.

PERIODIC STRUCTURES, GRATINGS, MOIRÉ PATTERNS AND DIFFRACTION PHENOMENA

Volume 240

SESSION 7

FABRICATION TECHNIQUES, APPLICATIONS III

Session Chairman
Terry L. Holcomb
Hughes Aircraft Company

Use of diffraction patterns as a regularity diagnostic for jittered arrays of periodic structures

C. Aime, F. Martin

Département d'Astrophysique de l'I.M.S.P., Université de Nice, E.R.A. 669 du C.N.R.S.
Parc Valrose, 06034 Nice Cedex France

Abstract

We describe a model which gives a general expression for the two-dimensional spectral properties of periodic structures of grains perturbed by random jitter. The theoretical formulation is valid no matter what may be the form of the grains of the probability density function of the jitter. The power spectrum of such arrays of grains is a mixed spectrum containing both a continuous term and a discrete line structure. As the intensity of the random jitter is increased, the relative intensity of the line components decreases. Experimental results, obtained with computer simulated arrays of grains, are in excellent agreement with theoretical model. Using Fourier techniques, experimental power spectra are obtained by an analogic optical method (Fraunhoffer diffraction) as well as by computer calculations (Fast Fourier Transform). The results presented here permit a quantitative determination of arrangement quality of periodic structures, randomly perturbed by jitter.

Jittered array of periodic structures

Let us define an underlying array of hexagonal cells arranged in a honeycomb structure. These cells are supposed all identical and the center of each one is at a fixed distance b from the center of each nearest neighbour, whose position vector is $\vec{\alpha}_{ij}$. The centers thus arranged form a double Dirac comb in the $\vec{r}(x,y)$ plane, for which we will use the notation $\underline{III}(\vec{r}/b)$ (see Bracewell[1]), and provide a reference array.

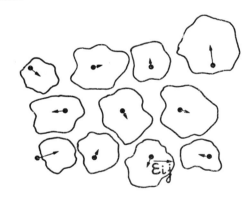

Figure 1. Example of the intensity distribution $D(\vec{r})$ for a spatial distribution of cells described by the model. The cells of variable amplitude and shape are at distances $\vec{\varepsilon}_{ij}$ from the Dirac comb.

We now suppose that the shape and the intensity distribution $B_{ij}(\vec{r})$ can change from one cell to another, and that the center of each cell is displaced by a random quantity $\vec{\varepsilon}_{ij}$ from $\vec{\alpha}_{ij}$ with a probability density function (pdf) $p(\varepsilon)$ for $\vec{\varepsilon}_{ij}$. In this case the total intensity distribution function, illustrated in figure 1, can be written :

$$D(\vec{r}) = \sum_i \sum_j B_{ij}(\vec{r} - \vec{\alpha}_{ij} - \vec{\varepsilon}_{ij}) \qquad (1)$$

The autocorrelation function $C(\vec{\rho})$ of $D(\vec{r})$ is computed in reference 2 and the spatial power spectrum $W(\vec{f})$ of $D(\vec{r})$ is obtained by taking the Fourier transform of $C(\vec{\rho})$ Wiener Kintchine theorem) ; $W(\vec{f})$ can be written :

$$W(\vec{f}) = S(\vec{f}) - S^{\star}(\vec{f})|\hat{\phi}(\vec{f})|^2 + \underline{III}(b\vec{f})S^{\star}(\vec{f})|\hat{\phi}(\vec{f})|^2 \qquad (2)$$

where $\hat{\phi}(\vec{f})$ is the characteristic function for $\vec{\varepsilon}_{ij}$, $S(\vec{f})$ and $S^{\star}(\vec{f})$ are respectively the average power spectrum* of the intensity distribution of cells. $\underline{III}(b\vec{f})$ is a double Dirac comb because the Fourier transform of a Dirac comb gives another Dirac comb. The general shape of the autocorrelation function $C(\vec{\rho})$ and the power spectrum $W(\vec{f})$ are shown on figure 2 and figure 3.

Expression (2) is formally identical with that given in the one-dimensional situation[3]. In both cases the power spectrum $W(\vec{f})$ is a mixed spectrum containing both a continuous term and a discrete line structure. As the randomness is increased the relative intensity of the line components decreases.

If we suppose that the cells are all identical, then equation (2) reduces to :

$$W(\vec{f}) = S(\vec{f})[1 - |\hat{\phi}(\vec{f})|^2 + III(b\vec{f})|\hat{\phi}(\vec{f})|^2] \qquad (3)$$

*and cross-power spectrum

where $S(\vec{f}) = \left| 2 \dfrac{J_1(\pi f d)}{\pi f d} \right|^2$ if the cells are also circular with diametre d (Airy function)

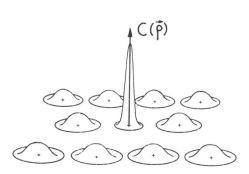

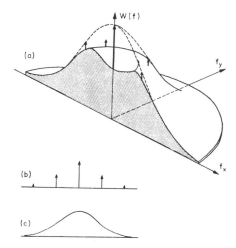

Figure 2. Autocorrelation function for the intensity fluctuations of a jittered array of cells. Centered at the origin is the average autocorrelation function of the individual cells. Outside the origin and centered on a double Dirac comb are peaks corresponding to the cross correlation function between the cells, convolved by the autocorrelation function of the pdf of the position fluctuations.

Figure 3. Spatial power spectrum corresponding to a jittered array of cells.
a) General case
b) Cut along the f_x axis in the non-jittered case.
c) Cut along the f_x axis for the case where the jitter is infinite (Poissonian repartition of cells).

Experimental results - Example of a particular application

The roles played by $S^{\star}(\vec{f})$ and $|\hat{\phi}(\vec{f})|^2$ are symmetrical and not easily dissociated. In order to obtain at least one of these functions independently we can replace each cell by a circular point of diameter d .

1) Analogical determination of the power spectrum

The power spectrum corresponding to a distribution of circular points is given by equation (3) ; the characteristics of this distribution can be obtained via the power spectrum. The experimental method used to obtain the two-dimensional power spectrum is based on classical optical arrangements, where the spectrum is observed as the Fraunhoffer diffraction pattern of the distribution[4,5]. In the case where the points are located at the center of hexagonal cells, the diffraction pattern is a pure line structure (figure 4a) as predicted by equation (3) where $|\hat{\phi}(\vec{f})|^2 = 1$. However, the lines are not true Dirac functions, because the spectral analysis window corresponding to the diameter of the point distribution is finite in size.

On the other hand, if we consider a totally random distribution of points corresponding to a Poissonian process, the diffraction pattern reduces to the central disk of the Airy function $S(\vec{f})$, as predicted by equation (3) ; this is shown in figure 4b.

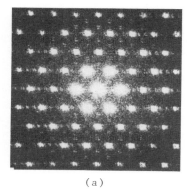

Figure 4.
Diffraction patterns given by point distributions corresponding respectively to

a) Hexagonal cells centers

b) Randomly distributed points (Poisson model)

(a)

(b)

Two intermediate cases of interest are shown in figure 5 and in figure 6. In these examples the reference lattice is composed of square cells ; although the calculation was developed by using a honeycomb reference lattice they remain valid for different types of periodic network. The diffraction pattern (figure 5b), corresponding to a weak jitter, shows the Dirac delta functions predicted by the term $S(\vec{f}).\underline{III}(b\vec{f})|\phi(f)|^2$ in expression (3). A light depletion zone, whose theoretical expression is given by $S(\vec{f})[1 - |\hat{\phi}(\vec{f})|^2]$, appears also in the central region.

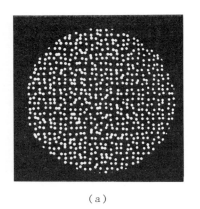

(a)

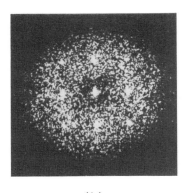

(b)

Figure 5. a) Array of points corresponding to a "weak jitter"
b) The corresponding diffraction pattern showing the Dirac peaks enlarged by the analysing spectral window, and the central light depletion zone.

If the jitter is increased (figure 6a) the peaks disappear in the diffraction pattern (figure 6b), but the light depletion zone, although narrower, is still visible. This decrease can be taken into account for the determination of the pdf of $\vec{\epsilon}_{ij}$ via $|\hat{\phi}(f)|^2$.

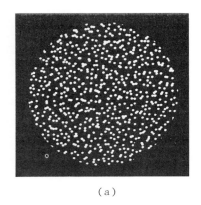

(a)

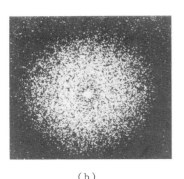

(b)

Figure 6. a) Array of points corresponding to a "strong jitter"
b) The diffraction pattern of (a). The Dirac peaks have disappeared and the central light depletion zone is still present but narrower than in figure 5b.

2) Numerical determination of the power spectrum : application to solar granulation

The model given here and its properties have been developed for the study of morphological properties of solar granulation[2,6] and for other applications[7,8]. In order to determine the characteristic function $\hat{\phi}(\vec{f})$ we used a photographic image of solar granules ; in this image each granule was replaced by a circular point positioned at the visually estimated granule center. $|\hat{\phi}(\vec{f})|^2$ and the pdf of intergranular distances were deduced[6].

This method presents two important inconveniences : i) The signal to noise ratio in the power spectrum is very poor ; to improve this ratio, processing of numerous photographs is necessary in order to obtain a better estimation of the power spectrum, ii) Determination of the granule centers is not objective. These two arguments imply a need for numerical processing.

All this numerical processing has been developed in cooperation with the "Centre de Dépouillement des Clichés Astronomiques" (Institut National d'Astronomie et de Géophysique) whose equipment consists of a Perkin Elmer Microdensitometer (PDS) and a Digital Equipment computer (PDP 1140).

We then studied, using digital techniques, a high quality photograph of granulation, obtained in 1978 with the 50 cm solar telescope of the Pic-du-Midi Observatory (France) at a wavelength of 5900 Å. The film covers an area of about 100" x 120". The cliché is divided in 49 areas of about 13" x 13". For each area, and after digitalisation, the following operations are performed : filtering, determination of granule boundaries using a negative Laplacian algorithm (figure 7a), and calculation of granule parameters, i.e. position of the centers of gravity (figure 7b), surface area, luminous flux integral and power spectrum via FFT. An example is given in figure 7c.

(a)

(b)

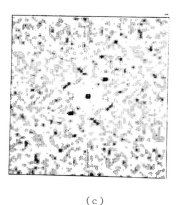

(c)

Figure 7.   a) Domains corresponding to granule boundaries obtained with a negative Laplacian algorithm
   b) Point distribution given by the gravity centers of the domains of a)
   c) Power spectrum of b) computed via an F.F.T. algorithm.

An improved statistical estimation of the power spectrum has been obtained by averaging the individual power spectra corresponding to the 49 areas. This is illustrated in figure 8; the figure shows a well defined decrease in energy in an annular zone centered at the origin. This very new and interesting result suggests local arrangements in the granulation

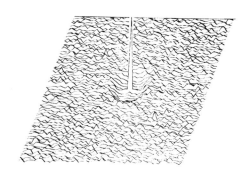

(a)

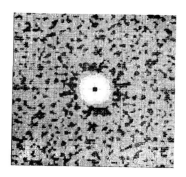

(b)

Figure 8.   Power spectrum estimation obtained by averaging 49 individual power spectra to improve the signal to noise ratio
   a) a relief representation
   b) a grey level representation

Conclusion

Morphological properties of two dimensional structures can be derived from the proposed model. We have shown that there is very good agreement between the theoretical formulation for the power spectrum, deduced from this model, and the corresponding diffraction patterns obtained by Optical Fourier Transform.

A sequential numerical processing technique using image segmentation has permitted objective determination of the power spectrum.

Other quantities related to the shape of the cells (contours, area; flux, mean size...) have also been determined but are not presented here.

References

1. Bracewell, R. The Fourier transform and its applications, McGrw Hill, 1965.
2. Aime, C., "A morphological interpretation of the spatial power spectrum of the solar granulation," Astron. Astrophys, Vol. 67, pp. 1-6, 1978.
3. White, D.R., Cha, M.Y., "Analysis of the 5 min. oscillatory photospheric motion. I - A problem in waveform classification," Solar Phys., Vol. 31, p. 23, 1973.
4. Goodman, J.W., Introduction to Fourier optics, McGraw Hill, 1968.
5. Stroke, G.W., An introduction to coherent optics and holography, Academic Press, 1969.
6. Aime, C., Martin, F., Grec, G., Roddier, F., "Statistical determination of a morphological parameter in solar granulation : Spatial distribution of granules, "Astron. Astrophys., Vol. 79, pp. 1-7, 1979.
7. Stark, M., "Diffraction patterns of nonoverlapping grains," J. Opt. Soc. Am., Vol. 67, p. 700, 1980.
8. Martin, F. Aime, C., "Some spectral properties of random arrays of grains," J. Opt. Soc. Am., Vol. 69, P. 1315, 1979.

Development of a compound interlaced grating for high energy laser systems

C. Chi, C. Smith
Hughes Aircraft Company, Culver City, California 90230

M. Ogan, C. O'Bryan
Air Force Weapons Laboratory, Kirtland AFB, New Mexico 87117

Abstract

In High Energy Laser (HEL) systems, there exist requirements for spectral beam splitting devices for purposes of beam control and diagnostics. This paper describes the development of one such aperture sharing component, called the Compound Interlaced (CI) diffraction grating. The CI grating is ruled such that a high frequency, visible to near IR wavelength grating is placed directly on top of a lower frequency, long wave IR grating. In a system configuration, the HEL beam is sampled with the long period ruling and directed toward a wavefront monitor or beam tilt sensor, while the specular reflection is sent through pointing optics to the target of interest. At the same time, shorter wavelength return energy from the target is sampled by the short period grating and is subsequently imaged by an appropriate sensor. Results of work completed to date show that indeed such a grating can be fabricated on a high power substrate and can yield enough efficiency for an operational environment. The paper addresses various design and fabrication issues including groove spacing and profile, blaze angles, apex angles, skew angles (of one ruling with respect to the other), substrate coatings and ruling techniques.

Introduction

A spectral beam splitting device is often necessary for High Energy Laser (HEL) system applications. Such a device, which allows direct coaxial viewing through the high power beam path, is referred to as an aperture sharing component. Aperture sharing can satisfy requirements of beam diagnostics, automatic focus of the beam on target, boresight error correction, hot spot tracking, and target imaging, among others. While beam diagnostics can be performed by a simple diffraction grating suitable for the HEL wavelength, the other functions normally require operation in other spectral bands. One solution to this problem is the use of a buried grating, which has particular fabrication complexity in terms of multilayer coatings and thermal effects. The component described in this paper provides an alternative solution and is called a Compound Interlaced (CI) diffraction grating. Beginning with a typical long wave IR grating, the aperture sharing is accomplished by ruling a high frequency grating (suitable for visible to near IR wavelengths) directly on top of the low frequency rulings.

When integrated into a complete optical control system, the HEL beam is sampled with the long period ruling while the specular reflection is sent through pointing optics to the target of interest. At the same time, shorter wavelength return energy is diffracted out of the high energy optical path by the short period grating and is subsequently imaged by an appropriate sensor. The particular center frequencies for the grating at hand are 10.6 μm (HEL beam), 3.7 μm (hot spot tracking or thermal imaging), and 0.61 μm (visible imaging for boresight correction or automatic focusing). Design and fabrication of the CI grating required analysis of the following issues: groove spacing and profile, blaze angles, apex angles, skew angles (of one ruling with respect to the other), substrate coatings and ruling techniques. Test results are presented which demonstrate current grating performance parameters (such as diffraction efficiency and scattering at the wavelengths of interest) which have been achieved.

Design Considerations

As described above, a Compound Interlaced grating is produced by superpositioning two diffraction gratings onto a common substrate. As an aperture sharing component, the CI grating is simple to fabricate and has excellent overall performance; however, it is useful only when the high energy laser (HEL) wavelength is at least two to three times longer than the diagnostic beam wavelength. The two superimposed grating designs have widely differing parameters. The b-grating (of period b) is designed to reflect the HEL beam with high efficiency while diffracting (sampling) it with low efficiency. It has a relatively "flat" profile and long period. The d-grating (of period d) is designed to diffract a much shorter wavelength with high efficiency. Typically, it works with visible (0.4 to 0.7 μm) radiation for alignment and middle infrared (3 to 5 μm) radiation for tracking. The d-grating has a short period and a "sharp" profile.

When the long wavelength HEL beam impinges on the CI grating it does not "see" the d-grating, as long as the grating period is shorter than 5.3 μm (one-half wavelength). The d-grating will, however, contribute to increased scatter since it does increase the effective surface roughness. When a short wavelength beam is incident onto the CI grating, it is diffracted by the d-grating in the manner and direction desired. Additionally, the short wavelength is diffracted by the b-grating, generating multiple ghost orders. The amplitudes of these ghost orders can be minimized by proper choice of groove shape, period and duty cycle. Various types of CI gratings are shown in Figure 1. A full CI grating is produced when the b-grating is ruled before the d-grating. This is usually desirable, since it yields the highest d-grating diffraction efficiency. A partial CI grating is produced by ruling the d-grating (fine) first and the b-grating (coarse) second, assuming the grating depths are comparable. The degenerate CI grating is a special case of the partial CI grating, wherein the high frequency grating is ruled intermittently. A skewed CI grating results when the grooves of the two superimposed gratings are not parallel. The skewed grating is often necessary to separate the return wave optics from the HEL diffraction orders. It has the additional advantage of separating the ghost orders of the b-grating from the primary diffraction orders.

Figure 2 is a simplified schematic of the central and diagnostic optics in an HEL system. It shows the manner in which the CI grating rhomb diffracts 0.5 to 0.7 μm and 3 to 5 μm, while simultaneously reflecting and sampling the 10.6 μm (HEL) beam. The outgoing and return waves each have a companion single frequency grating, arranged in a rhomb configuration. That is, the grating pairs are aligned and parallel. Light diffracted through the rhomb gratings emerges parallel to the input beam. The rhomb does shift various spectral components by differing amounts, so that the spatial extent of the output beam is larger than the input for broad band light. As shown in Figure 2 and described further below, the infrared and visible spectra can be controlled by the same rhomb pair, provided the diffraction orders for the desired wavelength regions are selected appropriately.

In developing the detailed grating design, several practical constraints had to be considered. First, a grating rhomb and subsequent diagnostic optics already existed for the 10.6 μm laser. It was most desirable not to interfere with the operation or location of this equipment. Second, a decision was made that the primary return sensor for the near term would be a SIT vidicon television camera. Investigation of the pointing optics showed that the peak visible transmission was at 0.62 μm, setting the approximate passband center. Finally, extensive examination of the diffracted and reflected high energy laser beams determined restrictions upon the location of the low power d-grating element. Consequently, a skew angle between the b- and d-grating directions was required. The diffraction grating computer design program of D. Maystre was used to determine the optimal grating profiles, which are shown in Figure 3. The skew geometry is depicted in Figure 4. Note that the HEL beam plane of incidence is not perpendicular to the axis of the b-grating. This causes the HEL diffraction orders to lie along an arc rather than a straight line.

Predicted spectral diffraction efficiencies of the visible grating profiles are shown in Figure 5 for rulings in either gold or silver. The gold grating can be used without overcoat. On the other hand, a silver grating would require a protective overcoating of ThF and either ZnSe or ZnS. The computer code was not used to determine diffraction efficiencies when non-planar coatings over the grating profile are applied. Figure 6 shows the diffraction efficiency as a function of groove apex angle. For all orders and materials investigated, the diffraction efficiency was very sensitive to reduction in apex angle down to 90°. Similar computer evaluations led to specification of the blaze angle to be 25°. The visible diffraction was blazed for the -6 order because this order would have the same diffraction angles as the -1 order of 3.7 μm.

Grating Fabrication

Two methods of ruling diffraction gratings are currently in use: planing and embossing. The planing method involves pushing a diamond tool across the substrate to "gouge out" the groove. Here, the substrate material is actually removed. Embossing involves "plowing" a groove by pulling the diamond tool across the surface. With this method, material is not actually removed, but is displaced. The ruling engines which produce diffraction gratings are of precision construction, and interferometrically controlled in a temperature stabilized and vibration isolated environment. This type of precision equipment is required to produce gratings with highly uniform period, low levels of scatter, and weak ghost orders.

To evaluate the feasibility and performance of the CI grating, full CI gratings were ruled on eight small (1.25 inches diameter) molybdenum witness samples. After evaluation of these samples, two large (10 inches diameter) test pieces were fabricated. One of these was on a cooled mirror substrate. Sample gratings were ruled in both gold and aluminum. A silver grating would consist of ruling in aluminum followed by a chrome coating and final silver overcoat. The aluminum or gold coatings are deposited to approximately 5 μm thickness onto the molybdenum substrates.

Two methods of depositing a rulable gold coating were investigated: sputtering and ion plating. The sputtering process uses a beam of charged particles (electrons or ions of inert gas) to eject particles of the desired material to the substrates. The coating has excellent adhesion with the surface because of the fairly high energy imparted to the ejected gold atoms. Ion plating requires an insulated substrate held at a high potential (typically 3 to 5 KV). The film material is evaporated from resistively heated coating "boats" operating at low voltage (∼5 volts) and high current (1000 to 1500 amperes). Before and during the evaporation of the film material, ions of an inert gas (argon, xenon or krypton) are admitted to the chamber. These ions are attracted towards the substrate by the potential difference and impart high amounts of energy (up to 1 KV) to the coating atoms, causing them to impact and adhere to the substrate. At the same time, coated film material is being removed from the substrate by the impact of high energy ions. Coating material must be evaporated from the boats faster than it is removed from the substrate to produce a coating. Properties of ion-plated gold differ greatly from those of sputtered gold. Ion-plated gold is usually soft (Knoop hardness ∼40) and cloudy. Sputtered gold is relatively hard (Knoop hardness ∼260) and shiny. The ion-plated gold ruled much better than sputter gold, due to its softness. A soft film produces a more predictable groove profile and results in less wear on the diamond tool.

The aluminum coatings are produced by evaporation. The evaporation process involves suspending the substrate over boats containing the coating material. The boats are heated, as in ion-plating, and the cloud of evaporating material flows over the substrate, creating a shiny surface with a Knoop hardness of 40. Evaporated aluminum produces a coating with good adhesion, but evaporated gold does not. Delamination of the film when attempting to rule into evaporated gold has been a widely reported experience. For this reason, evaporated gold was not used.

Test and Evaluation

During this effort, CI gratings were ruled into ion plated and sputtered gold, and evaporated aluminum. Several different methods and facilities were used to evaluate these samples. An efficiency measurement facility (Figure 7) was constructed for the 0.55 to 0.65 µm range. An existing scatter measurement facility (Figure 8) was used to measure scatter and diffraction efficiency at 10.6 µm. Laser calorimetry at 10.6 µm was employed to measure the absorption at the HEL wavelength. Typical absorption results are shown in Table 1.

TABLE 1. 10.6 µm ABSORPTION

Polarization	Absorption, Percent
TE	1.86
TM	2.36

Typical visible diffraction efficiency performance is presented in Figure 9. The relative spectral performance is similar to that predicted in Figure 5. Absolute efficiency is reduced from expected values because the desired apex angle of 90° was not produced by the ruling tool. Broadband average diffraction efficiency of 50% is, however, suitable for use in the system application.

One of the most important measurements made was the high energy diffraction and scattering at 10.6 µm. Comparison tests were made on both the low frequency grating alone as well as the full CI grating. The data presented in Figure 10 show minimal change in 10.6 µm performance when adding the high frequency grating. The data are normalized to the measured spectral reflectance.

Grating profiles were investigated with a Scanning Electron Microscope (SEM) and a Talystep Stylus Profilometer. Photographs similar to Figures 11 and 12 were obtained with the SEM. Edge profiles of the gratings were not taken because of elaborate means required, as well as the reluctance to destroy any sample gratings at this time. Figure 11 is a moderately low resolution photograph clearly showing the two grating frequencies ruled at a skew angle to each other. Figure 12 is a much higher resolution SEM photograph portraying the details of the d-grating profile. The Talystep profilometer is better suited for detailed evaluation of diffraction gratings. Dr. Jean Bennett of the China Lake Naval Weapons Center used a computer-interfaced Talystep to profile the CI grating witness samples. The scans were made using a 1 µm diameter diamond point stylus. Figures 13 and 14 show typical b- and d-grating profiles as measured by the Talystep. Figure 15 shows the modulation of the d-grating by the b-grating. Visible efficiency measurements of the first samples indicated that the desired groove profile was not being obtained (Figure 9). Profile measurements revealed that the apex angle produced was much broader than that of the optimal design. Computer analysis indicated that the lessened efficiency is due to the sub-optimal profile. Figure 6 showed that the sixth order diffraction efficiency is highly dependent on having a narrow apex angle.

All the results described have been measured from the witness samples. A typical sample is shown in Figure 16. At present, two large ion-plated gold coated substrates are being ruled with CI gratings. Gold is favored as a ruling film since it requires no protective overcoat. Protected silver coatings do not withstand normal environmental conditions well. After ruling, the gratings will be assembled into the rhomb configuration shown in Figure 2. The visible, 3 to 5 µm and 10.6 µm diffraction efficiency of the rhomb will be measured and the imaging quality of the rhomb system will also be evaluated. Figure 17 is a photograph of the first full size CI grating. Some areas of the gold film were destroyed during ruling. This is presumably due to a contamination problem encountered during coating.

Applications

The major impetus behind the development of the Compound Interlaced grating is its potential applicability to the Air Force Weapons Laboratory high energy laser program, specifically to experiments conducted on-board an airborne HEL testbed. The use of a jet aircraft as a platform for HEL technology implies adherence to requirements which are more stringent than those levied on a similar ground-based experiment. The need for proper boresighting (the ability to deliver energy on target at the precise aimpoint location) is in itself an extremely demanding task, particularly for off-axis sensors. The CI grating concept lends itself favorably to the boresighting requirement in that a visible sensor (for example), operating in conjuction with the high frequency grating may be aligned to see the exact spot at which the high power beam will strike. By keying on the shared aperture imagery, off-axis sensors may also be directly boresighted to the high power optical train, thereby improving initial pointing accuracies.

A second application for a visible imaging sensor whose field of view is coaxial with the HEL beam lies in the realm of autofocus techniques. Since aperture sharing allows simultaneous use of one set of optics, any beam defocus will be seen by the sensor as a defocused image of the target and can therefore be corrected. This one-to-one focus correspondence assumes that any initial HEL beam vergence is compensated for in the imager's fore-optics. An autofocus sensor in this configuration can minimize or delete the requirement for beam focusing via a rangefinder.

The fact that the short period ruling of the CI grating exhibits good efficiency at 3.7 μm allows for a third application in terms of hot spot tracking. In the present design, the first order diffraction of this wavelength is coincident with the sixth order visible radiation. A mid-IR sensor may therefore be placed near the visible imager with a dichroic beam splitter used to separate the two spectral bands. Again, because of the aperture sharing configuration, the initial boresight error should be minimized, resulting in improved hot spot tracking performance. In addition, this configuration allows the possibility of closing a control loop around a deformable mirror, thereby permitting adaptive optics correction. Other miscellaneous applications of the CI grating include a beam path safety check immediately prior to firing, a possible analysis of thermal effects on high power mirrors during a shot, and monitoring the target for damage effects.

The fabrication and testing of the CI grating described here demonstrates that such a concept is feasible. A design has been optimized for implementation in a specific optical configuration. Fabrication difficulties have been surmounted and performance goals have been accomplished. Future evaluations of the rhomb configuration and experiments in a system environment should establish the CI grating as a viable alternative to other aperture sharing components.

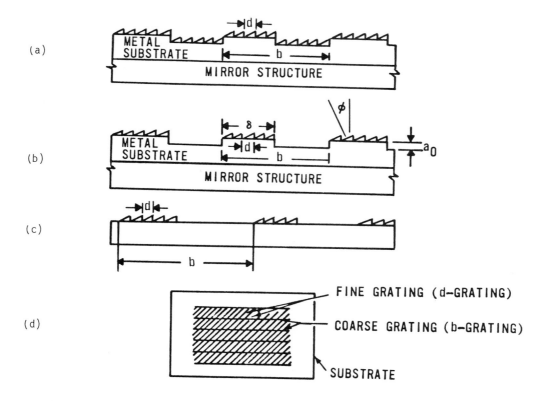

Figure 1. CI grating types: (a) Full, (b) Partial, (c) Degenerate, (d) Skewed

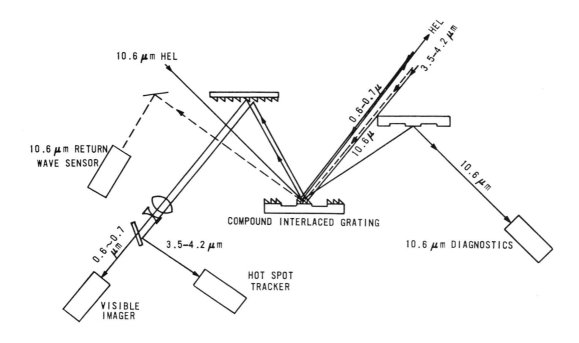

Figure 2. Schematic of CI grating rhomb system showing multi-spectral characteristics

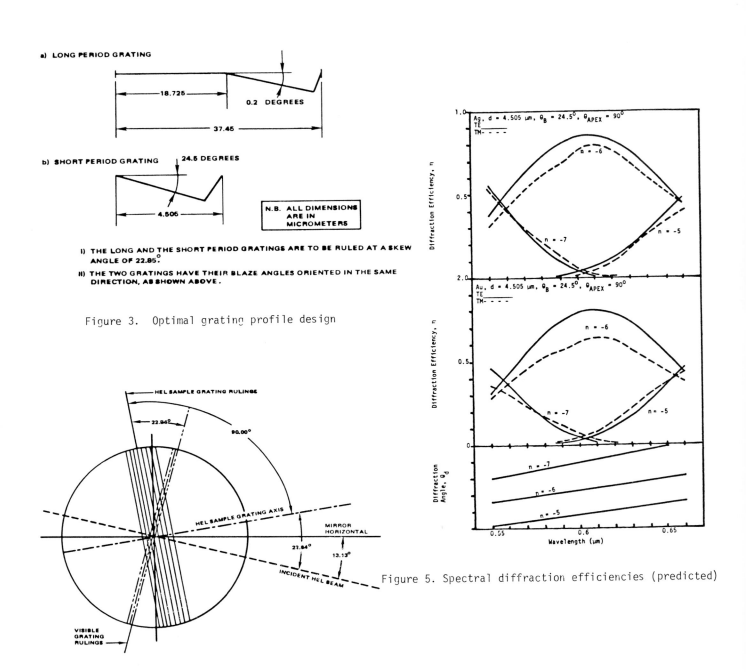

a) LONG PERIOD GRATING

18.725

0.2 DEGREES

37.45

b) SHORT PERIOD GRATING   24.5 DEGREES

4.505

N.B. ALL DIMENSIONS
ARE IN
MICROMETERS

I) THE LONG AND THE SHORT PERIOD GRATINGS ARE TO BE RULED AT A SKEW
ANGLE OF 22.85°.

II) THE TWO GRATINGS HAVE THEIR BLAZE ANGLES ORIENTED IN THE SAME
DIRECTION, AS SHOWN ABOVE.

Figure 3.   Optimal grating profile design

HEL SAMPLE GRATING RULINGS

22.84°

90.00°

HEL SAMPLE GRATING AXIS

MIRROR
HORIZONTAL

22.84°

13.13°

INCIDENT HEL BEAM

VISIBLE
GRATING
RULINGS

Figure 4.   CI grating ruling geometry

Ag, d = 4.505 μm, θ_B = 24.5°, θ_{APEX} = 90°
TE ———
TM — — — —

1.0

n = -6

Diffraction Efficiency, η

0.5

n = -7   n = -5

Au, d = 4.505 μm, θ_B = 24.5°, θ_{APEX} = 90°
TE ———
TM — — — —

2.0

Diffraction Efficiency, η

n = -6

0.5

n = -7   n = -5

0

Diffraction
Angle, θ_d

n = -7

n = -6

n = -5

0.55   0.6   0.65

Wavelength (μm)

Figure 5. Spectral diffraction efficiencies (predicted)

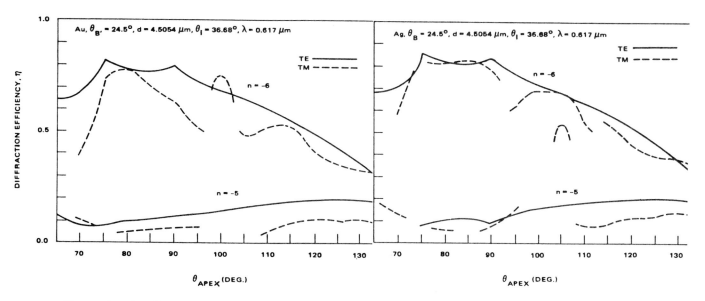

Figure 6. Predicted diffraction efficiencies as a function of apex angle for (a) Au and (b) Ag

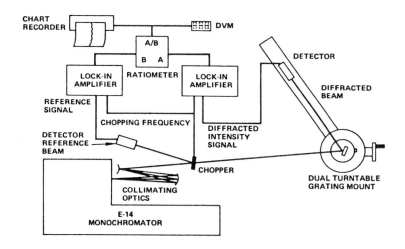

Figure 7. Efficiency measurement facility (visible wavelengths)

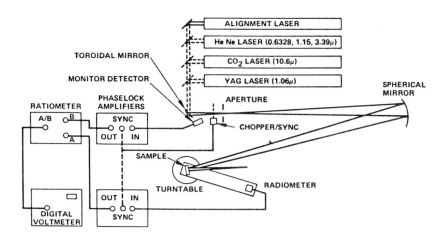

Figure 8. Scatter and diffraction measurement facility (10.6 μm)

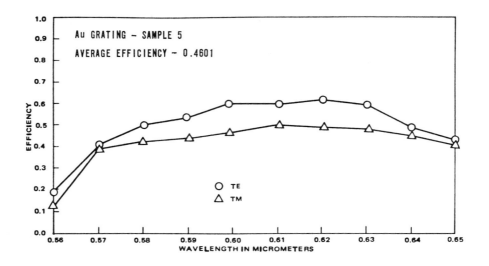

Figure 9. Visible diffraction efficiency (measured)

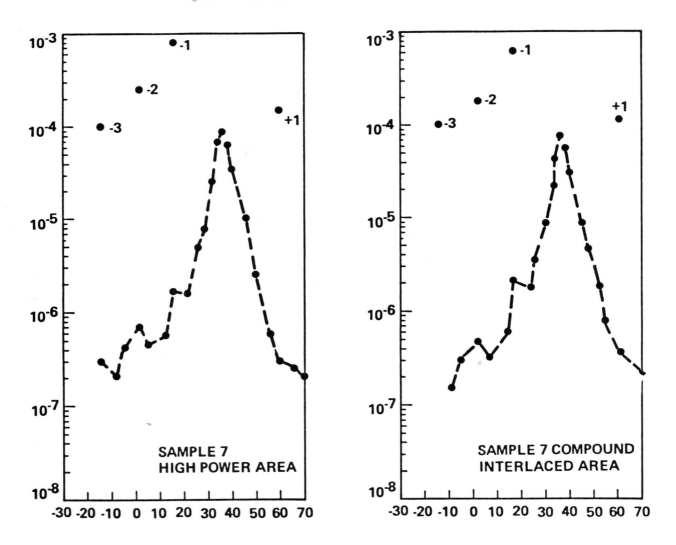

Figure 10. 10.6 µm scatter and diffraction efficiency for (a) long period ruling and (b) CI ruling

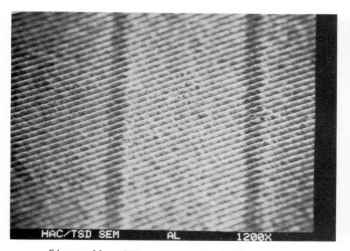

Figure 11. SEM photograph of CI grating

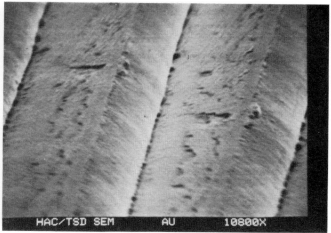

Figure 12. SEM photograph of short period grating

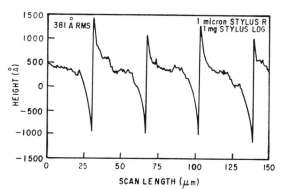

Figure 13. Talystep profile of b-grating

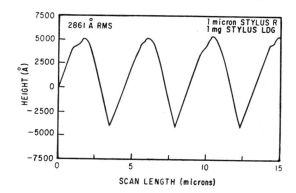

Figure 14. Talystep profile of d-grating

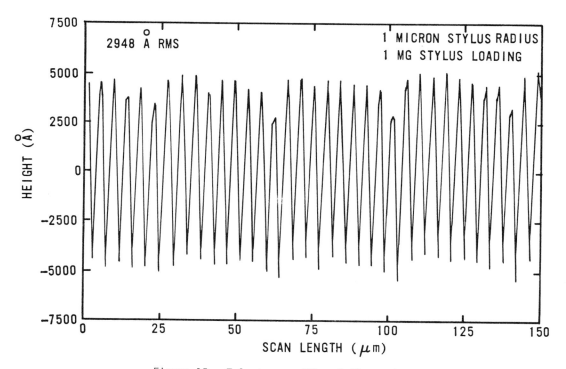

Figure 15. Talystep profile of CI grating

Figure 16. Photograph of CI grating witness sample

Figure 17. Photograph of full-size CI grating

Moiré topography for the detection of orthopaedic defects

Syed Arif Kamal

Department of Physics, Indiana University, Swain Hall, West 341, Bloomington, Indiana 47405, USA

Richard E. Lindseth

Department of Orthopaedic Surgery, School of Medicine, Indiana University, Indianapolis, Indiana 46223, USA

Abstract

Moiré topography is applied for the follow-up of scoliosis patients. The results are then compared with the X-rays. A special lamp and scale arrangement is utilized for patient alignment. It is suggested that this technique will be used for the detection of all orthopaedic defects.

During recent years there has been much activity in the field of moiré topography[1,2]. Moiré is a French word which means watered[3]. Moiré fringes are a series of interference fringes arising from the superposition of line grids, the lines of which are slightly inclined to each other, or are otherwise in register[4]. The width of the lines of the grid should be equal, to the space between them. The moiré effect also arises where there is an interference between a screen and its shadow which falls upon an object behind. In this case the various shadow lines - the contour lines - appear on the surface of the object at regular distances from the grid[5]. Takasaki[6,7] introduced contour moiré pictures of a full size living body. Adair et al.[8] applied this technique for the early diagnosis of scoliosis through a school screening program. Willner[5] has studied the correlation of the moiré fringe interval and the lateral deviation of the spine. Boyer and Goitein[9] have developed a new moiré camera. A method has been developed for the measurement of the angle of spinal curvature by moiré topographs[10]. In taking the moiré topographs there has been a problem of patient alignment[8]. This difficulty has now been resolved by introducing a new process for patient alignment[11,12].

With a little modification, this procedure is now being used at J.W. Riley Hospital for Children at Indianapolis for the follow-up of scoliosis patients. The screen is made up of fishing line of nylon, 0.5 mm thick, stretched vertically across a rectangular opening in a steel frame. The distance between the two threads is 0.5 mm. The frame has the horizontal support to make the screen standing. To align the patient parallel to the screen, two laser beams are used. A special scale is constructed with horizontal and vertical lines. One of the laser equipments is fixed right below the vertical scale. It is slightly tilted in the upward direction so that if a plane mirror is held parallel to the scale (which is in turn parallel to the screen) the light will be reflected back on the vertical scale. If the mirror is rotated along the horizontal axis, the light will move along the vertical scale. But if the mirror is rotated along the vertical axis, the light will no longer fall on the vertical scale. Therefore, this arrangement will show that the vertical plane in which the incident and reflected laser beams lie is perpendicular to the plane of the mirror as well as the plane of the scale (on which the light falls back) if the beam is reflected back on the vertical scale. Care should be taken in drawing the vertical scale. It would be preferable if a plumb line is used to draw the vertical scale. To insure that the mirror is not rotated along the horizontal axis also, a second laser beam is used. The incident and reflected beams from this second laser equipment lie in the horizontal plane. If the laser equipment is kept on the right side of the scale, the laser should be slightly tilted towards the left and vice versa. In this way the light reflected from the mirror will fall on the horizontal scale. If the mirror is rotated along the horizontal axis, the light will not fall on the horizontal scale, but above or below it.

To obtain the moiré topograph of the back, the patient stands behind the screen with his back close to the screen. The patient is asked to stand in a relaxed and normal mode, looking straight ahead. The mirror is held touching the stomach of the patient and adjusted in such a way that the light from the first laser falls on the vertical scale and from the second on the horizontal scale. If the patient twists in the horizontal direction, light from the first laser will not fall on the vertical scale and if the child twists in the vertical direction, light from the second lamp will not fall on the horizontal scale. Now the patient is ready to be photographed.

For photographing SX-70 (Sonar Focussing - One Step) Land Camera and SX-70 Land Film is used. The source of light used is 1000 watt lamp. The shadows of the fishing lines of nylon fall on the body and a pattern is formed depending on the depth and elevation of the surface. The distance h_n between the nth fringe, as projected on the body, and the screen is

$$h_n = nL(s_0^{-1}d - n)^{-1} \qquad (1)$$

where L is the horizontal distance from the light and the camera to the screen, d is the vertical distance between the camera and the light and s_0 is the screen interval[13]. In our experiment

$$L = 100 \text{ cm}; \quad d = 50 \text{ cm}; \quad s_0 = 1.00 \text{ mm}$$

Therefore

$$h_n = (100\ n)\ (500 - n)^{-1} \tag{2}$$

The difference between n^{th} and $(n + 1)^{th}$ fringe is

$$D = h_{n+1} - h_n = (n+1)L(s_0^{-1}d-n-1)^{-1} - nL(s_0^{-1}d-n) \tag{3}$$

This can be written as

$$D = Ls_0^{-1}d(n-s_0^{-1}d)^{-1}(n+1-s_0^{-1}d)^{-1} \tag{4}$$

Therefore

$$D = (100)\ (500)\ (n - 500)^{-1}(n + 1 - 500)^{-1} \tag{5}$$

If n = 1, the difference between the 1st and the 2nd fringe is 0.20 cm, whereas for n = 50, the difference is 0.24 cm.

The purpose of the experiment is to obtain moiré topographs of children having scoliosis at regular intervals and compare them with the previous moiré topographs. If there is a change observed in the moiré pattern, a radiographic examination may be taken. In this case the repeated taking of X-rays can be avoided. Only in the case of a change should X-rays be taken. This change will be noted from the change in the moiré pattern. It is also planned to find the angle of spinal curvature and compare it with the angle found by the X-rays. From moiré topographs the angle can be measured at regular intervals[10].

The method as described in ref.[10] consists of selecting a reference line on the moiré topograph. The line is drawn by joining the midpoint of the neck to the midpoint of the waist. From this line, the distances to the first visible moiré fringe on both sides are measured at different points. We adopt the convention that the distances on the right of the reference line are positive and those on the left are negative. Let d_1 be the position of the spine with respect to the reference line at the point of maximum asymmetry and d_2 be at the point of minimum asymmetry. D_1 is the distance between the points of maximum and minimum asymmetry on the reference line, then the angle of spinal curvature is given by[10]

$$\theta = 2 \tan^{-1}(D_1^{-1}|d_1 - d_2|) \tag{6}$$

If the spine is not curved at the midpoint, the above formula is modified as[10]

$$\theta = \tan^{-1}(D_1^{-1}d) + \tan^{-1}(D_2^{-1}d) \tag{7}$$

where D_1 is the distance between the point of maximum asymmetry and that of minimum asymmetry above it on the reference line and D_2 is the distance between the point of maximum asymmetry and that of minimum asymmetry below it. d is given by

$$d = \tfrac{1}{2} (|d_1 - d_2| + |d_1 - d_3|) \tag{8}$$

where d_1 is the position of spine with respect to the reference line at the point of maximum asymmetry and d_2 is at the point of minimum asymmetry above it, whereas d_3 is the position of the spine with respect to the reference line at the point of maximum asymmetry. At the time of initial examination, the points of maximum and minimum asymmetry can be judged with the help of comparison of initial X-rays and the moiré topograph, and in the following moiré topographs the same points may be used for measurements.

It is hoped that the set up will enable the medical personnel to reduce the number of X-rays taken, to keep a photographic record of the prognosis of scoliosis, and to record and measure the possible changes in the curve due to result of treatments (braces, exercises, surgery etc.). It is further hoped that the method will also be used for the follow-up of lordosis (topograph of front) and kyphosis (topograph of side with hand excluded) and many other orthopaedic defects. If the topographs of the full body of very young children are taken at regular intervals, it is unlikely that any orthopaedic disorder may remain undiagnosed at the very early stage[14].

References

1. P. S. Theocaris, Expt. Mech. 4, 153, 1964.
2. D. M. Meadows et al., Appl. Opt. 9, 942, 1970.
3. The World Book Encyclopedia, Field Enterprises, Chicago, 1964, Vol. 13.
4. J. Thewlis, Concise Dictionary of Physics, Second Ed., Pergamon Press, Oxford, 1979.
5. S. Willner, Acta. Orthop. Scand. 50, 295, 1979.
6. H. Takasaki, Appl. Opt. 9, 1467, 1970.

7. H. Takasaki, Appl. Opt. 12, 845, 1973.

8. I. V. Adair, M. C. van Wijk and G. W. D. Armstrong, Clinc. Orthop. 129, 165, 1977.

9. A. L. Boyer and M. Goitein, Med. Phys. 7, 19, 1980.

10. S. A. Kamal, Measurement of the angle of spinal curvature by moiré topographs, preprint, 1980, Bloomington, Indiana, U.S.A.

11. S. A. Kamal, A new process for patient alignment, 1980, Bloomington, Indiana, U.S.A. (submitted for patent).

12. S. A. Kamal and M. A. Khan, Moiré contour recorder, 19th Annual Science Conference, September 1979, Quaid-i-Azam University, Islamabad, Pakistan (in Urdu).

13. H. Takasaki, Moiré topography, Proceedings of the Symposium of Biostereometrics 1974, American Society of Photogrammetry, p. 590, 1974.

14. S. A. Kamal, The use of moiré topographs for the detection of orthopaedic defects in children of age group four to seven years, Research Project Proposal, June 1980, Bloomington, Indiana, U.S.A.

Angle-resolved light scattering from composite optical surfaces

J. M. Elson

Michelson Laboratory, Physics Division, Naval Weapons Center, China Lake, California 93555

Introduction

Described here is the derivation of an expression designed to predict the angular distribution of light scattered from composite surfaces. The scattering is assumed to be the result of microroughness, either periodic or random, present at each interface of the composite surface. The theory retains the vector nature of the fields and allows complex optical constants. The angles of incidence and scattering or diffraction are arbitrary. The number and thicknesses of the layers in the composite surface are arbitrary. The primary restriction on the validity of the theory is that the incident wavelength $\lambda \gg \delta$, where δ represents the root mean square (rms) roughness of any interface. This restriction therefore precludes the application of this theory to high-efficiency gratings. However, low-efficiency gratings, such as those designed for beam sampler applications, fall within the validity of the theory. Also, high quality optical components, such as laser gyro mirrors, typically have rms roughness values much less than a wavelength. Various aspects and applications of this theory have been published previously.[1]

Theory

Boundary conditions

The following derivation differs from that described previously.[1] We begin by using a starting point which has also been used by previous workers[2,3] for the case of scattering from uncoated surfaces. All aspects of this theory are limited to first order in δ/λ, where δ is the rms roughness of a given interface and λ is the incident wavelength. Shown in Figure 1 is the nomenclature regarding the composite surface. Each interface has a normal given by

$$\hat{n}_\ell = \hat{z} - \hat{x} \frac{\partial}{\partial x} (\zeta_\ell(\vec{\rho})) - \hat{y} \frac{\partial}{\partial y} (\zeta_\ell(\vec{\rho})) \quad , \tag{1}$$

where $\vec{\rho} = (x,y)$. The electric field in layers ℓ and $\ell+1$ may be expanded about the ℓth interface as

$$\vec{E}_j(\vec{\rho}, \tau_\ell + \zeta_\ell) \approx \vec{E}_j^{(0)}(\vec{\rho}, \tau_\ell) + \zeta_\ell(\vec{\rho}) \frac{\partial}{\partial z} (\vec{E}_j^{(0)}(\vec{\rho}, z))|_{z=\tau_\ell} + \vec{E}_j^{(1)}(\vec{\rho}, \tau_\ell) \quad , \tag{2}$$

where $j = \ell$ or $\ell+1$.

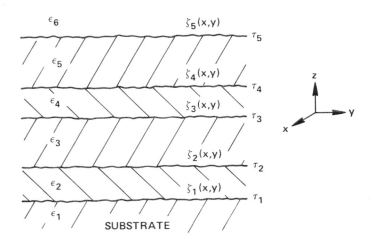

Figure 1. Schematic cross section of a four-layer composite surface showing terminology. The mean surface levels of the ith interface are denoted by τ_i, where $\tau_1 = 0$. The complex dielectric constants of the ith layer are written as ε_i. Each interface has a profile described by $\zeta_i(x,y)$ which may be either a random variable or periodic (i = 1-5).

The $\vec{E}^{(0)}$ is the zero-order field or that field which is calculated for the perfectly smooth situation where the $\zeta_\ell(x,y) = 0$ for all interfaces . The $\vec{E}^{(1)}$ is the first-order correction to the field which arises as a result of nonplanar interfaces. Consider first the boundary condition that the tangential components of the electric field given in Eq. (2) be continuous across an interface. This is written as

$$\hat{n}_\ell \times \Delta \vec{E}_\ell(\vec{\rho}, \tau_\ell + \zeta_\ell) = 0 \quad , \tag{3}$$

where $\Delta \vec{E}_\ell(\vec{\rho}, \tau_\ell + \zeta_\ell) = \vec{E}_{\ell+1}(\vec{\rho}, \tau_\ell + \zeta_\ell +) - \vec{E}_\ell(\vec{\rho}, \tau_\ell + \zeta_\ell -)$ is the difference of the $\vec{E}$ fields evaluated across the τ_ℓ boundary. Using Eqs. (1) and (2) in Eq. (3) yields

$$\Delta \vec{E}_{\ell x}^{(1)} = - \frac{\partial \zeta_\ell}{\partial x} \Delta \vec{E}_{\ell z}^{(0)} - \zeta_\ell \Delta \frac{\partial \vec{E}_{\ell x}^{(0)}}{\partial z} \tag{4a}$$

$$\Delta \vec{E}_{\ell y}^{(1)} = - \frac{\partial \zeta_\ell}{\partial y} \Delta \vec{E}_{\ell z}^{(0)} - \zeta_\ell \Delta \frac{\partial \vec{E}_{\ell x}^{(0)}}{\partial z} \tag{4b}$$

for discontinuities of the x and y components of $\vec{E}^{(1)}$ across the ℓth interfaces. The fields in Eqs. (4) have been evaluated at $z = \tau_\ell$. Let

$$\Delta \vec{E}_\ell^{(1)}(\vec{\rho}, \tau_\ell) = \int d^2 k \, \Delta \vec{e}_\ell^{(1)}(k, \tau_\ell) \exp(i\vec{k} \cdot \vec{\rho}) \tag{5}$$

and

$$\Delta \vec{E}_\ell^{(0)}(\vec{\rho}, \tau_\ell) = \Delta \vec{e}_\ell^{(0)}(k_o, \tau_\ell) \exp(ik_o x) \quad , \tag{6}$$

where the $\exp(-i\omega t)$ time dependence has been suppressed. Multiply both sides of Eq. (4a) by $\exp(-\vec{k}' \cdot \vec{\rho}) d^2 \rho$ and integrate. This yields

$$(2\pi)^2 \Delta e_{\ell x}^{(1)}(k', \tau_\ell) = -ik_x' \Delta e_{\ell z}^{(0)}(k_o, \tau_\ell) \zeta_\ell(\vec{k}_o - \vec{k}') \tag{7a}$$

$$(2\pi)^2 \Delta e_{\ell y}^{(1)}(k', \tau_\ell) = -ik_y' \Delta e_{\ell z}^{(0)}(k_o, \tau_\ell) \zeta_\ell(\vec{k}_o - \vec{k}') \quad , \tag{7b}$$

where

$$\zeta_\ell(\vec{k}_o - \vec{k}) = \int d^2 \rho \, \exp(i(\vec{k}_o - \vec{k}) \cdot \vec{\rho}) \zeta_\ell(\vec{\rho}) \quad , \tag{8}$$

which is the Fourier transform of the ℓth interface profile.

Similar equations hold for the other boundary conditions:

$$(2\pi)^2 \Delta h_{\ell x}^{(1)}(k', \tau_\ell) = i(\omega/c) \zeta_\ell(\vec{k}_o - \vec{k}') \Delta(\varepsilon_\ell e_{\ell y}^{(0)}(k_o, \tau_\ell)) \tag{9a}$$

$$(2\pi)^2 \Delta h_{\ell y}^{(1)}(k', \tau_\ell) = -i(\omega/c) \zeta_\ell(\vec{k}_o - \vec{k}') \Delta(\varepsilon_\ell e_{\ell x}^{(0)}(k_o, \tau_\ell)) \tag{9b}$$

$$(2\pi)^2 \Delta d_{\ell z}^{(1)}(k', \tau_\ell) = i\zeta_\ell(\vec{k}_o - \vec{k}')(k_x' \Delta(\varepsilon_\ell e_{\ell x}^{(0)}(k_o, \tau_\ell)) + k_y' \Delta(\varepsilon_\ell e_{\ell y}^{(0)}(k_o, \tau_\ell))) \tag{9c}$$

$$\Delta h_{\ell z}^{(1)}(k', \tau_\ell) = 0 \quad . \tag{9d}$$

To evaluate the right-hand sides of Eqs. (7) and (9), we need the zero-order fields.

Zero-order fields

The calculation of the zero-order fields is outlined here and follows Born and Wolf.[4] Let the incident beam be confined to the (x,z) plane and the fields in the ℓth layer be written as

$$\vec{E}_\ell^{(0)} = [\hat{x} f_\ell^{(+)}(z) + \frac{k_o}{q_\ell^{(0)}} \hat{z} f_\ell^{(-)}(z) - \hat{y} g_\ell^{(+)}(z)] \exp(ik_o x) \tag{10a}$$

and

$$\vec{H}_\ell^{(0)} = [\hat{x} q_\ell^{(0)} g_\ell^{(-)}(z) - \hat{z} k_o g_\ell^{(+)}(z) - \hat{y} f_\ell^{(-)}(z)(\omega/c)^2 \varepsilon_\ell / q_\ell^{(0)}] \exp(ik_o x)/(\omega/c) \quad , \tag{10b}$$

where the $\exp(-i\omega t)$ time dependence is suppressed. Note that in Eq. (10a), the $f_\ell^{(\pm)}$ refer to p-polarized incident light and the $g_\ell^{(+)}$ to s-polarized incident light. More explicitly, we have that in the ℓth layer

$$f_\ell^{(\pm)}(z) = a_\ell^{(-)} \exp(-iq_\ell^{(0)} z) \pm a_\ell^{(+)} \exp(iq_\ell^{(0)} z) \tag{11a}$$

$$g_\ell^{(\pm)}(z) = b_\ell^{(-)} \exp(-iq_\ell^{(0)} z) \pm b_\ell^{(+)} \exp(iq_\ell^{(0)} z) \quad , \tag{11b}$$

where $\varepsilon_\ell (\omega/c)^2 = k_o^2 + q_\ell^{(0)2}$ and where the $a_\ell^{(\pm)}$ and $b_\ell^{(\pm)}$ are to be determined. To do this, we concentrate on the p-polarized fields and define

$$Q_\ell(z) = \begin{bmatrix} f_\ell^{(+)}(z) \\ \dfrac{-f_\ell^{(-)}(z)\varepsilon_\ell(\omega/c)}{q_\ell^{(0)}} \end{bmatrix} \quad , \tag{12}$$

which are (except for $\exp(ik_o x)$ dependence) the x and y components of the electric and magnetic fields, respectively. Suppose a 4 x 4 matrix $T_\ell(\tau_\ell - \tau_{\ell-1})$ exists such that

$$Q_\ell(\tau_{\ell-1}) = T_\ell(\tau_\ell - \tau_{\ell-1}) Q_\ell(\tau_\ell) \quad , \tag{13}$$

where the T_ℓ matrix relates the fields at each boundary within layer ℓ. Calculation of the elements of T_ℓ is straightforward and yields

$$T_{\ell 11} = \cos\phi_\ell \tag{14a}$$

$$T_{\ell 12} = i\left(\frac{q_\ell^{(0)}}{(\omega/c)\varepsilon_\ell}\right) \sin\phi_\ell \tag{14b}$$

$$T_{\ell 21} = -i\left(\frac{(\omega/c)\varepsilon_\ell}{q_\ell^{(0)}}\right) \sin\phi_\ell \tag{14c}$$

$$T_{\ell 22} = \cos\phi_\ell \quad , \tag{14d}$$

where $\phi_\ell = q_\ell^{(0)}(\tau_\ell - \tau_{\ell-1})$. We note from Eq. (13) and the continuity of the tangential components of the electric and magnetic fields that

$$Q_{\ell+1}(\tau_\ell) = Q_\ell(\tau_\ell) \quad , \tag{15}$$

and, therefore, by successive application of Eqs. (13) and (15), we find that for a composite multilayer stack with L layers

$$Q_1(\tau_1) = T Q_{L+2}(\tau_{L+1}) \quad , \tag{16}$$

where T is a product of L matrices as

$$T = T_2(\tau_2 - \tau_1) T_3(\tau_3 - \tau_2) T_4(\tau_4 - \tau_3) \cdots T_{L+1}(\tau_{L+1} - \tau_L) \quad . \tag{17}$$

Note that in Eq. (16) the subscripts 1 and L+2 refer to the substrate and upper region, respectively (see Figure 1). Also, we know that there is no incoming beam in region (1) and thus

$$f_1^{(\pm)}(\tau_1) = a_1^{(-)} \quad , \tag{18}$$

since $a_1^{(+)} = 0$ and $\tau_1 = 0$ by definition. In region L+2 we know the strength of the incoming beam, and it follows that

$$f_{L+2}^{(\pm)}(\tau_{L+1}) = I\cos\theta_o \, \exp(-iq_{L+2}\tau_{L+1}) \pm a_{L+2}^{(+)} \, \exp(iq_{L+2}\tau_{L+1}) \quad , \tag{19}$$

where θ_o is the angle of incidence, and I is the incident field amplitude. Using Eqs. (18) and (19) in Eq. (16) yields

$$a_{L+2}^{(+)} = I\cos\theta_o \, \exp(-2iq_{L+2}\tau_{L+1}) \, [\frac{\alpha_- + \beta_-}{\alpha_+ + \beta_+}] \quad , \tag{20}$$

where

$$\alpha_\pm = (\frac{\omega}{c}) [T_{11}\varepsilon_1 q_{L+2}^{(0)} \pm T_{22}\varepsilon_{L+2}q_1^{(0)}] \tag{21a}$$

$$\beta_\pm = [T_{21}q_1^{(0)}q_{L+2}^{(0)} \pm \varepsilon_1\varepsilon_{L+2}(\omega/c)^2 T_{12}] \tag{21b}$$

and the T_{ij} are the elements of the T matrix of Eq. (17). Thus, the $a_{L+2}^{(+)}$ term may be calculated and all other $a_\ell^{(\pm)}$ terms may be generated by recursion using Eqs. (13) and (15). The $b_\ell^{(\pm)}$ may be calculated in a completely analogous way by using the y and x components of the electric and magnetic fields, respectively, in Eq. (12).

Having evaluated the zero-order fields, the right-hand sides of Eqs. (7) and (9) may be evaluated by using

$$\Delta e_{\ell z}^{(0)}(k_o, \tau_\ell) = - \frac{k_o f_\ell^{(-)}(\tau_\ell)}{q_\ell^{(0)} \varepsilon_{\ell+1}} \quad (\varepsilon_{\ell+1} - \varepsilon_\ell) \tag{22a}$$

$$\Delta(\varepsilon_\ell e_{\ell x}^{(0)}(k_o, \tau_\ell)) = f_\ell^{(+)}(\tau_\ell) \, (\varepsilon_{\ell+1} - \varepsilon_\ell) \tag{22b}$$

$$\Delta(\varepsilon_\ell e_{\ell y}^{(0)}(k_o, \tau_\ell)) = - g_\ell^{(+)}(\tau_\ell) \, (\varepsilon_{\ell+1} - \varepsilon_\ell) \quad . \tag{22c}$$

First-order fields

To use the boundary conditions given in Eqs. (7) and (9), we assume the following expressions for the first-order fields in the nth layer, i.e.,

$$\vec{E}_n^{(1)} = \int d^2k \, \exp(i\vec{k}\cdot\vec{\rho}) [\hat{k}\{p_n^{(+)} \, \exp(iq_n z) + p_n^{(-)} \, \exp(-iq_n z)\}$$

$$- \frac{\hat{z}k}{q_n} \{p_n^{(+)} \, \exp(iq_n z) - p_n^{(-)} \, \exp(-iq_n z)\}$$

$$+ \hat{k} \times \hat{z}\{s_n^{(+)} \, \exp(iq_n z) + s_n^{(-)} \, \exp(-iq_n z)\}] \tag{23a}$$

for the electric field, and

$$\vec{H}_n^{(1)} = \int d^2k \, \exp(i\vec{k}\cdot\vec{\rho}) \, [\hat{k}q_n\{s_n^{(+)} \exp(iq_nz) - s_n^{(-)} \exp(-iq_nz)\}/(\omega/c)$$

$$-\hat{z}k\{s_n^{(+)} \exp(iq_nz) + s_n^{(-)} \exp(-iq_nz)\}/(\omega/c)$$

$$-\hat{k} \times \hat{z} \, \frac{(\omega/c)\varepsilon_n}{q_n} \, \{p_n^{(+)} \exp(iq_nz) - p_n^{(-)} \exp(-iq_nz]$$

(23b)

for the magnetic field. Using Eqs. (23) in the boundary condition Eqs. (7) and (9), we may derive a recursion relation

$$P_{\ell+1} = M_{\ell+1}^{-1}(\tau_\ell)M_\ell(\tau_\ell)P_\ell - \frac{i}{(2\pi)^2} \zeta_\ell(\vec{k}_o - \vec{k})M_{\ell+1}^{-1}(\tau_\ell)S_\ell \quad , \tag{24}$$

where the P_ℓ and S_ℓ are column vectors defined as

$$P_\ell = \begin{bmatrix} p_\ell^{(+)} \\ p_\ell^{(-)} \end{bmatrix} \tag{25a}$$

and

$$S_\ell = k(\varepsilon_{\ell+1} - \varepsilon_\ell) \begin{bmatrix} f_\ell^{(+)}(\tau_\ell)\cos\phi - g_\ell^{(+)}(\tau_\ell)\sin\phi \\ \\ -\dfrac{k_o f_\ell^{(-)}(\tau_\ell)}{q_\ell^{(0)}\varepsilon_{\ell+1}} \end{bmatrix} \quad , \tag{25b}$$

where $\vec{k} = k(\hat{x}\cos\phi + \hat{y}\sin\phi)$. The matrix $M_n(\tau_\ell)$ has elements

$$(M_n(\tau_\ell))_{11} = \frac{k\varepsilon_n}{q_n} \exp(iq_n\tau_\ell) \tag{26a}$$

$$(M_n(\tau_\ell))_{12} = -\frac{k\varepsilon_n}{q_n} \exp(-iq_n\tau_\ell) \tag{26b}$$

$$(M_n(\tau_\ell))_{21} = \exp(iq_n\tau_\ell) \tag{26c}$$

$$(M_n(\tau_\ell))_{22} = \exp(-iq_n\tau_\ell) \quad . \tag{26d}$$

By successive application of the recursion relation in Eq. (24), it may be shown that

$$P_{L+2} = \sigma^{(L+1,1)} P_1 - \frac{i}{(2\pi)^2} \sum_{\ell=1}^{L+1} \zeta_\ell(\vec{k}_o - \vec{k})\sigma^{(L+1,\ell+1)}M_{\ell+1}^{-1}(\tau_\ell)S_\ell \quad , \tag{27}$$

where

$$\sigma^{(n,m)} = N_n N_{n-1} \cdots N_{m+1} N_m \tag{28a}$$

and

$$\sigma^{(n-1,n)} = \begin{pmatrix} 1 & 0 \\ 0 & 1 \end{pmatrix} \tag{28b}$$

by definition. Note that $N_n = M_{n+1}^{-1}(\tau_n)M_n(\tau_n)$, where

$$(N_n)_{11} = \frac{1}{2}(1 + \beta_n)\exp(-i\alpha_n^{(-)}) \tag{29a}$$

$$(N_n)_{12} = \frac{1}{2}(1 - \beta_n)\exp(-i\alpha_n^{(+)}) \tag{29b}$$

$$(N_n)_{21} = \frac{1}{2}(1 - \beta_n)\exp(i\alpha_n^{(+)}) \tag{29c}$$

$$(N_n)_{22} = \frac{1}{2}(1 + \beta_n(\exp(i\alpha_n^{(-)}) \tag{29d}$$

and

$$\beta_n = \frac{q_{n+1}\varepsilon_n}{q_n\varepsilon_{n+1}} \tag{30a}$$

$$\alpha_n^{(\pm)} = (q_{n+1} \pm q_n)\tau_n \quad . \tag{30b}$$

If we use the conditions that

$$P_{L+2} = \begin{bmatrix} p_{L+2}^{(+)} \\ 0 \end{bmatrix} \quad \text{and} \quad P_1 = \begin{bmatrix} 0 \\ p_1^{(-)} \end{bmatrix}$$

in Eq. (27), then it may be shown that

$$p_1^{(-)} = \frac{i}{2(2\pi)^2} \frac{1}{\sigma_{22}^{(L+1,1)}} \sum_{\ell=1}^{L+1} \zeta_\ell(\vec{k}_o - \vec{k}) \frac{q_{\ell+1}}{\varepsilon_{\ell+1}} \left[u_\ell \, \Gamma_{2-}^{(L+1,\ell+1)} + \frac{k\varepsilon_{\ell+1}v_\ell}{q_{\ell+1}} \, \Gamma_{2+}^{(L+1,\ell+1)} \right] , \tag{31}$$

where

$$\Gamma_{2\pm}^{(L+1,\ell+1)} = \sigma_{21}^{(L+1,\ell+1)} \exp(-iq_{\ell+1}\tau_\ell) \pm \sigma_{22}^{(L+1,\ell+1)} \exp(iq_{\ell+1}\tau_\ell) \tag{32a}$$

and

$$u_\ell = (\varepsilon_{\ell+1} - \varepsilon_\ell)[f_\ell^{(+)}(\tau_\ell)\cos\phi - g_\ell^{(+)}(\tau_\ell)\sin\phi] \tag{32b}$$

$$v_\ell = - \frac{k_o(\varepsilon_{\ell+1} - \varepsilon_\ell)f_\ell^{(-)}(\tau_\ell)}{q_\ell^{(0)}\varepsilon_{\ell+1}} \quad . \tag{32c}$$

Equation (31) is related to the fields transmitted through the composite stack, whereas the coefficient related to scattering into the L+2 medium is $p_{L+2}^{(+)}$. It is most convenient to derive $p_{L+2}^{(+)}$ by multiplying both sides of Eq. (27) by $[\sigma^{(L+1,1)}]^{-1}$. This yields

$$P_1 = [\sigma^{(L+1,1)}]^{-1} P_{L+2} + \frac{i}{(2\pi)^2} \sum_{\ell=1}^{L+1} \zeta_\ell(\vec{k}_o - \vec{k}) [\sigma^{(L+1,1)}]^{-1}\sigma^{(L+1,\ell+1)} M_{\ell+1}^{-1}(\tau_\ell)S_\ell , \tag{33a}$$

which may be rewritten as

$$P_1 = \mu^{(1,L+1)} P_{L+2} + \frac{i}{(2\pi)^2} \sum_{\ell=1}^{L+1} \zeta_\ell(\vec{k}_o - \vec{k}) \, \mu^{(1,\ell-1)} M_\ell^{-1}(\tau_\ell)S_\ell , \tag{33b}$$

where

$$\mu^{(1,n)} = [\sigma^{(1,n)}]^{-1} \quad . \tag{34}$$

The elements of $\mu^{(1,n)}$ may also be calculated from the matrix product

$$\mu^{(1,n)} = W_1 W_2 \ldots W_n \quad , \tag{35a}$$

where

$$(W_m)_{11} = \tfrac{1}{2}(1+\beta_m^{-1})\exp(i\alpha_m^{(-)}) \tag{35b}$$

$$(W_m)_{12} = \tfrac{1}{2}(1-\beta_m^{-1})\exp(-i\alpha_m^{(+)}) \tag{35c}$$

$$(W_m)_{21} = \tfrac{1}{2}(1-\beta_m^{-1})\exp(i\alpha_m^{(+)}) \tag{35d}$$

$$(W_m)_{22} = \tfrac{1}{2}(1+\beta_m^{-1})\exp(-i\alpha_m^{(-)}) \quad . \tag{35e}$$

Finally, from Eq. (33b) we find that

$$p_{L+2}^{(+)} = \frac{-i}{2(2\pi)^2 \mu_{11}^{(1,L+1)}} \sum_{\ell=1}^{L+1} \zeta_\ell(\vec{k}_0-\vec{k}) \frac{q_\ell}{\varepsilon_\ell} \left[u_\ell \Lambda_{1-}^{(1,\ell-1)} + \frac{k\varepsilon_\ell}{q_\ell} v_\ell \Lambda_{1+}^{(1,\ell-1)} \right] \quad , \tag{36a}$$

where

$$\Lambda_{1\pm}^{(1,\ell-1)} = \mu_{11}^{(1,\ell-1)}\exp(-iq_\ell\tau_\ell) \pm \mu_{12}^{(1,\ell-1)}\exp(iq_\ell\tau_\ell) \quad . \tag{36b}$$

Without going through any details, the corresponding expressions for the $s_1^{(-)}$ and $s_{L+2}^{(+)}$, which pertain to s-polarized scattered fields, are

$$s_1^{(-)} = \frac{i(\omega/c)^2}{2(2\pi)^2 Y_{22}^{(L+1,1)}} \sum_{\ell=1}^{L+1} \zeta_\ell(\vec{k}_0-\vec{k}) \left[\frac{(\varepsilon_{\ell+1}-\varepsilon_\ell)[g_\ell^{(+)}(\tau_\ell)\cos\phi + f_\ell^{(+)}(\tau_\ell)\sin\phi]K^{(L+1,\ell+1)}(\tau_\ell)}{q_{\ell+1}} \right], \tag{37a}$$

where

$$K^{(L+1,\ell+1)}(\tau_\ell) = Y_{21}^{(L+1,\ell+1)}\exp(-iq_{\ell+1}\tau_\ell) - Y_{22}^{(L+1,\ell+1)}\exp(iq_{\ell+1}\tau_\ell) \tag{37b}$$

and

$$Y^{(n,m)} = X_n(\tau_n)X_{n-1}(\tau_{n-1})\ldots X_m(\tau_m) \tag{38a}$$

with the $y^{(n,m)}$ having properties as in Eq. (28b) and

$$(X_n(\tau_n))_{11} = \tfrac{1}{2}(1+\eta_n)\exp(-i\nu_n^{(-)}) \tag{38b}$$

$$(X_n(\tau_n))_{12} = \tfrac{1}{2}(1-\eta_n)\exp(-i\nu_n^{(+)}) \tag{38c}$$

$$(X_n(\tau_n))_{21} = \tfrac{1}{2}(1-\eta_n)\exp(i\nu_n^{(+)}) \tag{38d}$$

$$(X_n(\tau_n))_{22} = \tfrac{1}{2}(1+\eta_n)\exp(i\nu_n^{(-)}) \quad . \tag{38e}$$

Finally,

$$\eta_n = \frac{q_n}{q_{n+1}} \tag{39a}$$

and

$$\nu_n^{(\pm)} = (q_{n+1} - q_n)\tau_n \quad . \tag{39b}$$

Also,

$$s_{L+1}^{(+)} = \frac{-i(\omega/c)^2}{2(2\pi)^2 B_{11}^{(1,L+1)}} \sum_{\ell=1}^{L+1} \frac{\zeta_\ell(\vec{K}_0 - \vec{k})(\epsilon_{\ell+1} - \epsilon_\ell)}{q_\ell} \cdot \psi^{(1,\ell)}(\tau_\ell)[g_\ell^{(+)}(\tau_\ell)\cos\phi + f_\ell^{(+)}(\tau_\ell)\sin\phi], \tag{40a}$$

where

$$\psi^{(1,\ell)}(\tau_\ell) = B_{11}^{(1,\ell-1)}\exp(-q_\ell\tau_\ell) - B_{12}^{(1,\ell-1)}\exp(iq_\ell\tau_\ell) \tag{40b}$$

and

$$B^{(1,n)} = V_1(\tau_1)V_2(\tau_2)\ldots V_n(\tau_n) \tag{41a}$$

with Eq. (41a) reducing to the identity matrix for n=0 and

$$(V_n(\tau_n))_{11} = \frac{1}{2}(1 + \eta_n^{-1})\exp(i\nu_n^{(-)}) \tag{41b}$$

$$(V_n(\tau_n))_{12} = \frac{1}{2}(1 - \eta_n^{-1})\exp(-i\nu_n^{(+)}) \tag{41c}$$

$$(V_n(\tau_n))_{21} = \frac{1}{2}(1 - \eta_n^{-1})\exp(i\nu_n^{(+)}) \tag{41d}$$

$$(V_n(\tau_n))_{22} = \frac{1}{2}(1 + \eta_n^{-1})\exp(-i\nu_n^{(-)}) \quad . \tag{41e}$$

Considering only scattering in the upper hemisphere (L+2 region), we may let n = L+2 in Eqs. (23) and use Eqs. (36) and (37). This yields

$$\vec{E}_{L+2}^{(1)} = \int d^2k \; e^{i\vec{k}\cdot\vec{\rho}} \left[p_{L+2}^{(+)}(\hat{k} - \hat{z}k/q_{L+2}) + s_{L+2}^{(+)} \hat{k} \times \hat{z} \right]\exp(iq_{L+2}z) \tag{42a}$$

$$\vec{H}_{L+2}^{(1)} = \int d^2k \; e^{i\vec{k}\cdot\vec{\rho}} \left[-\frac{\epsilon_{L+2}(\omega/c)}{q_{L+2}} (\hat{k} \times \hat{z})p_{L+2}^{(+)} + \frac{(\hat{k}q_{L+2} - \hat{z}k)}{(\omega/c)} s_{L+2}^{(+)} \right]\exp(iq_{L+2}z) \quad . \tag{42b}$$

Angle-resolved scattering

To obtain the expression for the angle-resolved scattering, consider the time-averaged Poynting vector $\vec{S}^{(1)} = \frac{c}{8\pi} \text{Re}(\vec{E}^{(1)} \times \vec{H}^{(1)*})$. To separate out the total power radiated away from the surface, we calculate

$$P = \int d^2\rho \; \vec{S}^{(1)} \cdot \hat{z} \quad , \tag{43}$$

where the integration is over the surface. This yields

$$P = \frac{\pi c}{2} \int d^2k \, \exp(-2 \operatorname{Im}(q_{L+2})z) \left[|p_{L+2}|^2 \frac{\varepsilon_{L+2}(\omega/c)}{q_{L+2}} + |s_{L+2}|^2 \frac{q_{L+2}}{(\omega/c)} \right] \quad .$$

Note that $q_{L+2} = [\varepsilon_{L+2}(\omega/c)^2 - k^2]^{1/2}$, and if ε_{L+2} is real then q_{L+2} is pure imaginary if $k > \frac{\omega}{c} \sqrt{\varepsilon_{L+2}}$. In this case, the radiation decays exponentially away from the surface. If $k < \frac{\omega}{c} \sqrt{\varepsilon_{L+2}}$, then $\operatorname{Im}(q_{L+2}) = 0$, and radiation emanates from the surface unimpeded. Thus, we are only interested in situations where $k \leq \frac{\omega}{c} \sqrt{\varepsilon_{L+1}}$, and we therefore write $k = \frac{\omega}{c} \sqrt{\varepsilon_{L+1}} \sin\theta$, $k_x = k \cos\phi$, $k_y = k \sin\phi$, and $d^2k = (\omega/c)^2 \varepsilon_{L+1} d\Omega$, where $d\Omega = \sin\theta \, d\theta \, d\phi$. Also, the total power incident on the surface may be shown to be $P_0 = \frac{c}{8\pi} L^2 I^2 \sqrt{\varepsilon_{L+2}} \cos\theta_0$, where L^2 is the area of illumination. Incorporating the changes above in Eq. (44) and normalizing with respect to P_0 yields

$$\frac{1}{P_0} \frac{dP}{d\Omega} = \frac{4\pi^2 (\omega/c)^2 \varepsilon_{L+2}}{L^2 \cos\theta_0} \left[|p_{L+2}^{(+)}|^2 + |s_{L+2}^{(+)}|^2 \cos^2\theta \right] \quad . \tag{45a}$$

The corresponding equation for angle-resolved scattering in the region below the multilayer stack (the substrate) may be written as

$$\frac{1}{P_0} \frac{dP}{d\Omega} = \frac{4\pi^2 (\omega/c)^2 \varepsilon_1^{3/2}}{L^2 \sqrt{\varepsilon_{L+2}} \cos\theta_0} \left[|p_1^{(-)}|^2 + |s_1^{(-)}|^2 \cos^2\theta \right] \quad , \tag{45b}$$

where the angle θ is interpreted as the polar scattering angle in the substrate. The substrate in this case can be $\varepsilon_1 = (1., 0.)$. Clearly, the ε_1 must be real and nonnegative ($\operatorname{Im}(q_1) = 0$, $q_1 = (\omega/c)(\varepsilon_1 - \sin^2\theta)^{1/2}$) for transmission and scattering in the region below the stack to be observed. These equations were derived on the basis that the profile at each interface was known _a priori_. We now discuss the treatment of periodic and random roughness.

Periodic Roughness

Consider as a practical example a beam sampler which has a rectangular profile at each interface. Assume that each profile is identical in shape and that the grating amplitude $H \ll \lambda$. Following Eq. (8) we transform the profile shown schematically in Figure 2. This yields

$$\zeta(\vec{k}_0 - \vec{k}) = 2H\delta(k_y) \sum_{m=-\infty}^{\infty} \delta(k_0 - k_x) + \frac{2\pi m}{d}) \exp(\pi i m\alpha) \frac{\sin(\pi m\alpha)}{m} \quad . \tag{46}$$

Using Eq. (46) in Eq. (45a) with Eqs. (36a) and (40a) shows that the δ-functions $\delta(k_0 - k_x + 2\pi m/d)$ and $\delta(k_y)$ are squared. Thus, we make use of the relations

$$[\delta(k_0 - k_x + 2\pi m/d)]^2 = \pi L \, \delta(k_0 - k_x + 2\pi m/d) \tag{47}$$
$$[\delta(k_y)]^2 = \pi L \, \delta(k_y)$$

as $L \to \infty$. With the Dirac δ-functions, the integration of Eq. (45a) is straightforward and yields

$$\frac{P_n}{P_0} = \frac{4\pi^4}{\cos\theta_0 \cos\theta_n} \left[|p_{L+2}^{(+)}|^2 + |s_{L+2}^{(+)}|^2 \cos^2\theta_n \right] \quad , \tag{48}$$

where $\sin\theta_n = \sin\theta_0 + n\lambda/d$. The diffracted power p_n is the result of diffraction from each

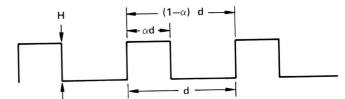

Figure 2. Schematic representation of a rectangular profile showing the associated parameters.

interface. Note that all terms of Eq. (48) involving θ and ϕ are evaluated at the angles $\phi_n = 0$ or π, depending on the direction of diffraction, and θ_n as given following Eq. (48). Cases where the direction of the grating grooves are not perpendicular to the plane of incidence have been discussed elsewhere.[6]

Random Roughness

To use Eq. (45) for cases where the interface profiles $\zeta_\ell(\vec{\rho})$ are not known, we may determine an average by using statistical properties of the $\zeta_\ell(\vec{\rho})$. Taking the ensemble average $\frac{1}{P_o} \langle dP/d\Omega \rangle$ of Eq. (45) indicates that we need to consider $\langle |p_{L+2}^{(+)}|^2 \rangle$ and $\langle |s_{L+2}^{(+)}|^2 \rangle$ since these expressions contain the random variables $\zeta_\ell(\vec{k}_o - \vec{k})$. If we write

$$p_{L+2}^{(+)} = \sum_{\ell=1}^{L+1} C_\ell \, \zeta_\ell(\vec{K}) \tag{49}$$

for brevity, then we see that

$$\langle |p_{L+2}^{(+)}|^2 \rangle = \sum_{\ell=1}^{L+1} \sum_{n=1}^{L+1} C_\ell C_n^* \langle \zeta_\ell(\vec{K}) \, \zeta_n^*(\vec{K}) \rangle \quad . \tag{50}$$

In Eq. (50) with Eq. (8) we see that

$$\langle \zeta_\ell(\vec{K}) \zeta_n^*(\vec{K}) \rangle = \int d^2\rho \, d^2\rho' \, \exp(i\vec{K} \cdot (\vec{\rho} - \vec{\rho}')) \, \langle \zeta_\ell(\vec{\rho}) \zeta_n(\vec{\rho}') \rangle \quad . \tag{51}$$

By a change of variables $\vec{\rho}' - \vec{\rho} = \vec{\tau}$, we have

$$\frac{\langle \zeta_\ell(\vec{K}) \zeta_n^*(\vec{K}) \rangle}{L^2} = \int d^2\tau \, \exp(-i\vec{K} \cdot \vec{\tau}) G_{\ell n}(\vec{\tau}) \quad , \tag{52}$$

where $G_{\ell n}(\vec{\tau})$ is the correlation function between the ℓth and nth interfaces. Thus, we see that the choice of correlation properties between interfaces $G_{\ell n}(\vec{\tau})$ and along interfaces $G_{\ell\ell}(\vec{\tau})$ play an important role in evaluating $\langle |p_{L+2}^{(+)}|^2 \rangle$ and $\langle |s_{L+2}^{(+)}|^2 \rangle$. More details on these questions have been discussed elsewhere.[5] We note that if all interfaces are identical, then $G_{\ell n}(\vec{\tau}) = G_{\ell\ell}(\vec{\tau})$ and all interfaces are correlated. If there is no correlation between interfaces, then $G_{\ell n}(\vec{\tau}) = 0$ for $\ell \neq n$.

Conclusion

A theory for angle-resolved scattering from composite optical surfaces has been derived. The theory is valid for cases where the rms roughness is much less than the wavelength. The cases of periodic and random roughness were considered. Comparisons between theory and experiment have been discussed elsewhere[7] and continue to be ongoing efforts.[8]

References

1. Elson, J. M., Low Efficiency Diffraction Grating Theory, Air Force Weapons Laboratory, Kirtland AFB, N.M., Technical Report AFWL-TR-75-210. 1976; Elson, J. M. Appl. Opt., Vol. 16, pp. 2872-2881. 1977; Scott, M. L., and Elson, J. M., Appl. Phys. Lett., Vol. 32(3), pp. 158-161. 1978; Elson, J. M., J. Opt. Soc. Am., Vol. 69, pp. 48-54. 1978.
2. Kröger, E., and Kretschmann, E., Z. Physik, Vol. 237, pp. 1-15. 1970.
3. Vlieger, J., and Bedeaux, D., Physica, Vol. 82A, pp. 221-246. 1976; Physica, Vol. 85A, pp. 389-398. 1976.
4. Born, M., and Wolf, E., Principles of Optics, Pergamon Press, New York 1965. Pp. 66-70.
5. Elson, J. M., and Bennett, J. M., J. Opt. Soc. Am., Vol. 69, pp. 31-47. 1979.
6. Elson, J. M. Bennett, H. E., and Bennett, J. M., "Scattering from Optical Surfaces," in Applied Optics and Optical Engineering, Academic Press 1979. Pp. 191-244.
7. Elson, J. M., Rahn, J. P., and Bennett, J. M., Appl. Opt., Vol. 19, pp. 669-679. 1980.
8. Experiment and theory comparisons have been made between surfaces covered with one and two layers of dielectric material deposited on opaque silver. These results are still being analyzed and will be published.

Some ideas on ruling engine control systems

Donald N. B. Hall

Kitt Peak National Observatory*, P.O. Box 26732, Tucson, Arizona 85726

Abstract

The magnitudes of various effects which may lead to wavefront errors in gratings ruled on the Harrison 'C' engine, and which are not corrected by the translation servo loop, are evaluated. It appears that thermal expansion is not a problem with the current level of temperature control ($\pm$ 5 x 10^{-3} C) and that variations in the refractive index of air could be compensated to the 10 nm level. However an interferometer which permits direct monitoring of the spacing between the diamond carriage assembly and the reference fused silica straightedge indicates larger wavefront errors may arise because of the present diamond carriage support.

The 'C' engine represents the culmination of a twenty five year effort by the late Dean Harrison[1,2] to develop facilities to rule large, high quality diffraction gratings at the Massachusetts Institute of Technology. Early in 1974 the engine was moved to the Kitt Peak National Observatory's headquarters in Tucson and it has since been used to rule gratings for astronomical applications; during this period the engine has been left in essentially the same configuration as when it left MIT. However we are now planning to upgrade the translation servo system to take advantage of technology developed at Kitt Peak for the control of Fourier Transform Spectrometers. Because of the greatly improved precision anticipated with this system, and also because of wavefront errors which have shown up in recent attempts to rule two large gratings (a 632 line/mm x 30 cm groove length x 40 cm ruled width x 56° blaze angle ruling intended for solar spectroscopy and a 452 groove/mm x 8 cm groove length x 38 cm ruled width x 68° blaze angle echelle to be used in the Space Telescope's High Resolution Spectrograph), we are presently carrying out an exhaustive evaluation of factors outside the translation interferometer servo control loop which might result in wavefront errors. The present paper summarizes progress to date in this area.

The 'C' engine rules continously. The blank is translated at constant speed under interferometric control while the diamond reciprocates across the blank, being raised on the forward stroke and lowered to rule a groove of the required length on the return. Almost all of the potential causes of wavefront errors not corrected by the translation servo loop arise in the area of the diamond carriage; this part of the engine is shown schematically in Figure 1. Beams in the two arms of a Michelson interferometer are reflected from mirrors which are attached respectively to the end of the blank being ruled and to the back of an optically polished fused silica bar which serves as a reference straightedge for the grooves. The diamond carriage is suspended from a Nitralloy monorail and rides on four graphitar bearing surfaces. The entire diamond carriage assembly is free to rotate about the center of the monorail and its location in this axis is defined by a Rulon button which slides along the fused silica bar. The diamond itself is supported by a parallelogram flex mount which is used to lower and raise it. The actual diamond carriage assembly is shown in Figure 2.

The replacement translation servo-control interferometer is inherently capable of very high accuracy. In Fourier Transform Spectrometer Applications[3,4] phase is determined with an rms error significantly less than a millifringe (λ=6328 Å) in a bandpass of several kilohertz. On the other hand, we are presently seeing slow (~ several days) wavefront drifts with amplitudes up to a fringe! As a compromise we have adopted the criterion that wavefront errors due to sources outside the translation servo loop should be held below a thirtieth of a fringe at λ=6328 Å i.e., an amplitude $\leq$ 10 nm. Those processes which could, in principal, yield errors of this magnitude are listed in Table 1 together with the precision to which the controlling parameter must be held to reduce them below 10 nm and potential steps to achieve this.

* Operated by the Association of Universities for Research in Astronomy, Inc., under contract with the National Science Foundation.

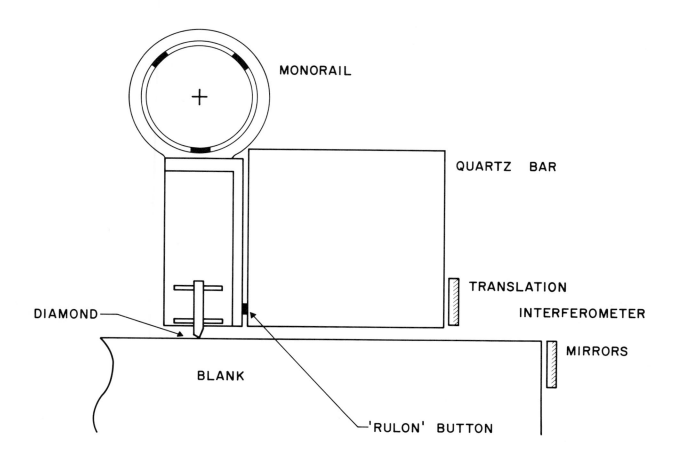

MONORAIL

QUARTZ BAR

TRANSLATION

INTERFEROMETER

DIAMOND

MIRRORS

BLANK

'RULON' BUTTON

HARRISON 'C' ENGINE — DIAMOND CARRIAGE ASSEMBLY

Figure 1. Schematic drawing of principal elements of the diamond carriage assembly.

Thermal expansion of the blank, the fused silica bar and metal spacer between the diamond and the bar have all long been recognized as potential sources of wavefront error. It is evident from Table 1 that all of these effects require temperature excursions at least several times the $\pm 5 \times 10^{-3}$ C range within which the engine is typically held during a ruling. In any case they can be further reduced by introducing compensation outside the other arm of the interferometer or they could be calibrated and open loop corrected.

Because the engine rules in air, changes in atmospheric refractive index would result in wavefront errors away from the interferometer's white light fringe unless corrected. In the present 'C' engine control system atmospheric pressure is monitored and its changes are compensated, although not to the precision given in Table 1. It is further evident from Table 1 that changes in precipitable water content can cause substantial errors although the effect of change in atmospheric refractive index with temperature is acceptable for temperature control of $\pm 5 \times 10^{-3}$ C. All three of these refractive index effects can be precisely compensated if the atmospheric refractive index is directly determined by measuring phase shift in a temperature insensitive, unequal path interferometer mounted inside the engine enclosure.

Potentially more serious sources of wavefront errors arise at the Rulon button and monorail bearing. Any changes in the thickness of the Rulon button translate directly into the wavefront and so its temperature, compression and wear rate must remain constant throughout the ruling. The lever arm between the Rulon button and the monorail reduces the effect of any translation of the latter in the wavefront by an order of magnitude but our

Figure 2. Photograph of the diamond carriage assembly.

criterion still requires the horizontal location of the carriage on the monorail to be
constant to 100 nm.

In order to monitor these effects, we have installed a compact interferometer which
monitors the distance between the diamond carriage and the fused silica bar. An 8 mm
diameter, optically flat surface is mounted in the diamond carriage, parallel to the fused
silica bar and only about a millimeter away from it. The axis is in line with the vertical
plane containing the diamond and the button and is the same distance above the button as
the ruled surface is below it. A collimated laser beam is fed to the carriage parallel to
the fused silica bar; a 90° prism deflects it into the interferometer where fringes are
produced between the flat and the surface of the bar. The output of the interferometer is
fed to a TV camera and displayed on a video monitor. A time lapse camera can be set to
photograph the fringe pattern, at the same point in the stroke, each time a preset number of
grooves have been ruled.

The interferometer has been used to measure a stiffness constant of 135 gm/fringe for
the present button. The 10 nm criterion then requires that the preload on the button,
typically ~ 300 gm for fine rulings, remain constant to ± 5 gm. This in turn requires that
friction between the graphitar bearing surfaces and the monorail be kept to substantially
lower levels than they may have been during previous long rulings. We are presently
installing a continuous oiling system to achieve this.

Table 1. Estimates of the Magnitude of Factors Outside the Translation Control Loop

Factor	Variable	Change for 10 nm Jump in Spacing	Remedy
Expansion of aluminum diamond carriage (22 mm)	Engine temperature	$\Delta T = 22 \times 10^{-3}$ C	Control engine temperature to $\pm 5 \times 10^{-3}$ C
Expansion of fused silica (100 mm)	Engine temperature	$\Delta T = 280 \times 10^{-3}$ C	Control engine temperature to $\pm 5 \times 10^{-3}$ C
Expansion of blank (60 cm maximum)	Engine temperature	$\Delta T = 48 \times 10^{-3}$ C	Control engine temperature to $\pm 5 \times 10^{-3}$ C
Alignment of laser beams (60 cm maximum)	Angular collimation	$\Delta \theta = 0.5$ "	Stable interferometer optics
Variation of refractive index of air with pressure (60 cm unequal path)	Refractive index (pressure)	$1:6 \times 10^{7}$ ($\Delta p = 0.025$ mm Hg)	Monitor refractive index directly
Variation of refractive index of air with temperature (60 cm unequal path)	Refractive index (engine temperature)	1.6×10^{7} ($\Delta T = 50 \times 10^{-3}$ C)	Control engine temperature to $\pm 5 \times 10^{-3}$ C
Variation of refractive index of air with pp H_2O (60 cm unequal path)	Refractive index (pp H_2O)	$1:6 \times 10^{7}$ (Δpp $H_2O = 0.3$ mm Hg)	Monitor refractive index directly
Compression of Rulon button	Preload force	$\Delta F = 5$ gm	Maintain constant preload or actively compensate
Thermal expansion of Rulon button	Button temperature	$\Delta T = 30 \times 10^{-3}$ C	Maintain constant temperature or actively compensate
Wear of Rulon button	Wear rate	?	Maintain constant or actively compensate
Translation of carriage on monorail	Translation	$\Delta x = 100$ nm	Maintain constant or actively compensate

The interferometer has also allowed us to measure the compression of the Rulon button when the diamond is lowered to rule a groove. In the present rulings the total fringe shift is ≤ 0.2 and this should be constant across a uniformly coated blank.

Thermal expansion of the Rulon button is another potential problem. Calculations indicate the button temperature must be held constant to ± 0.03 C to meet the 10 nm criterion. Although the frictional heat dissipation at the button is ≤ 30 mwatt, changes in frictional characteristics with time could produce such effects. We intend to monitor the fringe shift when ruling is stopped and the button cools down.

By monitoring fringe shift over an extended period of ruling, it should be straightforward to utilize the interferometer to measure the wear rate of the Rulon button. The button travels over 100 miles during a large ruling and its wear is reflected directly in the groove spacing. A linear wear rate does not affect the wavefront as it only results in an undetectable change in groove spacing. However non-linearities in the wear rate will distort the wavefront.

An obvious solution to all of the problems arising from use of the Rulon button is to actively monitor and control the spacing between the diamond carriage and the fused silica bar. This would also open the possibility of actively curving grooves or varying their spacing. However this approach will only work if the bearing surfaces on the monorail are horizontally defined to sufficient precision (≤ 100 nm). Seperate monitoring of button and monorail induced effects could be accomplished by installation of a second interferometer near the top of the fused silica bar.

To summarize, it appears that thermal expansion and refractive index induced wavefront errors can be reduced to acceptably low levels. Monitoring of the separation between the reference straightedge and the diamond carriage during actual ruling should allow those sources of wavefront error associated with the Rulon button and the monorail to be identified and either eliminated or actively compensated.

Acknowledgments

Most of the credit for the successful implementation and alignment of the test interferometer belongs to Ian Gordon; he has also carried out most of the subsequent tests. I am grateful to L. Barr, I. Ghozeil, E. Loewen, D. Schrage and R. Wiley for their advice and support during the course of this project.

References

1. Harrison, G. R., Thompson, S. W., Kazukonis, H. and Connell, J. R., J. Opt. Soc. Am., 62, p. 751. 1972.
2. Harrison, G. R., Applied Optics, 12, p. 2039. 1973.
3. McCurnin, T. W., Conference on Laser and Electro-Optical Systems, Topical Meeting on Inertial Confinement Fusion held in San Diego, CA, February 26-28, 1980, proceedings to be published.
4. Hall, D. N. B., Ridgway, S., Bell, E. A. and Yarborough, J. M., SPIE, Vol. 172, p. 121. 1979.

Optimization of Fourier gratings

Magnus Breidne*, Daniel Maystre**

*Department of Physics II and Institute of Optical Research, Royal Institute of Technology
S-100 44 Stockholm, Sweden
**Laboratoire d'Optique Electromagnétique, E.R.A. au C.N.R.S. n° 597
Faculté des Sciences et Techniques, Centre de St-Jérôme, 13397 Marseille Cedex 4, France

Abstract

During recent years, the technique to blaze holographic diffraction gratings has been developed considerably. In particular, reproducible methods using successive exposures enable one to superpose several harmonic to the fundamental. So, it is possible to obtain what we call Fourier gratings. We show that by using only one harmonic, it is possible to find groove profiles having efficiency curves very close to - or even better than those of ruled gratings.

Introduction

Attempts to make unsymmetrical holographic gratings were made not long after the first holographic diffraction gratings had been produced, around 1967. This desire to make unsymmetrical gratings is due to the low efficiency of symmetrical (sinusoidal) gratings when more than two orders propagate.

Today, several techniques to blaze holographic gratings are known. [1-3] Different blaze techniques have been compared[4] and the Fourier blaze technique suggested by Schmahl[2] was found superior. The idea is to make a quasi-triangular profile by "simply" adding a number of harmonics to the fundamental frequency, i.e. to make a Fourier synthesis of the desired profile.

All blaze techniques have in common the intention to make quasi-triangular grooves, i.e. to imitate the profile of the echelette grating. The primary reason for this is the well known blaze effect of the echelette grating. Furthermore it has been shown[5] that for shallow gratings used in the scalar region (i.e. wavelength to period ratio $\lambda/d < 0.2$) the triangular profile can be considered as the optimal one.

In the resonance region, however, no such analytical solution to the problem of the optimal profile can be found. Nevertheless, several optimization studies [6-8] have been made in this region where the possible choice of groove profile has been restricted (local optimization). Recently, a global optimization study[9] using an inverse scattering method has been published. That study is, however, restricted to the case where the electric vector is parallel to the grooves, TE-case.

Here we present some results from an optimization study in the resonance region, using a direct integral method[10] and calculating both the TE- and TM-efficiency. The idea behind this optimization study is to investigate the performances of the groove profiles possible to manufacture with experimental set up at the Institute of Optical Research[4] (Stockholm).

The important result is that there exist groove profiles giving efficiency curves much better than the efficiency curves of echelettes.

Fourier gratings

By a Fourier grating we mean a grating with profile :

$$y(x) = \sum_{n=1}^{j} h_n \sin(nKx + \phi_n) \ , \tag{1}$$

with $j > 1$ and $K = 2\pi/d$, d being the period of the grating.

In practice it is very difficult to do more than two consecutive exposures and as the influence of the higher order harmonics decreases rapidly we have chosen to study exclusively the family of Fourier gratings expressible in the form :

$$y(x) = h_1 \sin Kx + h_2 \sin(2Kx + \phi) .$$ (2)

For an echelette grating with period d, blaze angle α, and apex angle 90°, we have :

$$h_n = \frac{d}{(n\pi)^2} (\tan \alpha + \cot \alpha) . \sin (n\pi \cos^2 \alpha) ,$$

$$\phi_n = 0 \qquad \forall n .$$

For $\alpha < 20°$, $h_2 \simeq - h_1/2$, this means that (2) could be written :

$$y = h_1 \left[\sin Kx + \frac{1}{2} \sin(2Kx + 180°) \right] .$$

We will see that $\phi = 180°$ will give the best efficiency for shallow Fourier gratings. The ratio $|h_2/h_1|$ will, however, differ from 1/2.

Method of optimization

To be able to use the words "optimal" and "best" (or similar words) properly one has to define a criterion. Different users of gratings have (often implicit) criteria that varie considerably. Ours has been to maximize the integral under the efficiency curve over a λ/d-interval between 0.2 and 1.2. That is, loosely speaking, to have as high efficiency over as wide wavelength-to-period interval as possible, without strong anomalies.

We started with a sinusoidal grating with groove depth to groove period ratio equal to h/d ($h_1 = h/2$; $h_2 = 0$). Adding a first harmonic with a small amplitude h_2 we calculated the efficiency curve as ϕ was changed from 0° to 315° in steps of 45°. Then h_2 was increased and the procedure was repeated (sometimes with a smaller increment). The value of h/d was varied between 0.10 and 0.42, that is from shallow gratings to very deep gratings.

Presentation of the results

The values of h_1 , h_2 and ϕ have been varied to obtain the best possible efficiency curve. From the calculated efficiency curves [11] it is possible to distinguish three different intervals of the ratio h/d, where the behaviour of the gratings in each of these intervals is similar.

h/d < 0.2

In this interval the optimal Fourier grating is found for $\phi = 180°$ and a value of $h_2 \simeq 0.7 h_1$. An example of gratings in this region is shown in Fig.1. Here the sinusoidal grating (Fig.1 a) with h/d = 0.18 is compared to the optimal Fourier grating (Fig.1 b) with $h_1 = h/2$ and with the corresponding echelette grating (i.e. having the same fundamental) (Fig.1 c). The optimal Fourier grating is much better than the corresponding sinusoidal grating and almost as good as the corresponding echelette.

$0.2 \leqslant \frac{h}{d} < 0.3$

In this transition interval the differences between the sinusoidal, the corresponding echelette and the optimal Fourier grating are small. Still $\phi = 180°$ gives the best efficiency. This is examplified in Fig.2 where h/d = 0.26.

$0.3 \leqslant \frac{h}{d} < 0.4$

In this region we will find the optimal Fourier gratings for $\phi \simeq 330°$. They have very high efficiency and large blazewidths, and are better than the corresponding echelette grating. In Fig.3 the best Fourier grating we have found is compared to the corresponding sinusoidal and echelette grating respectively.

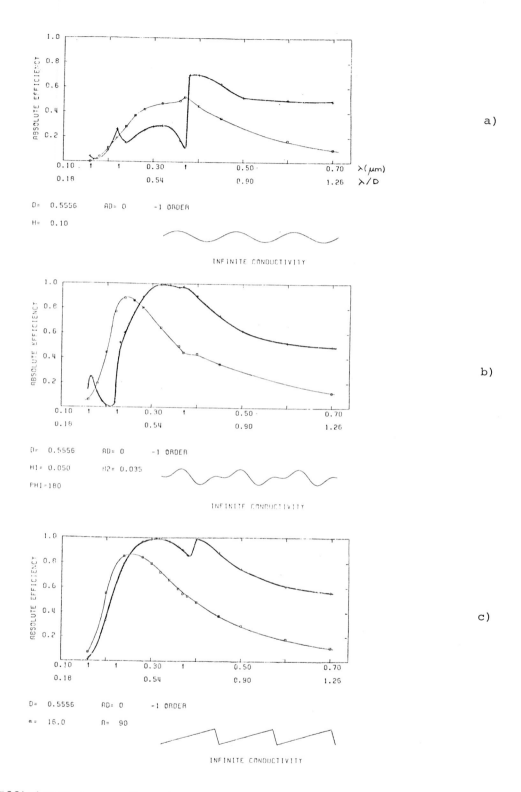

Figure 1. Efficiency curves for sinusoidal (a), Fourier (b) and echelette (c) gratings (with the same fundamental), as functions of the wavelength λ (upper scale) or λ/d (lower scale) for Littrow mount (deviation AD = 0), with h/d = 0.18. The heavy line corresponds to the TM case of polarization and the light line to the TE case.

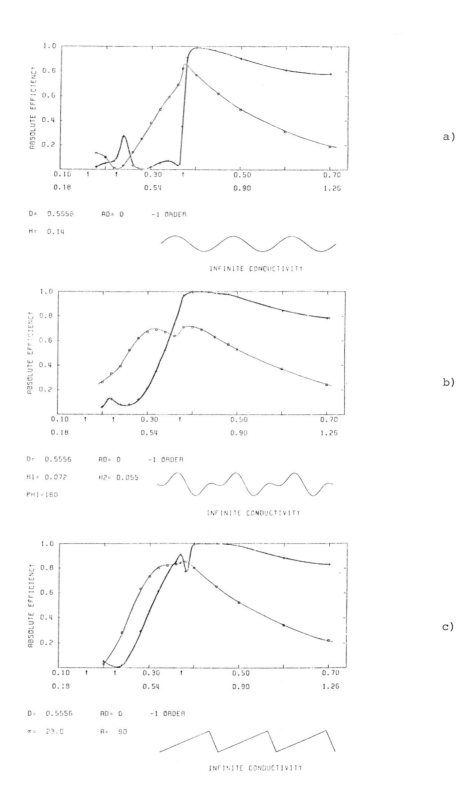

Figure 2.   Same as Fig.1, for the ratio h/d = 0.26.

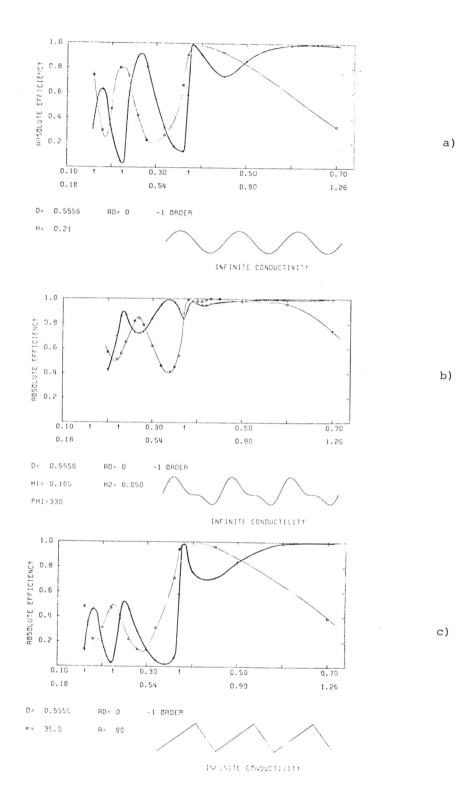

Figure 3.   Same as Fig.1, for the ratio h/d = 0.34

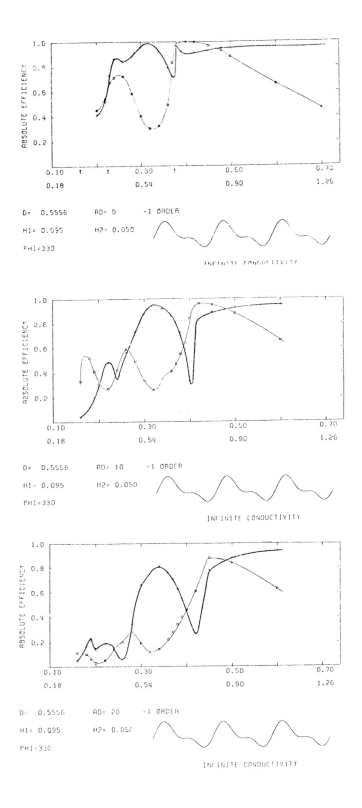

Figure 4.   Efficiency curves for a Fourier grating for various deviations

Thus, in the resonance domain there exist groove profiles superior to the triangular profile.

Finally, concentrating on the interval $0.3 < h/d < 0.4$, the behaviour in non-Littrow configurations and for finite conductivity is shown in Figs.4 and 5 for the optimal Fourier grating $h/d = 0.34$.

Increasing the angular deviation enhances the anomaly at $\lambda/d = 0.67$ for TM-efficiency, but otherwise there are no drastic changes when the angular deviation is increased for either TE- or TM-efficiency. The influence of finite conductivity aluminum coating is shown in Fig.5.

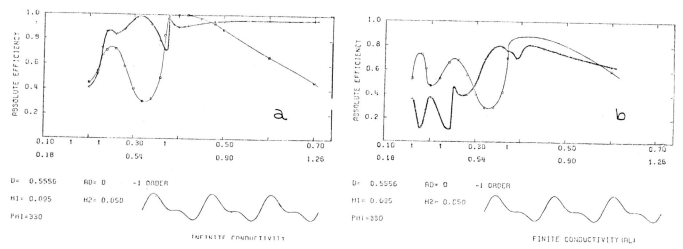

Figure 5. Efficiency curve for a Fourier grating. a) Infinite conductivity b) Aluminum

The TM-efficiency curve is the most sensitive of the two curves to a change to finite conductivity, especially at the Rayleigh wavelengths. The TE-efficiency curve essentially keeps the same shape but is lowered by a factor equal to the reflectance of aluminum.

Conclusions

We have shown that, in the resonance domain, there exist groove profiles giving a better efficiency curve than the triangular profile of the echelette grating. For holographic gratings the only method able to construct these profiles is the Fourier blaze method. However, the best gratings are very deep and have steep slopes. The first attempts to produce such gratings have led to evaporation problems. However, we think that these problems can be surmounted. Furthermore, the Fourier blaze technique is of interest even for shallow gratings. Here the performance of Fourier gratings is better than that of the corresponding sinusoidal ones and almost that of the corresponding echelette.

Acknowledgements

The authors wish to thank K. Biedermann and S. Johansson for fruitful discussions. The research work described in this paper was performed with the financial support of the Centre National de la Recherche Scientifique.

References

1. N.K. Sheridon, Appl. Phys. Lett. 12, 316 (1967)
2. G. Schmahl, J. Spectrosc. Soc. Japan 23, Suppl.1,3 (1974)
3. I.J. Wilson, Thesis, University of Tasmania (1977)
4. M. Breidne, S. Johansson, L.E. Nilsson, H. Åhlen, Optica Acta 26, 1427 (1979)
5. E.G. Loewen, M. Nevière, D. Maystre, J. Soc. Am. 68, 3476 (1978)
6. D. Maystre, R. Petit, Opt. Commun. 4, 25 (1971)
7. R.C. Mc Phedran, M.D. Waterworth, Optica Acta 20, 177 (1973)
8. R.C. Mc Phedran, I. J. Wilson, M.D. Waterworth, Opt. Commun. 7, 331 (1973)
9. A. Roger, Opt. Commun. 32, 11 (1980)
10. D. Maystre, J. Opt. Soc. Am. 68, 490 (1978)
11. M. Breidne, D. Maystre, Institute of Opt. Research Techn. Report (1980)

&
& &

Interference gratings for ultraviolet spectroscopy

M. C. Hutley

National Physical Laboratory, Teddington, Middlesex, United Kingdom

Abstract

The development of interferographic techniques has led to an improvement in the quality of gratings available for spectroscopy and has permitted the manufacture of entirely new types of gratings. Unfortunately different aspects of performance have frequently been studied in isolation and it is not always appreciated that they are interrelated. A dramatic improvement in say, the focal properties may be at the expense of reduced efficiency or dispersion. The relative importance of these factors and the extent of any overall improvement will depend strongly upon the design of the instrument and the purpose for which it is built. The purpose of the present paper is to review the contribution of interference grating technology to spectroscopy in the whole of the ultraviolet region. Special emphasis will be given to the properties of blazed gratings both made in photoresist and ion etched into the substrate. Plane and concave gratings may be made with blaze wavelengths between 300 nm and 3 nm. The efficiency is generally equal or superior to those of ruled gratings and this can sometimes, but not always, be combined with good holographic correction of aberrations.

Introduction

It has been evident since the late 1960s that gratings made by recording interference fringes in photoresist offer several advantages over ruled gratings.[1,2] Over the past twelve years they have become accepted by spectroscopists, but perhaps less rapidly than was originally expected. This delay was due at least partly to the fact that some of the advantages of interference gratings proved to be accompanied by less attractive features. Many of the disadvantages have since been overcome but there still remain constraints on the types of grating that can be produced and an improvement in one aspect may be bought at the expense of some other feature. For example, steps to increase efficiency may result in a limited spectral range, or near perfect focussing properties may in some cases only be achieved in relatively coarse and inefficient gratings. The three features of a grating which are of most interest to spectroscopists are the level of stray light, the resolution or focal properties, and the efficiency. Unfortunately the last two have often been studied in isolation by different workers and the interdependence of the various features is not always fully appreciated.

The relative merits of ruled and interference gratings for use in the visible are now well understood and have been reviewed on several occasions[3,4,5]. It is perhaps for ultraviolet spectroscopy that interference gratings have been able to offer most benefits and the purpose of this paper is to summarise the various contributions that have been made to the manufacture of gratings for wavelengths ranging from the near UV to hard X-rays: from say, 300 nm to below 1 nm.

Stray light

Because the positions of the grooves on an interference grating are determined by the conditions of interference there need be no short term errors in position, no "ghosts" or "grass", and the spectrum is significantly cleaner than for a ruled grating. The scattering from the grating can be so low that the level of stray light in an instrument is determined by diffraction or by scattering from other components. It must be remembered, though, that the nature of the scattering is different in the two types of grating. The angular dependence is certainly different and the wavelength dependence is probably different although this has yet to be confirmed experimentally. With an interference grating the scattering is diffuse and caused by surface roughness whereas with a ruled grating it is dominated by "grass" which is concentrated in the plane of dispersion. Most measurements of stray light are performed, for convenience, with visible light and show interference gratings to be significantly better than ruled ones. However, one would expect the intensity of ghosts and grass to increase as $1/\lambda^2$ and diffuse scattering to increase at $1/\lambda^4$. It may happen therefore, that an improvement which is measured in the visible may be significantly eroded in the ultraviolet. On the other hand, particularly with gratings which are ion etched into the substrate, the surface texture of interference gratings is often much smoother than that of a ruled grating and the level of stray light is still significantly reduced.

It is, in fact, difficult to make a meaningful measurement of stray light because this will depend upon the geometry of the optical system. For diffuse scattering the measured flux is proportional to the _area_ of the exit slit whereas for "grass" it is proportional to the width of the slit. Furthermore a given amount of flux will be more deleterious in a photographic instrument if it is concentrated along the line of dispersion than if it is evenly diffused. This is because a photographic plate responds to the irradience, (i.e. the flux density) rather than to the total flux.

If the quality of the grating is such that the instrumental stray light is limited by diffraction then it is necessary to take into account the _focal_ properties of the grating. One of the effects of aberrations is to reduce the intensity of the main peak and redistribute the energy in the wings of the instrumental profile where it will appear as stray light. Thus, even if the grating is to be used at less than its theoretical resolving power, it may still be necessary to look for a diffraction limited performance.

In the ultraviolet it is usually necessary to employ concave gratings which both focus and disperse the radiation. This significantly increases the luminosity of the instrument but it usually introduces aberrations which greatly exceed those encountered with plane gratings. In an interference grating the aberrations may be reduced or even eliminated if the fringes are generated by carefully chosen spherical wavefronts so that the grooves are curved and not of constant spacing. The "holographic" correction of aberrations has long been appreciated as one of the advantages of interference gratings, but instrument designers usually concentrate on reducing the size of the image or in reducing the instrumental linewidth. It would appear that studies of the far field diffraction pattern would yield valuable information concerning the apparent stray light.

Focal properties

According to Fermat's principle, the condition for a focus is that the derivative of the light path function should be zero. In the case of a grating this may be expanded in a Taylor series in terms of the coordinates on the grating aperture, the position of the grooves and the coordinates of the source and image. In this expression the first terms describe the grating equation and the focal curve and the higher order terms represent various aberrations. For a conventional ruled grating these describe the usual geometrical aberrations such as astigmatism, coma and spherical aberration but in the case of an interference grating the terms also contain a "holographic" component which is dependent upon the recording geometry. These parameters may be chosen so that for a given aberration the holographic component cancels the geometrical component and the aberration is eliminated. This technique has attracted a great deal of attention[6,7] particularly with regard to the reduction of astigmatism in mountings such as the Seya Namioka which is particularly convenient for ultraviolet spectroscopy but for which the astigmatism of conventional gratings is particularly severe. The holographic correction of aberrations is a very powerful technique but the following points must be borne in mind.

1) It is not generally possible to cancel all aberrations simultaneously so that in reducing one, another may become more significant.

2) The correction generally applies only to one wavelength. It may happen that at different wavelengths the aberrations are worse than they otherwise would have been.

3) The calculation applies only to a point source. If an extended source, e.g. an entrance slit is used then the correction will not apply to points off the meridional plane. Slit curvature cannot be avoided[6] and the overall improvement may be less than that predicted on the basis of a point source.

Once the grating has been designed and the recording parameters have been chosen the complete focal properties may be studied by exact raytracing. A spot diagram can be produced which shows the size and shape of the image for various wavelengths and for sources of various shapes so that a check can be made on the three points listed above.

The extent to which the reduction of aberrations improves the performance of the instrument depends upon the type of instrument and the way it is to be used. For example, in a monochromator with an extended slit a reduction of astigmatism by a factor of ten may only increase the luminosity by a factor of, say, two because in effect much of the light is simply passing through a different part of the exit slit. On the other hand, if an image of a point source is to be recorded photographically then a reduction of image size by ten will increase the irradiance at the plate by the same factor. In view of the wide variety of instruments used in the ultraviolet, it is impossible to quantify the advantages to be gained by the holographic correction of aberrations. We can, however, emphasise the need for an exact raytracing analysis in order to assess the performance of a grating before it, and the rest of the instrument, are made.

A very instructive example of the analysis of the focal properties of an aberration corrected grating was performed by Namioka, Noda and Seya[8]. In this, not only did they compute the shape of the image at various wavelengths but they measured the flux passing through the exit slit. They showed that for gratings of equal efficiency there was, in their case, a gain of luminosity of a factor of about three for a grating with reduced coma. For a grating with reduced astigmatism the gain was approximately two even though the astigmatism for a point source was reduced by a far greater amount. It is, however, interesting to note that efficiencies of the holographic gratings were between 10% and 20%. A well blazed grating with a sawtooth groove profile may well have had an efficiency of 70% so that the gain in luminosity achieved by holographic correction could well be offset by the reduction of efficiency.

Efficiencies of only 10% or 20% are typical of gratings with a quasi sinusoidal groove profile used in a configuration where several diffracted orders are allowed. It is possible to achieve higher efficiencies but only if the grating constant is approximately equal to the wavelength and only one diffracted order is allowed. Unfortunately this condition is precluded by the recording geometry which has been chosen to reduce aberrations. This is therefore a good example of the way in which two features, the efficiency and the focal properties, are interrelated and it illustrates the danger of studying them in isolation. This is particularly relevant in the ultraviolet first because it is often essential to use concave gratings, and second because the wavelength is shorter than that used to record the grating and the single order condition is more difficult to fulfil.

Efficiency

In most cases the efficiency of a grating is the most important aspect of its performance, particularly in the vacuum ultraviolet where the reflectance of metals is low. It is in fact the luminosity of the whole instrument which is of ultimate importance and this will be influenced by other aspects of the grating such as the focal properties and the dispersion, but the diffraction efficiency is still very important. Since it is not usually practical in the ultraviolet to use the grating in conditions where high efficiencies can be obtained from a symmetrical groove profile, a blazed sawtooth profile should be adopted whenever possible.

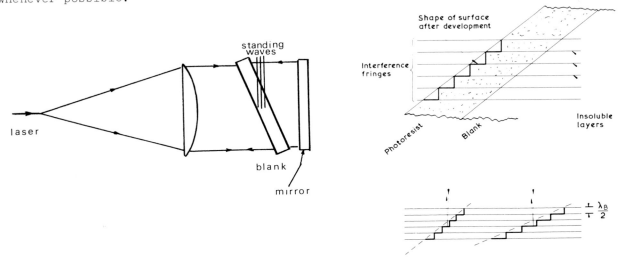

Figure 1. A simple method for the manufacture of blazed interference gratings

Interference gratings with such a profile can be made in photoresist by inclining the blank to a standing wave interference pattern as shown in figure 1. The nodal planes of the standing waves do not affect the resist which remains insoluble in these regions and when the resist is developed they define the facets of the grating grooves. The spacing of the nodal planes of the standing waves is $\lambda_0/2n$ where λ_0 is the laser wavelength and n is the refractive index of the resist. One would therefore expect the Littrow blaze wavelength to be λ_0/n but in practice it is rather lower than this. The groove profile compares very favourably with that of a ruled grating and so do the blaze properties which are summarised in figures 2 and 3. At the blaze peak, relative efficiencies in excess of 70% are not uncommon.

The laser wavelengths used at NPL in making these gratings are 458 nm (argon) 351 nm (krypton) and 257 nm (frequency doubled argon) and the blaze wavelengths obtained are 220 nm, 160 nm and 100 nm respectively. They are therefore very well suited to spectroscopy in the ultraviolet between 300 and 100 nm. This range can be extended by ion etching the grating through the resist and into the substrate. If the etching is controlled in such a way that the substrate etches more slowly than the resist the groove depth and hence the blaze wavelength will be reduced by the ratio of the etch rates. This ratio may range from 1 to 10 but values between 3 and 6 are commonly used. By this means the range of blaze wavelengths may be extended down to 20 nm or less which is the effective limit for "normal incidence" spectroscopy.

For shorter wavelengths it is necessary to resort to grazing incidence optics. The blaze wavelength of a grating away from the Littrow condition is given by

$$\lambda_B = \lambda_{B(Litt)} \sin \phi$$

where ϕ is the <u>grazing</u> angle of incidence of radiation on the facet of the groove. To achieve a reasonable efficiency for X-rays ϕ must be less than the critical angle for total external reflection. According to simple theory the critical angle at a given wavelength is given by

$$\sin \phi_c = K\lambda$$

where K depends upon the atomic properties of the material. Therefore under limiting conditions

$$\lambda_B = \lambda_{B(Litt)} K \lambda_B$$

and the minimum value of $\lambda_{B(Litt)}$ is 1/K which in the case of gold is 15.4 nm and platinum 14.7 nm. Thus the shortest Littrow blaze wavelength that is ever likely to be required is about 15 nm and this can be produced by ionetching a photoresist grating. It can therefore be claimed that with a combination of these different techniques it is possible to cover the needs of the whole of the ultraviolet region, and it is NPL's experience that the efficiencies of such gratings are as high or higher than those of the equivalent ruled gratings, and that the level of stray light is significantly lower.

The optical system shown in figure 1 may be modified to accommodate concave substrates as shown in figure 4. In this case the standing waves are spherical and centred on the centre of curvature of a convex mirror. If this point lies on the Rowland circle then the focal curve of the grating will be the same as that of a conventional ruled grating and the two types are interchangeable. There is an element of holographic correction for gratings made in this way, but one has very little control over it. The focus is perfect for λ_0 in the Eagle mounting and for λ_0 and $\lambda_0/2$ in a Paschen mounting but for the Wadsworth and Seya Nanioka mountings the improvement is insignificant. In choosing a grating of this type one therefore has to balance the relative merits of high efficiency (perhaps 2 to 5 times higher) and whatever gain may be achieved through the holographic correction of aberrations.

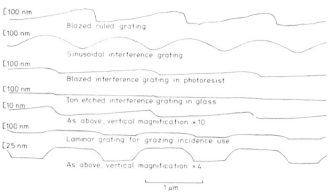

Figure 2. Various forms of grating groove profile

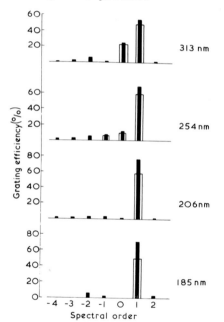

Figure 3. The distribution of light among various diffracted orders of a blazed interference grating. 1 solid line relative efficiency, open line absolute efficiency.

In addition to high efficiency, gratings made according to figure 4 have the advantage over ruled gratings that the blaze remains practically constant over the aperture. For ruled gratings the facets are all parallel and since, with a concave grating the angles of incidence, diffraction and the substrate normal vary across the aperture, the facets become progressively mis-oriented. With an interference grating the facets lie on a series of concentric spheres and correspond much more closely to the ideal case depicted in figure 5. Measurements made at the Naval Research Laboratory[9], Washington have confirmed this and in the case of one 1 m, 1200 gr/mm grating the efficiency of a ruled grating varied by a factor of 3.7 across a 27 mm aperture whereas that of the equivalent interference grating varied by only 1.2.

Non spectroscopic properties

So far we have considered only the spectroscopic properties of interference gratings for the ultraviolet. Because interference gratings are usually easier, faster and cheaper to make than ruled ones it is often as practical to use a master as it is to use a replica. Master interference gratings, particularly those etched into the substrate, are more robust than replicas. They can withstand far greater power densities which is useful for synchrotron radiation studies. They can be baked and are therefore compatible with ultra high vacuum systems and if in the course of (mis)use they become contaminated, they may be cleaned and recoated. This is not a feature which is strictly unique to interference gratings because laminar gratings for grazing incidence work were originally derived from ruled gratings. Nowadays, however, even laminar gratings are generated interferographically.

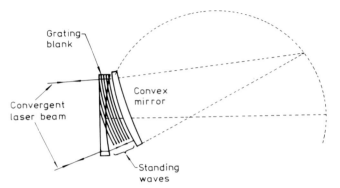

Figure 4. The manufacture of concave blazed gratings

Summary

We have seen that the various spectroscopic properties of a grating are interrelated and that in choosing a grating it is essential to consider the influence of all of these properties on the complete system. In the case of aberration corrected gratings a very thorough raytrace analysis is to be recommended.

Interference techniques have made various contributions to the manufacture of diffraction gratings in general, and the present technology is such that only in rare cases should one need to consider ruling a grating for ultraviolet use.

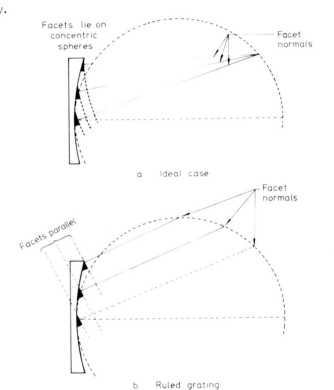

Figure 5. a) The ideal orientation of facets facets for a concave grating.

b) The orientation of facets on a ruled grating.

References

1. Labeyrie, A. and Flamand, J., Optics Commun, Vol. 1, pp 5-8, 1969
2. Rudolph, D. and Schmahl, G., Optik, Vol. 30, pp 475-487, 1970
3. Cordelle, J., Laude, J.P., Petit, R. and Pieuchard, G., "Réseaux classiques - Réseaux holographiques" Nouv. Rev. d'Opt Appliquée Vol. 1, pp 149-159, 1970
4. Hutley, M.C., "Interference (holographic) diffraction gratings" J. Phys. E, Vol. 9, pp 513-520, 1976
5. Schamahl, G. and Rudolph, D. "Holographic Diffraction Gratings" Progress in Optics XIV,
6. Pouey, M. "Imaging properties of ruled and holographic gratings" J. Spect. Soc. (Japan) Vol. 23, supp 1 pp 67-81, 1974
7. Namioka, T., Noda, H. and Seya, M. "Possibility of using the holographic concave grating in vacuum Ultraviolet Monochromators" Science of Light, Vol. 22, pp 77-99, 1973
8. Namioka, T., Noda, H. and Seya, M. "Performance of Aberration-Reduced Holographic Concave Gratings Designed specifically for Seya-Namioka Monochromators" J. Spect. Soc. (Japan), Vol. 23, Supp 1, pp 29-35, 1974
9. Hutley, M.C. and Hunter, W.R. "The variation of blaze of concave diffraction gratings". To be published.

Periodic structure defects in mechanical surfaces diagnosed through diffraction phenomena analysis

E. H. Soubari*, C. Liegeois, P. Meyrueis, C. Gateau*

Louis Pasteur University, GREPA, 3, rue de l'Université, 67000 Strasbourg, France
*BRGM, B.P. 6009, Orléans, France

Abstract

Periodic surface defects in mechanical industry, caused for instance by cavitation, are very difficult to analyze when the size of the defects is too small for conventional methods of analysis. The use of coherent light with grazing angle allows us to process the diffraction phenomena through a Fourier transform device, to obtain data concerning the size, distribution, and shape characteristics of periodic effects. The results are very promising in geophysics for the study of natural periodic materials (surface and particle) and in the study of damage caused by cavitation in hydraulic equipment for nuclear power and in the petroleum industry. The data are analyzed with very simple equipment. Other applications of this process are possible. We are working on a coupling of computers for very fast and large scale automatized control.

Introduction

We use the possibilities of the optical Fourier transform for characterization studies of surfaces with polishing defects. The method consists of a non-destructive technique utilizing the coherence properties of laser light and it enables us to have information on defects. This process was tested on cavitation damage to a polished surface.

Theory

Analysis of particle distribution and surface roughness

The search of the above-mentioned parameters is achieved by the directional and frequential spectral analysis. The energy spectrum gives information about size, shape and lengthening of the structures included in the image and that is why we analyse textures by optical Fourier transform. For example, the circular symmetry of a diffraction pattern means that most of the textures are circular, the halo effect means a concentration of large circular textures. Also, the lenthening of a texture in a determined direction is visualized in the diffraction pattern by a lenthening perpendicular to the real one.

After these few remarks, one can begin with size analysis of defects by quantitive on-line evaluation. This can be achieved electronically by examining the spectral intensity which is directly related to shape, size and lengthening of the defects. The method is extremely efficient in the case of almost spherical defects (cavitation defects). The distance of the first minimum in the Fourier spectrum gives us their diameter (Fig. 1).

The equation for a circular hole is:

$$S(x,y) = Circ\left(\frac{r}{a}\right) = \begin{vmatrix} 1 \text{ if } r \leq a \\ 0 \text{ if } r > o \end{vmatrix} \tag{1}$$

$$S(u,v) = \frac{\pi a^2}{i\lambda f} \frac{2J_1(Z)}{Z} \qquad \text{fig. 1}$$

where

$$Z = 2\pi\rho a$$

$$\rho = (u^2 + v^2)^{1/2} = \left(\frac{\xi^2 = \eta^2}{\lambda f}\right)^{1/2}$$

a = radius = constant
r = radial distance fromany point M in the plane
$J_1(Z)$ is the Bessel function 1st kind and 1st order

The intensity of the energy spectrum is:

$$I_i = \frac{(\pi a_i^2)^2}{\lambda^2 f^2} \quad \frac{2J_1(2\pi\rho\ a_i)^2}{2\pi\rho\ a_i)} \tag{2}$$

$I_{io} = (\pi a_i^2)^2$ is the square of the surface of the particle which has a cross section of radius a_i

$$Z_i = 2\pi\rho\ a_i = \frac{2\pi}{\lambda f}\ (\xi^2+\eta^2)^{1/2} a_i = k.\theta.a_i \quad \text{with tg } \theta = \frac{d}{f} \simeq \theta$$

$$I_i\ (\theta) = \frac{\pi^2 a_i^4}{\lambda^2 f^2}\ \frac{2J_1\ (k\theta\ a_i)}{k\theta\ a_i}^2 \tag{3}$$

which is the Airy function for a circular defect of radius a_i

For example

$a_i = 1 \text{ um}$
$Z = k\theta a_i = 3,83$
$\lambda = 633 \text{ nm}$
$k = \frac{2\pi}{\lambda}$

we have $\theta = 1,22 \frac{\lambda}{2a_i} = 22,12°$ $\tag{4}$

and for $\theta = 44,24° \rightarrow a_i = 0,5 \text{ um}$ (5)

The number of defects is given by the analysis of the intensity profile. So, we see that every defect size is given by the abscissa and the number by the ordinates of the optical energy spectrum.

For a set of defects of variable diameter we obtain:

$$I\ (\theta) = \sum_{i=0}^{\infty}\ I_i\ (\theta)\ \Delta a_i \tag{6}$$

where Δa_i represents the variations of the diameter $(a_i + \Delta a_i - a_i)$

$$\Delta a_i \rightarrow 0 : \quad I\ (\theta) = \int_0^\infty I_i\ (\theta)\ da_i$$

$$I\ (\theta) = c \int_0^\infty a_i^4\ \left[\frac{2J_1\ (k\theta a_i)}{k\theta a_i}\right]^2\ da_i \tag{7}$$

where c is a constant depending on the optical system.

It should be noted that the spatial invariance by translation of defects in their plane is assured by the properties of the coherent optics systems; the diffraction pattern does not change and stays motionless, but the amplitude is multiplied by a phase factor which is directly related to the displacement of a defect in the entrance plane:

$\text{Exp}\ (i\ \frac{2\pi}{\lambda}\ \sigma_n)$ where σ_n is the optical path difference introduced by the displacement.

For N identical defects randomly distributed, we have:

$$I_n\ (u,v) = |\sum_{\sigma_n=1}^{N}\ \text{Exp}\ (ik\sigma_n)\ s\ (u,v)|^2 \tag{8}$$

$$I_n\ (u,v,) = N.\ |\ S\ (u,v)|^2$$

which effectively means that the intensity of the optical spectrum is multiplied by N.

Usually we introduce a statistical weight into the equation, hence:

$$I(\theta) = c \int_0^\infty (a_i^4) \, P(a_i) \left[\frac{2J_1(k\theta a_i)}{(k\theta a_i)}\right]^2 da_i \qquad (9)$$

where $P(a_i)$ is the statistical weight.

This formula enables us to calculate the defect size and the relative number of defects starting from the optical energy spectrum by measuring the physical and geometrical parameters.

To each θ_j belongs an intensity $I(\theta_j)$ given by equation (9) which corresponds to the diffraction of the defects of the same diameter $2a_i$ for an angle θ_j for which the measured intensity is $I_j = I(\theta_j)$

In making several measures of θ_j one can solve the problem for all the existing defects in the entrance signal in the following way:

Formula (7) can be written as a matrix:

$$I(\theta_j) = I_j = \sum_{i=1}^{m} P(a_i) \, A_{ij} \qquad (10)$$

$$\text{where } A_{ij} = c(a_i)^4 \left[\frac{2J_1(k\theta_j a_i)}{(k\theta_j a_i)}\right]^2 \qquad (11)$$

$$P(a_i) = \sum_{j=1}^{n} A_{ij}^{-1} \cdot I_j \qquad (12)$$

where A_{ij}^{-1} is the inverse matrix of A_{ij}.

n is the number of values θ_j and I_j to be measured to determine the a_i's, so it is recommended to take n>m (usually n=2m).

The function $P(a_i)$ can be obtained easily by the solving of a system of linear equations if one makes as much I_j measures as chosen possibilities for a_i (i=j=10), if not one has the possibility of utilizing the least square method when j>i where the number of I_j measures is superior to the number of unknown a_i. But it is impossible to determine $I(\theta)$ for all the continuous values of θ_j, so we have to make a sampling of discrete values for θ and $I(\theta)$, which results in the diameter sampling of the defect which is indicated on the frequency spectrum by numerical methods (Fig.2).

We can avoid the sampling by using optical spatial filtering methods (in triple optical diffraction) (Fig. 3), where it would be interesting to look for a continuous frequency spectrum in order to have the total information on all the possible existing sizes. This avoids the computing time of the system (12) by the improvement of the number of samples of measures.

Fig. 3 gives us a spatial filtering method avoiding the inconvenience of large size defects which give rise to a non-diffracted light near the optical axis. With a high-pass filter, one restores an image of the small defects of the sample, and then realizes a third Fraunhofer diffraction and, with the help of a photomultiplier, makes a scanning to obtain a continuous spectrum infrequency of the defects.

This operation requires an anti-vibrating bench of a minimum length of 6f, where f is the focal length of the chosen lenses.

For example, the numerical result given in (5) requires large diameter lenses (Ø = 80mm, f = 300 mm).

When the defects are very fine, the diffraction pattern is on the Fournier sphere of radius f (Figs. 4 and 5), and the photomultiplier's scanning operation becomes very difficult since it has to rotate on an axis of radius f.

Experiments

We tried the method on transparent and opaque objects and applied the results to mechanical parts which had been damaged by cavitation. Fig. 13 shows the frequency spectrum for such a sample.

Conclusion

We have proved that optical Fourier transform can be a rapid, economic and efficient means of metrology of surface defects caused by cavitation and other processes.

Developments are in process for commercial products with industrial partners.

Since this method also works in the case of moving particles, we will develop a system for the metrology of moving particles.

References

1. HERVE Ph. (1977): "Influence de l'Etat de Surface sur le Rayonnement Thermique des Matériaux Solides". Thèse de Doctorat d'Etat, Paris VI

2. HARVEY L. Kasdan (1979): "Industrial Applications of Diffraction Pattern Sampling", Optical Engineering, Vol. 18, No. 5, September-October 1979

3. SAWATARI T. and MUELLER R.K. (1977) "Surface Flaw Detection using Optical Filters". International advances in non-destructive testing: Vol. 5, pp. 1-15

4. LIND M.A., HARTMAN J.S. and HAMPTON H.L. (1978): "Specularity Measurements for Solar Materials", SPIE, Vol. 161, Optics applied to solar energy IV

5. THOMPSON B.J. (1977): "Hybrid Processing Systems - An Assessment", Optical Computing, Proceedings of IEEE, Vol. 65, No. 1, pp. 62-76

6. GOODMAN J.W. (1968): "Introductions à l'optique de Fourier et à l'Holographie", Masson et Cie, Paris

7. SOUBARI E.H., MEYRUEIS P., LIEGEOIS C. (1979): "Application of Optical Filtering to Geophysics Mapping and Remote Sensing", SPIE, Vol. 201, Optical Pattern Recognition

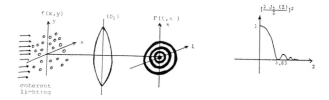

Fig. 1a: Optical diffraction by spherical defects

Fig. 1b: Intensity profile in the optical energy spectrum

a. Profile of the diffraction pattern   b. Frequency spectrum

Fig. 2: From the diffraction pattern to the frequency pattern

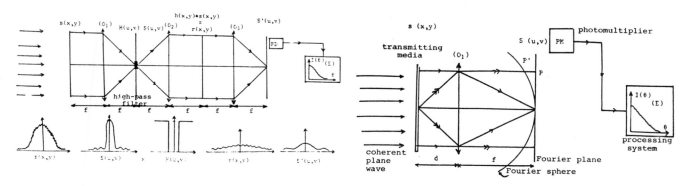

Fig. 3: Determination of size of defect or roughness of a surface by optical filtering

Fig. 4: Diffraction of coherent plane wave by a transmitting material

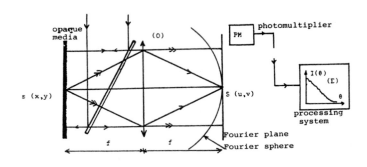

Fig. 5: Diffraction of a coherent plane wave by a reflecting opaque object

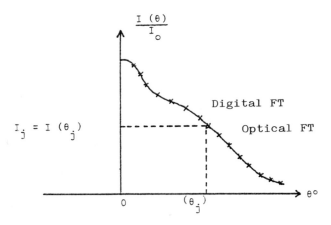

a. Diffraction figure of a surface

Fig. 6: Surface defect measurement by the non-destructive method (Fourier optics)

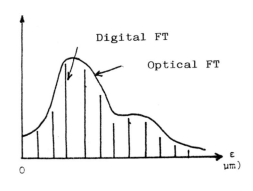

b. Frequency spectrum of surface defects by optical method and method of numerical analysis of the diffraction figure

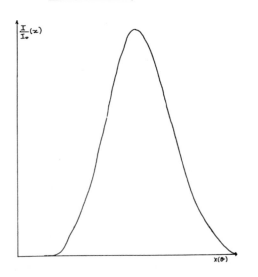

Fig. 7

Profile of the laser beam

Filtered and magnified

Fig.8: Diffraction of a mirror

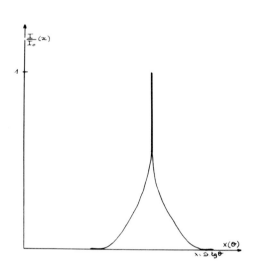

Fig 9: Diffraction figure of mechanical object with defects

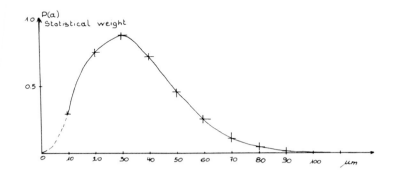

Fig.10 Statistical weight of a mechanical object with defects

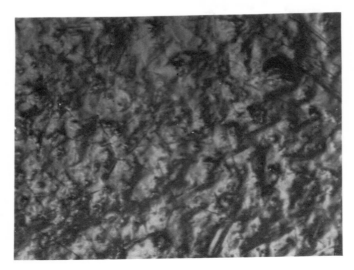

Fig. 11: Cavitation **defect** on **a** mechanical surface   - $\simeq$ 92,000 strokes - average size 20

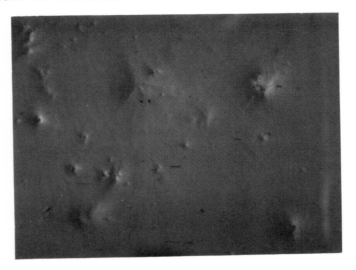

Fig. 12: Cavitation defect on a mechanical surface   - $\simeq$ 15 strokes - average size 20

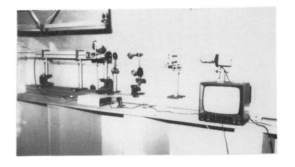

Fig. 13: Set up

AUTHOR INDEX

SUBJECT INDEX